# Cells, Forces, and the Microenvironment

# Cells, Forces, and the Microenvironment


edited by

Charles M. Cuerrier

Andrew E. Pelling



CRC Press
Taylor & Francis Group
Boca Raton London New York

CRC Press is an imprint of the
Taylor & Francis Group, an **informa** business

First published 2015 by Pan Stanford Publishing Pte. Ltd.

Published 2022 by CRC Press
Taylor & Francis Group
6000 Broken Sound Parkway NW, Suite 300
Boca Raton, FL 33487-2742



ISBN-13: 978-981-4613-36-1 (hbk)
ISBN-13: 978-981-4613-37-8 (ebk)



**Visit the Taylor & Francis Web site at**
**http://www.taylorandfrancis.com**

**and the CRC Press Web site at**
**http://www.crcpress.com**

**British Library Cataloguing-in-Publication Data**
A catalogue record for this book is available from the British Library.

**Cells, Forces, and the Microenvironment**

# Contents

# Preface

While you are reading this book, comfortably seated in a quiet and relaxing environment, your body is producing dramatic and dynamic mechanical phenomena. Indeed, your heart is beating, your lungs are inflating and deflating, blood courses through your vasculature, your stomach and intestines are moving to digest your last meal and all of these movements generate mechanical forces that have important consequences, even at the cellular level. Furthermore, tissues and organs are composed of a variety of scaffold proteins that are organized in a myriad of ways resulting in a wide range of cellular microenvironments in vivo. Together, scaffolding proteins not only control the biochemistry of the microenvironment but also its mechanical properties, which vary depending on the tissue and organ type. The mechanical forces and properties that arise in the cellular microenvironment play a role not only in mammalian cell biology but also in plants and bacteria, which are well known to possess distinct biochemical mechanisms to sense and respond to the diversity of physical cues in their microenvironment.

For decades, cellular responses to their extracellular environment have been directly understood in terms of their genetics and biochemistry. The development of new technical approaches has facilitated quantitative investigations and insights into how physical cues in the microenvironment regulate cell biology. Integrating these new concepts has prompted many new questions about cell biology, such as: What physical information in the cellular microenvironment affects cell biology? How do cells detect and respond to changes in the physical properties of the microenvironment? Can we re-engineer the purpose of living systems using physical cues? It is now well known that many parameters such as cell differentiation, morphology, and migration, to name only a few, can be controlled and modulated by physical cues. These findings have emphasized the importance of physical phenomena in cell biology in the context of health and disease.

In this book, chapters written by leading experts in their respective research fields provide an overview of how mechanical forces and the physical properties of the microenvironment affect cell and tissue homeostasis. The book addresses current knowledge on how prokaryotic and eukaryotic cells are able to sense and integrate physical signals. In the organism, cells are exposed to a variety of forces, material properties, and topographical features that influence a diversity of biological functions. The chapters in this book detail many of the modern methods that have been utilized to develop a better understanding of how these physical cues influence cell biology. These methods include the development of traction force microscopy, atomic force microscopy, micropost methods, microfluidic devices, deformable membrane arrays, and novel biomaterials at the single cell to whole organism level. All chapters point out future perspectives and limitations of these technologies and discuss the significance or application of the research at hand. In addition to research on animal systems, several chapters highlight and review the approaches that have been employed to shed new light on the importance of physical phenomena in governing bacteria and plant cells.

This review volume is dedicated to a broad range of inter-disciplinary scientists (biologists, physiologists, pharmacologists, engineers, physicists, etc.) who are interested in these new and important physical aspects of cell biology. It is intended to provide an extremely broad overview of the role of physical cues in governing diverse aspects of cell biology across kingdoms and in physiological and pathological processes. To conclude, we would like to thank and acknowledge the time and great efforts made by all of the contributors who participated in this enjoyable project.

**Charles M. Cuerrier**
**Andrew E. Pelling**

# Chapter 1

# Micromechanical Cues Converging on Fibroblasts, Cardiac Myocytes, and Stem Cells

**James R. Pinney,[a] Golnar Doroudian,[b] Patrick Chew,[a] Tejal A. Desai,[c] and Brenda Russell[d]**

[a]*Bioengineering Department, University of California San Francisco, 1700 4th Street, San Francisco, CA 94158, USA*
[b]*Bioengineering Department, University of Illinois at Chicago, 835 South Wolcott Ave, Chicago, IL 60612, USA*
[c]*Bioengineering and Therapeutic Sciences, University of California San Francisco, 1700 4th Street, San Francisco, CA 94158, USA*
[d]*Physiology Department, University of Illinois at Chicago, 835 South Wolcott Ave, Chicago, IL 60612, USA*

russell@uic.edu

## 1.1 Introduction

Renewed interest in how cells are affected by their physical surroundings has begun to reveal the intimate interplay between the mechanical microenvironment and cell phenotype and behavior. Recent research has provided an insight into the role of biochemical and mechanical cues in two and three dimensions to

*Cells, Forces, and the Microenvironment*
Edited by Charles M. Cuerrier and Andrew E. Pelling

ISBN 978-981-4613-36-1 (Hardcover), 978-981-4613-37-8 (eBook)
www.panstanford.com

effect significant changes at the cell and tissue level [4,10,27, 28,38,95,99–101]. Cells are sensitive to passive and active physical stimuli via external forces (eliciting outside-in signaling) and by forces generated in the cell (inside-out signaling) [61]. External forces occur in vivo through shear stress, extension, or compression [19,64,134]. Internally detected forces, which can be generated and transmitted by motor proteins or sliding filaments within the cytoskeleton, can reorganize the cytoskeleton in response to external stiffness, surface topography, and ligand density [113,151].

The mechanisms underlying these cell responses to physical cues are thought to depend on mechanotransduction, which is defined as translating mechanical forces into biochemical events. Study of the mechanotransductive and cytoskeletal pathways that control cell traction and matrix deposition continues to elucidate the mechanisms of cellular response to the microenvironment, which may overcome the high hurdles found in many chronic clinical conditions. Despite historical enthusiasm for broad-based pharmacologic solutions or the introduction of cellular replacements for injured tissue [22,63,72,74,135], there is significant potential in using intrinsic responses to local micromechanical changes to reduce maladaptive repair processes and promote more effective healing [16,22,62,99,100,135]. This has driven recent interest in harnessing the potential of materials-based micromechanical cues to dictate the behavior of cell populations to mitigate pathologic responses. This chapter focuses on three cell types: fibroblasts, cardiac myocytes, and stem cells.

## 1.2 Fibroblasts

### 1.2.1 Microenvironmental Forces and Fibroblastic Remodeling

Fibroblasts play an active part in both physiologic and disease states through mechanisms of cell-to-cell communication, signal propagation, and periods of growth and recession, in addition to building and depositing extracellular matrix (ECM) proteins [59,112]. Although these phenomena can be instigated by a range of biological, chemical, or mechanical signals, much of this activity can be attributed to the interaction of fibroblasts with the local microenvironment of the surrounding tissue (Fig. 1.1).

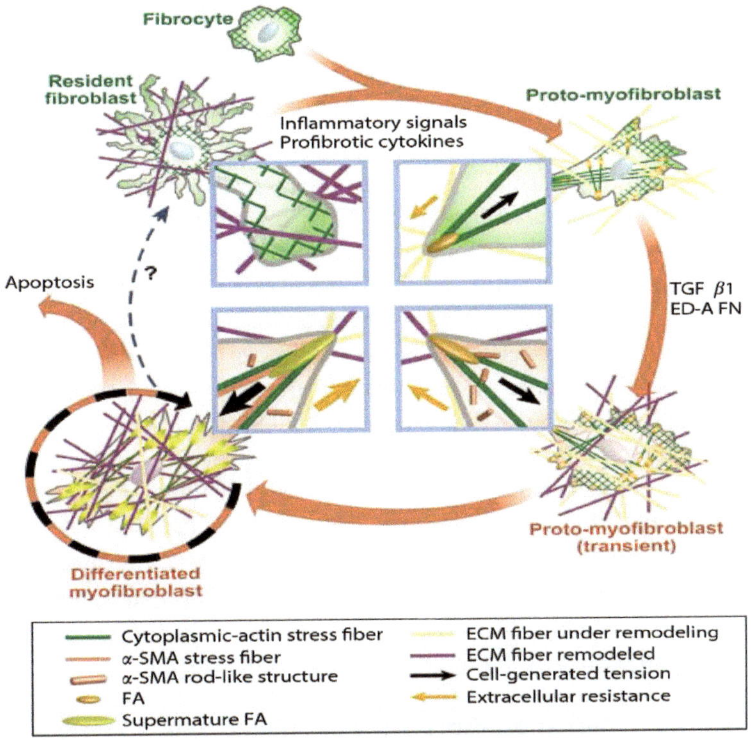


**Figure 1.1** In the setting of hypoxic cardiac injury, up-regulation in inflammatory signals and profibrotic cytokines, as well as breakdown of the native ECM by invading immune cells, induces fibroblasts to change their junctional attachments with ECM components and alter their cytoskeletal mechanics. This leads to the differentiation of highly contractile myofibroblasts, which generate stiff ECM and scar tissue in a runaway positive feedback response. Reprinted with permission from Ref. [59].

The cardiac environment serves as a prime example of the multifaceted characteristics of resident fibroblasts in tissue. Previously thought to be senescent, it is now recognized that cardiac fibroblasts are vital to cardiac tissue, playing an inimitable role in structure, function, and disease [13,133]. Despite accounting for a minority of the total tissue volume of the heart muscle, cardiac fibroblasts greatly outnumber cardiomyocytes, making up nearly 70% of all cells [16]. In addition to depositing the intricate and durable ECM required for effective heart function, cardiac fibroblasts couple both chemically [13,63] and electrically [2,17] to cardiomyocytes and are integral in the maintenance and limited regenerative capacity of the heart muscle [147]. Nonetheless, regulation of fibroblast quantity and phenotype by local forces and the microenvironment is still under investigation.

After myocardial infarction (MI), a population of highly contractile fibroblasts, referred to as myofibroblasts, appears in the wake of inflammatory matrix metalloproteases that degrade the injured tissue. Although the origin of this cell type has been the source of controversy, it is characterized by accelerated ECM production, increases in signaling factor release, and high contractility that is responsible for the scar tissue found after MI [5,14,33,59,60]. This runaway cellular response results in maladaptive tissue stiffening and may prevent cardiomyocyte progenitor cells from repopulating the area, leading to progressive heart wall thinning at the expense of cardiac function [59]. This shift in the fibroblast's phenotype illustrates the responsiveness of the cell type to mechanical changes in the microenvironment and provides insight into novel methods of dictating the behavior of cells and tissues by manipulating the surrounding milieu.

#### 1.2.1.1 Mechanosensors in the cell membranes

Fibroblasts sense and respond to local microenvironmental changes that cause them to undergo phenotypic transformations into contractile myofibroblastic cells [59]. Transmission of force throughout the tissue can integrate stretch activation of ion channels in the cell membrane with the gap junctions between the cells [93]. The opening of mechanosensitive calcium ion channels has been linked to short-term increases in the production of alpha smooth muscle actin (αSMA) in fibroblasts, promoting the contractile myofibroblastic transformation [2]. Changes in the gap junctional protein profile has also been indicated in this transformation as upregulation of connexins 43 and 45 in the peri-infarct zone link post-injury cardiac fibroblasts to one another [17]. This type of communicable reaction to micromechanical stimuli helps to explain how local changes in the extracellular environment, such as those found in the scarred region due to tissue stiffening post-infarct, can set in motion fibroblastic responses across the entire organ [132].

#### 1.2.1.2 Focal adhesion and integrin-mediated mechanotransduction

In addition to sensors in the membranes, mechanotransducers are located in focal adhesions and the cystoskeletal network, reviewed elsewhere [27,118,121]. Integrin-mediated perception of external stress plays a vital role in influencing how fibroblasts behave in the cardiac environment [59]. Myosin-based contractility from within the fibroblast affects integrin binding to the ECM and focal adhesion formation [14,78]. The integrin complexes have other mechanotransductive proteins such as paxillin and focal adhesion kinase (FAK), which can transmit signals downstream through intracellular mediators, connecting directly to the actin network [75,118,121]. After myocardial infarct, an increase in supporting intracellular proteins such as vinculin and tensin coincides with a simultaneous rise in integrins αvβ3 and α5β1, leading to the development of robust and pronounced focal adhesion complexes [37,58].

The mechanosensors and downstream effectors permit fibroblasts to detect and respond to the stiffness of the environment. After an infarct the matrix becomes more rigid due to excess collagen deposition, and as fibroblasts continue to pull against their bound ligands, they experience increasing mechanical resistance from the stiffened tissue of the dense scar [13,133]. This is illustrated by a rise in the complexes of integrin α3 and collagen VI [14], which in turn affects tensegrity-based signaling mechanisms in the fibroblast leading to myofibroblastic transformation and increased ECM production and deposition [60]. As a whole, the pathologic response can be partly attributed to increased local tensile stresses on fibroblasts that are no longer able to offload mechanical stress by attachment to the now degraded and transformed network of ECM proteins. A positive feedback cycle develops as the fibroblasts continue to adapt to the new character of their microenvironment, bypassing the productive apoptotic signals that would normally allow for healthy resolution and retraction of scar tissue.

#### 1.2.1.3 Cryptic ECM signaling domains

A third mechanism proposed to contribute to the ability of fibroblasts to respond to changes in their mechanical environment highlights the complex mechanochemical nature of the ECM. The native ECM is rich with potent signaling molecules secreted by the resident cells, such as latent TGF-β complexes [59]. As the matrix is broken down after injury or put under stress by the contractile nature of the new fibroblastic population, these proteins can be physically stretched or separated from the structural proteins of the matrix, revealing activating cleavage sites or previously cryptic binding sites that may offer enhanced affinity or bioactivity [143]. In the case of TGF-β, this propagates a positive feedback mechanism as now activated TGF-β ligands promote the myofibroblastic transformation, including increased production of αSMA and deposition of collagens [13,59].

### 1.2.2 Fibroblasts and Topographical Cues

One strategy to study the response of fibroblasts to changes in their microenvironment is to introduce local artificial anomalies in the extracellular space that serve as instructive cues for the cells. Topographical cues play a well-established role in the migration and matrix deposition characteristics of fibroblasts. A number of mechanisms to explain this interaction have been proposed to attempt to bridge the extremely sensitive and variable behavior displayed on substrates of differing geometry and stiffness. Cells sense topographical features and adapt by a phenomenon called contact guidance [115]. Spreading fibroblasts between posts forms arc-like shapes that present an inward curvature that increases with time. The importance of cell tension to this response can be appreciated through its reversal in the presence of a myosin II inhibitor [50]. As such, fibroblasts alter their growth and migration poles to orient themselves with a direction that maximizes adhesion potential and traction-force generation [1,6,25,79,84,109]. This behavior corroborates observed microscale topographical feature-size minimums and responses to external forces in the presence of patterned topography, shown in Fig. 1.2 [10,28,73,99].

Observations of how fibroblasts react to angled or constrained geometries that put a stress on their ability to bind to and contract against ligands in their microenvironment are also instructive. In these environments, fibroblasts increase their production of ECM components such as collagens and hydroxyproline, highlighting the sensitivity of even single cells to their local mechanical environment [5,73]. A critical dependence on myosin-II generated traction forces is cited as an important player in this response to topography and pattern density, as the sensitivity to physical environmental cues can be abrogated by specific inhibition of this internal force generation [18,35]. Extrapolation to contact-dependent upstream signaling pathways of Rho-associated kinase or MAP kinase, among others, are fruitful in understanding fibroblastic interactions with microscale topographical substrates [10,49,107,137]. Altered internal cell dynamics in response to microtopography, including decreased alpha smooth muscle actin fibers (αSMA), mitigated response to stimulators of the myofibroblastic transformation such as TGF-β, and changes in integrin surface molecule expression are also indicated [5,97]. Grooved topography has been shown to affect matrix-metalloprotease production by fibroblasts, which, coupled with the effects on cell tension networks, could serve to limit downstream activation and recruitment of fibrotic processes after injury [15].

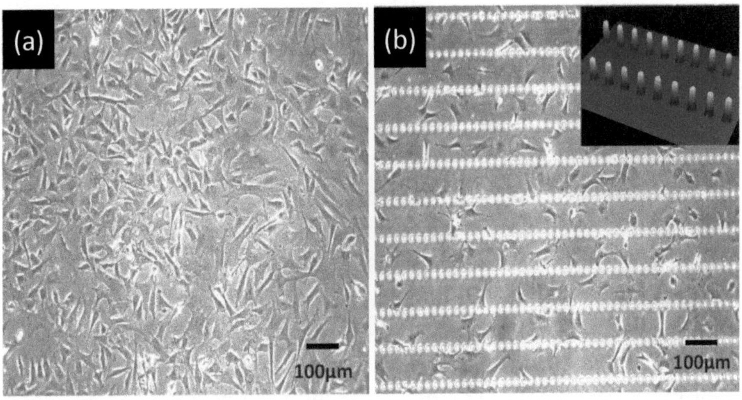


**Figure 1.2** Fibroblasts grown on a flat membrane for five days (a) show higher proliferation and growth characteristics than those grown on a membrane patterned with microposts (b) Microposts are 10 μm tall and spaced 30 μm apart, with 100 μm between rows (inset). Reprinted with permission from Ref. [10].

Improved bioengineering methods create three-dimensional culture environments, which enable observations of cell behavior in their native microenvironment surrounded by ECM on all sides. These three-dimensional substrates, changes in cell motility, aspect ratio, and growth characteristics suggested that cells were able to sense the mechanical nature of their environment even when not in contact with the surface [99]. Sequential patterning organized multilayered constructs that exhibit improved strength and function [55]. Complex geometries generated by electrospinning techniques or porous scaffold systems control and direct fibroblastic growth, which will be critical to future efforts to grow multicellular, structurally robust cell constructs for tissue engineering applications [70,85,140].

Discrete microstructures of defined density and stiffness can be included in a three-dimensional gel of fibroblasts and are able to mitigate fibrotic matrix deposition, integrin character, and proliferation rates [4,101]. Injecting microstructures into post-infarct injury zones has shown changes in the fibrotic makeup and improvements in cardiac function over time from our group (unpublished). These types of applications may pave the way for the use of micromechanical therapies that seek to take advantage of the widespread influence of environmental characteristics on tissue function to address clinical issues of tissue recovery, regeneration, and controlled growth.

### 1.2.3 Fibroblasts in Tissue Engineering

As a whole, the growing collection of data on the effect of introduced topography and micromechanical cues continues to expand our understanding of the role and response of fibroblasts in maintaining tissue structure in a dynamic fashion. Providing topographical and discretized mechanical cues serves to instruct the growth and migration of fibroblasts in a manner that alters cell-cell and cell-matrix contacts. Furthermore, adjusting the spacing, shape, and patterning of micromechanical cues dictates how and where fibroblasts are able to bind or interact with their environment. This plays a role in their ability to contract, generate force, and activate mechanotransductive pathways, which contribute to fibrotic activation, senescence, or apoptosis. Lastly, examination of the interaction of fibroblasts with micromechanical cues in three-dimensional culture environments provides insight into the stretch and ligand-spacing dependent mechanisms that can reveal or hide cryptic binding sites in the ECM or permit or disallow the assembly of cell-surface binding complexes. Coupled with new research into the effects of dynamic micromechanical changes [78] and the cooperative use of pharmacologic approaches, these novel micromaterial strategies promise to remain exciting.

## 1.3 Cardiomyocytes

### 1.3.1 Micromechanical Forces and Cardiomyocytes

No discussion of micromechanical forces would be complete without consideration of the beating heart cell: the cardiomyocyte. From early in the embryo throughout life, cardiomyocytes generate dynamic cyclical forces to pump blood through the body. Cardiomyocytes act as a functional syncytium, propagating electrical signals to synchronize their contractions. After myocardial infarction, many cardiomyocytes die, leaving the survivors to work harder. Compounding their load, fibroblast and ECM content increases, scarring and stiffening the heart and interrupting normal communication from cell to cell [14], eventually culminating in heart failure. Muscle injury and efforts to repair heart tissue have been well described [11]. However, less is known about how muscle cells convert mechanical cues from the ECM into intracellular signals [65–67,87,92], and the necessary micromechanical cues needed to engender cardiomyocyte interconnectivity and effective function [33,93].

#### 1.3.1.1 Matrix stiffness and contraction

Cardiomyocytes are able to match the work required of them to meet external demands by regulating muscle mass. Work is greater when the stiffness of the matrix in which they reside is increased; therefore, this is a critical component in health and disease. During morphogenesis, epithelial sheets are folded with physical forces playing a crucial role in the formation of the complex body plan of the embryo. Moreover, mesoderm tension is less stiff than the ectoderm and endoderm layers of early embryos and, interestingly, cells require internal contractility rather than adhesivity to sort to the correct location [80]. The formation of the heart tube thus depends on local forces, which are tied to the contractile force that can be generated against a matrix of variable stiffness [142], and these mechanical loads can influence gene expression patterns during development [150]. Studies on rodent heart tissue demonstrate a several fold increase in stiffness of the heart tissue from embryonic to post-natal periods coincident with increased strength of contraction and gap-junction formation. Sarcomeric character also changes during this time including sarcomeric lengthening and an increase in less compliant isoforms of titin [69].

Artificial manipulation of substrate stiffness has a significant impact on the growth and function of cultured neonatal cardiomyocytes. Increasing the stiffness of gel substrates increases the resting sarcomere length, which, along with intracellular calcium levels, leads to greater twitch power [114]. As stiffness exceeds normal adult myocardium levels of 10–50 kPa, sarcomeric shortening continues to decrease and as this threshold is crossed, the cardiomyocyte is no longer able to establish significant anisotropy parallel to the direction of contraction. This leads to poor contractile capability, the effect of which can be directly observed in the post-infarct heart stiffened by reactive fibrosis [27,43,68].

#### 1.3.1.2 Costameres and focal adhesions

Like fibroblasts, cardiomyocytes depend on a network of intracellular, transcellular, and extracellular proteins to integrate with the extracellular environment and transduce their contractile forces from the myofibrils to the tissue as a whole. Much of this interactivity occurs at the specialized focal adhesion complexes, called costameres, which surround the cardiomyocyte peripherally at the Z-discs. Costameres transduce mechanical signals from extracellular ligands via integrins across webs of communicating proteins including talin, vinculin, paxillin, FAK, dystrophin, α-actinin, and muscle LIM protein (MLP), thereby activating a variety of downstream signaling cascades [27,121]. These signaling cascades are numerous and beyond the scope of this discussion, but generally include a variety of MAP kinase and JAK/STAT related protein kinases that can activate changes in calcium transients and hypertrophic responses [105,138].

#### 1.3.1.3 Cyclic and anisotropic strain and the sarcomere

Forces of both longitudinal and transverse strain are a characteristic feature of the cardiomyocyte's natural environment. Although the study of these phenomena is complicated by internal stresses generated through the actively contracting network of the heart tissue, several important principles have been identified that characterize the role of mechanical straining in the adaptive response of the cardiomyocyte.

Both static and cyclic strain applied externally to cardiomyocytes in culture results in increases in expression and phosphorylation of focal adhesion components such as FAK and ERK that are directly proportional to an increase in cyclic strain frequency [126]. However, the geometry of the cardiomyocyte is characterized by anisotropic longitudinal and lateral structures [9,24,131]. Distinct directional pathways are implicated by findings that static transverse and longitudinal loading activates stress induced-MAP kinase [83] and anisotropic mechanosensing FAK phosphorylation [126]. MLP has anisotropic myofibrillar distribution that may contribute to the changes in cell shape, spreading, and elongation seen in response to cyclic straining [118]. Longitudinal strain contributes to the insertion of new sarcomeric units [154] through the activity of active isoforms of FAK, protein kinase Cε (PKCε), and others. These signaling molecules may serve to partially dissociate Z-disc complexes, enabling capping proteins such as CapZ to add actin subunits to the thin filament, restructuring the cell's cytoskeletal network [90,118]. Transverse strain, on the other hand, is shown to have a greater effect on focal adhesion relocation, parallel addition of filaments and other mechano-sensor/signaling molecules found at the Z-disc, such as MLP [10,27,118].

Cyclic strain may also affect cardiomyocyte progenitor cells. Both mesenchymal stem cells (MSCs) and embryo-derived cardiac progenitors grown on cyclically strained membranes show increases in cardiac differentiation markers such as atrial natriuretic factor, sarcomeric α-actinin, cardiac troponin-I, tropomyosin, and α-myosin heavy chain [57,125]. Clearly, dynamic mechanical forces are important to the growth and function of cardiomyocytes, yet are still incompletely defined.

### 1.3.2 Microtopography

Muscle strength in the heart is determined by the fiber cross-sectional area and the ability to transmit useful force depends on the longitudinal orientation of the contractile filaments, which can be mimicked in culture by microtopography (Fig. 1.3). Microposts provide a perpendicular attachment equivalent to the intercalated disc and the attached cardiomyocytes have increased myofibrillar height, indicating more effective contractile machinery [96]. As a whole, extracellular microtopography may act to permit larger focal adhesions for anchorage of perpendicular myofibrils that organize sarcomeric architecture and promote more physiologic and efficient contractile mechanisms [8,10,31,95,96,108,117,126]. Microgrooves facilitate cellular alignment and also regulate levels of cadherin and connexin expression, recapitulating the organization and interaction of cardiomyocytes seen in vivo [95] (Fig. 1.3). The importance of microtopographical cues is also found in embryonically derived cardiomyocytes, as beating rates increase, proliferation falls, and tension generation is increased [8].

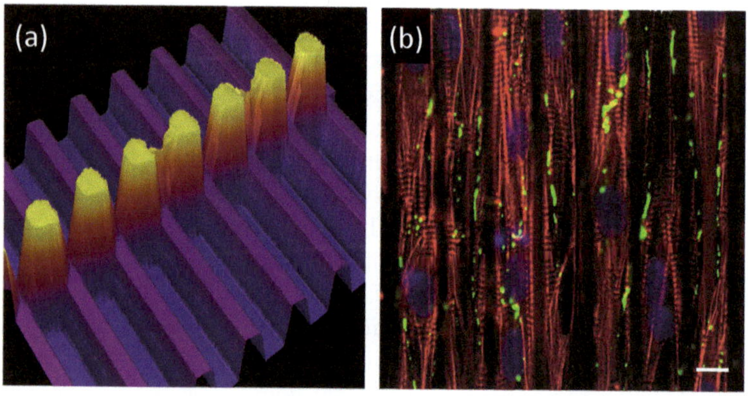


**Figure 1.3** Profile of microfabricated grooves with microposts (a) that facilitate alignment of actin fibers (b, red) of cardiomyocytes and physiologic connectivity through connexin 43 (b, green), (nuclei = blue in b). Scale bar = 10 μm. Reprinted with permission from Ref. [95].

Significant phenotypic changes occur in myocytes cultured with microstructures, or microrods, in 3D, including hypertrophy and increased spontaneous beating rates [27,29]. Microtopographical cues promote structural anisotropy and hypertrophy through a variety of mechanisms, including a critical dependence on the RhoA/ROCK family downstream of integrin and PKC activity [29]. Microarray gene ontology analyses of these cell cultures also highlights potential roles for components of signal transduction pathways and mediators of stimulus response. Genes affected by microrods include several proteins which interact with focal adhesion-associated proteins such as FAK and paxillin. Expression of both tyrosine and serine kinases known to bind FAK and paxillin and contribute to actin cytoskeleton reorganization are indicative of the widespread effects introduced microtopographical cues can induce [3,27,144]. Similarly, GTPase activating proteins that co-localize with FAK or the Z-discs and contribute to actin assembly and the response to overloading are indicated [12,27]. In all, the functional relationship of this gene subset reinforces the notion that any mechanisms which contribute to microrod-triggered myocyte hypertrophy are dependent on the inherent activity of focal adhesions and actin assembly. Focal adhesion kinase, specifically, has been identified as an early load-induced mediator of hypertrophy, as phosphorylation of the protein affects compensatory responses of cardiac myocytes [32,139].

### 1.3.3 Micromechanical Forces for Tissue Engineering

Applications of microtopographical manipulation of cardiomyocyte growth to tissue engineering are extremely promising. Current cell replacement and supportive therapies in the setting of infarction, for example, are limited by challenges presented by the pathologic scarred environment that may be too stiff for effective recruitment of cardiac progenitor cells. However, integrating proper microscale forces and micromechanical cues may enable the design of more appropriate niches for the establishment of healthy cardiac tissue [26,28]. Combining structure with proper biomechanical cues and supportive stiffnesses may lead to more effective cell engraftment and improved cardiac function. Understanding the interaction of all of these forces will be vital to the development of tissue engineering solutions for heart disease [21,81,104,128].

## 1.4 Stem Cells

### 1.4.1 Stem Cells and the Environment

The replication and differentiation potential of stem cells make them an extremely attractive subject of research in tissue engineering and regenerative medicine. Their function depends not only on chemical factors but also physical cues of the microenvironmental niche. The niche is defined by a combination of structural and cellular components that vary from one tissue to another to control proliferation and differentiation [130]. Traditionally, stem cell differentiation has been controlled through the use of biochemical and genetic mediators, including growth factors and transfected DNA or RNA elements, many of which are problematic for clinical applications. As a result, an increasing amount of research has focused on the role of micromechanical factors at the scale of local cellular microenvironments in regulating and modulating stem cell differentiation. Nonetheless, precisely how mechanical parameters in the physical microenvironment affect stem cell function is not yet well understood. Modulation of stem cell fate to better understand these mechanisms is now being approached by design of specialized substrates and the application of dynamic forces.

### 1.4.2 Substrate Stiffness

The physical interaction of stem cells with the surrounding substrate differentially controls lineage commitment. Seminal work by Engler et al. demonstrated that human MSCs respond to matrix stiffness on a physiologic scale, which has continued to be supported by other work in the field [38,40,153]. When cultured on softer surfaces, stem cells differentiate toward a neurogenic lineage, while those on rigid gels develop into osteogenic precursors, with intermediate stiffnesses leading to myogenic precursors. Control of matrix stiffness can also be used to maintain stem cells in a quiescent state. Soft, fibronectin-coated polyacrylamide gels have been shown to delay differentiation and instead promote quiescence of MSCs. This ability to maintain stem cells in an undifferentiated state may allow long term maintenance of stem cell pluripotency [149]. The interactions between cells and the matrix substrate are complex but it is clear that substrate stiffness can play a significant role in directing stem cell fate [47,119,146].

### 1.4.3 Microtopography

Although substrate stiffness can affect terminal differentiation of MSCs, topography and dimension also play a lesser role in directing cell differentiation [89]. Microfabrication techniques now allow precise control of microscale features, creating complex microtopographical substrates that can mimic the natural niches of stem cells. The environment surrounding a cell contains a complex variety of topographical features such as pores, ridges, and grooves at the micro- and nanoscale [77,88]. These features have been shown to have an influence on cell proliferation, apoptosis, and in the case of stem cells, differentiation [41,82].

Neural stem cells are particularly sensitive to the effects of substrate topography. Adult neural stem cells cultured on PDMS substrates patterned with microchannels show a channel-width-dependent decrease in differentiation and dendrite growth while also directing stem cell migration and dendrite orientation [7]. This level of topographical control could be critical for integrating stem cell technologies into therapeutic applications. Silicon substrates patterned with linear and circular micropatterns of varying densities have been shown to increase neuronal differentiation by 10% as compared to flat controls or smaller "dot"-patterned substrates. In contrast, astrocyte differentiation is reduced by 10–12% on linear and circular wafers while smaller dot patterns show no difference to unpatterned controls [111]. Furthermore, differentiation to oligodendrocytes is increased on all three types of patterned features compared to the control, suggesting that microstructural cues can preferentially induce differentiation towards specific neural lineages [111].

Mesenchymal stem cells have multipotent potential and can differentiate into osteogenic, adipogenic, and chondrogenic lineages, which can be influenced by topographical features [30]. For example, MSCs in engineered 3D constructs increase expression of SMA and cardiac markers [53,71,98]. Micropatterned hydrophilic titanium substrates induce the expression of nearly three times the level of markers for osteoblastic differentiation, such as RUNX2 and osteocalcin, suggesting that it may be advantageous to incorporate stem cells into implants with microtopographic features for therapeutic treatments [122]. The introduction of discrete, stiff microstructures controls local micromechanical cues directly, as described for fibroblasts and myocytes above. Stiff microrods with MSCs in 3D gel have increased proliferation and clustering, and instigated marked changes in gene expression leading to decreased osteogenic differentiation (Fig. 1.4) [23].

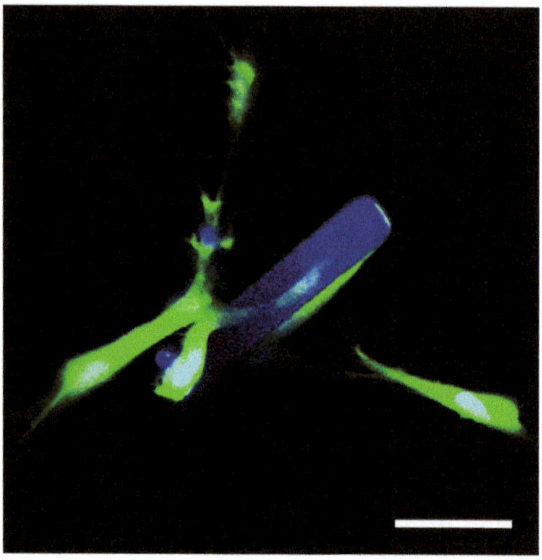

**Figure 1.4** Human mesenchymal stem cells (green: Calcein AM staining cell cytoplasm) clustering around a microrod (Blue: Hoechst staining nuclei and also SU-8 microstructures). Scale bar = 50 μm. Reprinted with permission from Ref. [23].

The mechanisms underlying micromechanical regulation of lineage commitment are not yet understood. When MSCs are cultured on adhesive islands under conditions promoting both adiopogenesis and osteogenesis, cells in the interior of the features express adipogenic lineage specific PPARγ while those at the edges stain positive for osteogeneic cbfa-1/runx2, which has led to the suggestion that this is due to gradients from the higher stresses at the edge of the cell aggregates to the lower interior stresses. Similar segregation of differentiation potential from the outside to the center is seen in three-dimensional MSC cell clusters [116].

### 1.4.4 Applied Micromechanical Forces

Significant interest has also focused on the role of mechanotransduction on stem cell behavior by using forces such as uniaxial stress, tension, and pressure to induce and modulate lineage commitment [94]. Strain amplitude has been varied from 1% to 20%, with the high end having the maximum effect on various stem cells [46,48,129]. Markers considered to be specific for vessels or early cardiomyocytes, such as "αSMA," increase in a strain-dependent manner for MSCs [48,76]. In particular, the morphology, proliferation, and differentiation of MSCs are affected by mechanical stimuli [46,71, 98,148]. Also, embedded embryonic stem cells (ESCs) strained at 2 and 3 Hz increase other cardiac markers [56], while those at the low frequency of 0.2 Hz at 10% strain reduced differentiation of human ESCs (hESCs), maintaining an undifferentiated state [120].

Cyclic strain regulates differential fates on MSCs when applied equiaxially versus uniaxially. While equiaxial cyclic strain results in decreased expression of smooth muscle markers, uniaxial cyclic strain, such as that exerted by blood flow through the vasculature, leads to transient upregulation of α-actin and SM-22α [106]. The effects of mechanotransduction depend on the orientation of cells with respect to the strain axis as well [86]. Cyclic strain of cultured MSCs induces myofibroblast-like phenotypes, which may explain the benefits found in repair and matrix remodeling after injury [123]. This behavior mirrors that seen in fibroblasts and cardiomyocytes, as discussed above, highlighting the shared mechanisms of mechanical response across disparate cell populations. For example, after 2 days of stretch (20% strain at 1 Hz), more f-actin labeled fibroblasts with random orientation were found in environments of cyclic biaxial strain compared to anisotropically stimulated or statically anchored controls, emphasizing the importance of mechanical forces in cell phenotype [51].

### 1.4.5 Stem Cells and Focal Adhesions

Focal adhesion formation can be induced through application of external force by strain and microtopography [45,113,124,127]. Mediators such as RhoA/ROCK and FAK regulate mechanical strain-induced actin fiber formation and cytoskeleton reorganization of MSCs [152]. The focal adhesion complex is the anchor for the cytoskeletal organization of cells plated on microposts (Fig. 1.5), and depends on both FAK and the contractile forces generated by myosin II [44,50,136]. Meanwhile, RhoA regulates actin polymerization in topography-induced focal adhesion formation [127]. Combining cyclic strain with microtopographical cues results in a synergistic effect on human MSC cytoskeletal remodeling and interaction with the environment. Although the effects of cyclic strain appear to be dominant, the inclusion of microtopographical cues results in increases in the expression of MMP13 and focal adhesion associated proteins such as paxillin and vinculin [34].

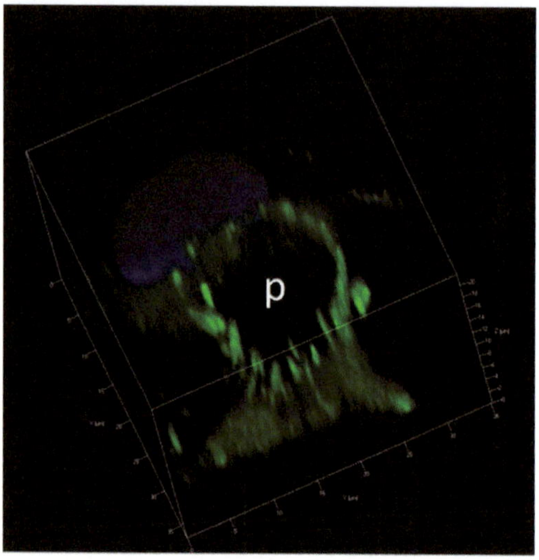


**Figure 1.5** Mesenchymal stem cell interaction with a PDMS micropost (p) (green: paxillin; blue: DAPI staining nucleus). Reprinted with permission from Ref. [34].

### 1.4.6 Proliferation

Proliferation varies with stem cell type, cell culture, and topography. Human MSC proliferation increased in culture when microstructures were added to 3D gel [23], whereas division was inhibited in mouse ESCs near a post [8]. Mechanical strain also affects cell division and depends on the duration, frequency, amplitude and direction of strain. Proliferation of MSCs increases with 10% uniaxial strain at either 1 Hz after 4 h or 0.26 Hz after 3 days [48,71]. Uniaxial strain in concert with microgrooved topography increased MSC proliferation while imposing cell alignment parallel to the strain. However, the importance of the orientation of anisotropy was demonstrated when MSC proliferation was not affected by aligned micropatterns that were perpendicular to the axis of mechanical strain [86].

### 1.4.7 Stem Cells in the Cardiac Microenvironment

Stem cell treatments for myocardial infarction have previously been challenged by low efficiency, low retention and integration, and inconsistent differentiation. Micromechanical factors may be able to influence how stem cells interact with the local cardiac tissue, allowing researchers to dictate cell fate to regenerate healthy tissue [52].

The cardiac ECM is generated by the rich population of cardiac fibroblasts and is composed of a variety of proteins, including collagens type I, III, and VI, fibronectin, and vitronectin providing both mechanical and biochemical cues capable of directing stem cell differentiation [54,91]. Human ESCs cultured on gels with higher ratios of native decellularized ECM show greater expression of cardiac troponin T as compared to gels with lower ECM ratios, suggesting the makeup and structure of the native ECM plays a role in encouraging cardiogenic differentiation [36]. Conversely, infarcted tissues demonstrate limited potential to instruct stem cells to differentiate into cardiomyocytes, perhaps due to the altered extracellular composition and stiffness. Indeed, half as many differentiated cardiomyocytes were found in the infarcted tissue compared to hESCs injected into healthy cardiac tissue [20]. Embryonic cardiomyocytes cultured on stiff substrates, similar to post-infarct fibrotic tissues, are not able to maintain beating and overstrain themselves while soft substrates allow cells to beat, but less efficiently [39].

Microtopographical cues have also been shown to influence cardiogenic differentiation of mouse ESCs. Micropost arrays can attenuate proliferation of murine ESCs in a manner that is reversed by inhibiting Rho kinase and myosin light chain kinase—proteins that play a role in maintaining stress fibers in the cytoskeleton. These microstructural cues interact with tension generating mechanotransductive networks of stem cells, affecting their ultimate differentiation and proliferation potential. Cardiomyocytes derived from ESCs and cultured on these substrates also show attenuated proliferation but with an increase in beating rate, further indicating that microtopographical elements affect cell differentiation and potential through tension generating networks [8]. Fluid flow and hydrostatic pressure on stem cells has also proven to be a potent differentiating force. Titration of pulsatile, radial, and shear forces induces adipose-derived stem cells into chondrogenic lineages [110] or MSCs into smooth muscle cell types [102]. Similarly, aortic stenosis in humans causes cardiac hypertrophy and enhancement of the differentiation of cardiac stem cells, suggesting that these endogenous stem cells amplify and commit to the myocyte lineage in response to an increased workload [141]. Improved cardiomyocyte differentiation of murine ESCs has also been shown by combining embryoid body formation with shear stress in a rotary orbital shaker resulting in greater expression of mesodermal and cardiac differentiation markers [122]. It is clear that further understanding of how various types of structural cues and forces influence stem cell development and differentiation potential will be invaluable to bridge to effective stem cell therapies.

## 1.5 Summary and Conclusions

Micromechanical forces regulate tissue structure, healthy function, and disease. Developing a stronger understanding of how these cues integrate with different cell types will become increasingly important as tissue engineering solutions become more prevalent. A variety of forces, environmental factors, and topographical features contribute to the overall micromechanical environment that ultimately influences cell morphology and function and an appreciation of each of these factors is necessary to fully understand the intricate and dynamic interplay between cells and their environment. The last decade has seen an accelerating interest in the effect of environmental forces on mechanically sensitive cells such as fibroblasts, cardiomyocytes, and more recently, stem cells. Deeper insight into how micromechanical cues, such as stiffness, topography, and applied forces, trigger mechanotransduction and downstream signaling cascades is needed. These may pave the way for improved design and implementation of materials-based therapies and tissue-engineered applications.

### Acknowledgments

The authors would like to acknowledge funding support from the NIH (HL 62426, HL 090523). The work was also made possible by a grant from the California Institute for Regenerative Medicine (grant number TG2-01153). The contents of this publication are solely the responsibility of the authors and do not necessarily represent the official views of CIRM or any other agency of the State of California.

## References

1. Al-Haque, S., Miklas, J. W., Feric, N., Chiu, L. L., Chen, W. L., Simmons, C. A., Radisic, M. (2012). Hydrogel substrate stiffness and topography interact to induce contact guidance in cardiac fibroblasts. *Macromol. Biosci.*, **12**(10), 1342–1353.
2. Asazuma-Nakamura., Y., Dai, P., Harada, Y., Jiang, Y., Hamaoka, K., Takamatsu, T. (2009). Cx43 contributes to TGF-beta signaling to regulate differentiation of cardiac fibroblasts into myofibroblasts. *Exp. Cell Res.*, **315**(7), 1190–1199.
3. Avraham, H., Avraham, S., Taniguchi, Y. (2000). Receptor protein tyrosine phosphatases in hematopoietic cells. *J. Hematother. Stem Cell Res.*, **9**(4), 425–432.
4. Ayala, P., Lopez, J., Desai, T. (2010). Microtopographical cues in 3D attenuate fibrotic phenotype and extracellular matrix deposition: Implications for tissue regeneration. *Tissue Eng. A.*, **16**(8), 2519–2527.
5. Ayala, P., Desai, T. A. (2011). Integrin alpha3 blockade enhances microtopographical down-regulation of alpha-smooth muscle actin: Role of microtopography in ECM regulation. *Intgr. Biol. (Camb)*., **3**(7), 733–741.
6. Baker, D. W., Liu, X., Weng, H., Luo, C., Tang, L. (2011). Fibroblast/fibrocyte: Surface interaction dictates tissue reactions to micropillar implants. *Biomacromolecules*, **12**(4), 997–1005.
7. Béduer, A., Vieu, C., Arnauduc, F., Sol, J. C., Loubinoux, I., Vaysse, L. (2012). Engineering of adult human neural stem cells differentiation through surface micropatterning. *Biomaterials*, **33**(2), 504–514.
8. Biehl, J. K., Yamanaka, S., Desai, T. A., Boheler, K. R., Russell, B. (2009). Proliferation of mouse embryonic stem cell progeny and the spontaneous contractile activity of cardiomyocytes are affected by microtopography. *Dev. Dyn.* ,**238**(8), 1964–1973.
9. Bloch, R. J., Gonzalez-Serratos, H. (2003). Lateral force transmission across costameres in skeletal muscle. *Exerc. Sport Sci. Rev.*, **31**(2), 73–78.
10. Boateng, S. Y., Hartman, T. J., Ahluwalia, N., Vidula, H., Desai, T. A., Russell, B. (2003). Inhibition of fibroblast proliferation in cardiac myocyte cultures by surface microtopography. *Am. J. Physiol. Cell Physiol.*, **285**(1), C171–C182.

11. Boheler, K. R., Czyz, J., Tweedie, D., Yang, H. T., Anisimov, S. V., Wobus, A. M. (2002). Differentiation of pluripotent embryonic stem cells into cardiomyocytes. *Circ. Res.*, **91**(3), 189–201.
12. Borges, L., Bigarella, C. L., Baratti, M. O., Crosara-Alberto, D. P., Joazeiro, P. P., Franchini, K. G., Costa, F. F., Saad, S. T. (2008). ARHGAP21 associates with FAK and PKCzeta and is redistributed after cardiac pressure overload. *Biochem. Biophys. Res. Commun.*, **374**(4), 641–646.
13. Brown, R. D., Ambler, S. K., Mitchell, M. D., Long, C. S. (2005). The cardiac fibroblast: Therapeutic target in myocardial remodeling and failure. *Ann. Rev. Pharmacol. Toxicol.*, **45**(1), 657–687.
14. Bryant, J. E., Shamhart, P. E., Luther, D. J., Olson, E. R., Koshy, J. C., Costic, D. J., Mohile, M. V., Dockry, M., Doane, K. J., Meszaros, J. G. (2009). Cardiac myofibroblast differentiation is attenuated by alpha(3) integrin blockade: Potential role in post-MI remodeling. *J. Mol. Cell. Cardiol.*, **46**(2), 186–192.
15. Brydone, A. S., Dalby, M. J., Berry, C. C., Dominic Meek, R. M., McNamara, L. E. (2011). Grooved surface topography alters matrix-metalloproteinase production by human fibroblasts. *Biomed. Mater.*, **6**(3), 035005.
16. Camelliti, P., Borg, T. K., Kohl, P. (2005). Structural and functional characterisation of cardiac fibroblasts. *Cardiovasc. Res.*, **65**(1), 40–51.
17. Camelliti, P., Green, C. R., Kohl, P. (2006). Structural and functional coupling of cardiac myocytes and fibroblasts. *Adv. Cardiol.*, **42**, 132–149.
18. Chang, S. S., Guo, W. H., Kim, Y., Wang, Y. L. (2013). Guidance of cell migration by substrate dimension. *Biophys. J.*, **104**(2), 313–321.
19. Chen, K. D., Li, Y. S., Kim, M., Li, S., Yuan, S., Chien, S., Shyy, J. Y. (1999). Mechanotransduction in response to shear stress: Roles of receptor tyrosine kinases, integrins, and Shc. *J. Biol. Chem.*, **274**, 18393–18400.
20. Chen, Y. R., Li, Y., Chen, L., Yang, X. C., Su, P. X., Cai, J. (2011). The infarcted cardiac microenvironment cannot selectively promote embryonic stem cell differentiation into cardiomyocytes. *Cardiovasc. Pathol.*, **20**(2), 77–83.
21. Christman, K. L., Fok, H. H., Sievers, R. E., Fang, Q., Lee, R. J. (2004). Fibrin glue alone and skeletal myoblasts in a fibrin scaffold preserve cardiac function after myocardial infarction. *Tissue Eng.*, **10**(3–4), 403–409.
22. Christman, K. L., Vardanian, A. J., Fang, Q., Sievers, R. E., Folk, H. H., Lee, R. J. (2004). Injectable fibrin scaffold improves cell transplant survival, reduces infarct size, and induces neovasculature formation in ischemic myocardium. *J. Am. Coll. Cardiol.*, **44**(3), 654–660.

23. Collins, J. M., Ayala, P., Desai, T. A., Russell, B. (2010). Three Dimensional culture with stiff microstructures increases proliferation and slows osteogenic differentiation of human mesenchymal stem cells. *Small*, **6**(3), 355–360.

24. Collinsworth, A. M., Zhang, S., Kraus, W. E., Truskey, G. A. (2002). Apparent elastic modulus and hysteresis of skeletal muscle cells throughout differentiation. *Am. J. Physiol. Cell Physiol.*, **283**(4), C1219–C1227.

25. Cortese, B., Riehle, M. O., D'Amone, S., Gigli, G. (2013). Influence of variable substrate geometry on wettability and cellular responses. *J. Colloid Interface Sci.*, **394**, 582–589.

26. Curtis, M. W., Russell, B. (2009). Cardiac tissue engineering. *J. Cardiovasc. Nurs.*, **24**(2), 87–92.

27. Curtis, M. W., Sharma, S., Desai, T. A. (2010). Hypertrophy, gene expression, and beating of neonatal cardiac myocytes are affected by microdomain heterogeneity in 3D. *Biomed. Microdevices*, **12**(6), 1073–1085.

28. Curtis, M. W., Russell, B. (2011). Micromechanical regulation in cardiac myocytes and fibroblasts: Implications for tissue remodeling. *Pflugers Archiv. Eur. J. Physiol.*, **462**(1), 105–117.

29. Curtis, M. W., Budyn, E., Desai, T. A., Samarel, A., Russell, B. (2013). Microdomain heterogeneity in 3D affects the mechanics of neonatal cardiac myocyte contraction. *Biomech. Model Mechanobiol.*, **12**, 95–109.

30. Delaine-Smith, R. M., Reilly, G. C. (2012). Mesenchymal stem cell responses to mechanical stimuli. *Muscles Ligaments Tendons J.*, **2**(3), 169–180.

31. Deutsch, J., Motlagh, D., Russell, B., Desai, T. A. (2000). Fabrication of microtextured membranes for cardiac myocyte attachment and orientation. *J. Biomed. Mater. Res.* (*Appl. Biomater*)., **53**(3), 267–275.

32. DiMichele, L. A., Doherty, J. T., Rojas, M., Beggs, H. E., Reichardt, L. F., Mack, C. P., Taylor, J. M. (2006). Myocyte-restricted focal adhesion kinase deletion attenuates pressure overload-induced hypertrophy. *Circ. Res.*, **99**(6), 636–645.

33. Dobaczewski, M., Bujak, M., Li, N., Gonzalez-Quesada, C., Mendoza, L. H., Wang, X. F., Frangogiannis, N. G. (2010). Smad3 signaling critically regulates fibroblast phenotype and function in healing myocardial infarction. *Circ. Res.*, **107**(3), 418–428.

34. Doroudian, G., Curtis, M. W., Gang, A., Russell, B. (2013). Cyclic strain dominates over microtopography in regulating cytoskeletal and focal adhesion remodeling of human mesenchymal stem cells. *Biochem. Biophys. Res. Commun.*, **430**(3), 1040–1046.

35. Doyle, A. D., Kutys, M. L., Conti, M. A., Matsumoto, K., Adelstein, R. S., Yamada, K. M. (2012). Micro-environmental control of cell migration-myosin IIA is required for efficient migration in fibrillar environments through control of cell adhesion dynamics. *J. Cell Sci.*, **125**(Pt 9), 2244–2256

36. Duan, Y., Liu, Z., O'Neill, J., Wan, L. Q., Freytes, D. O., Vunjak-Novakovic, G. (2011). Hybrid gel composed of native heart matrix and collagen induces cardiac differentiation of human embryonic stem cells without supplemental growth factors. *J. Cardiovasc. Trans. Res.*, **4**(5), 605–615.

37. Dugina, V., Fontao, L., Chaponnier, C., Vasiliev, J., Gabbiani, G. (2001). Focal adhesion features during myofibroblastic differentiation are controlled by intracellular and extracellular factors. *J. Cell Sci.* **114**(Pt 18), 3285–3296.

38. Engler, A. J., Sen, S., Sweeney, H. L., Discher, D. E. (2006). Matrix elasticity directs stem cell lineage specification. *Cell*, **126**(4), 677–689

39. Engler, A. J., Carag-Krieger, C., Johnson, C. P., Raab, M., Tang, H. Y., Speicher, D. W., Sanger, J. W., Sanger, J. M., Discher, D. E. (2008). Embryonic cardiomyocytes beat best on a matrix with heart-like elasticity: Scar-like rigidity inhibits beating. *J. Cell Sci.*, **121**(Pt 22), 3794–3802.

40. Eroshenko, N., Ramachandran, R., Yadavalli, V. K., Rao, R. R. (2013). Effect of substrate stiffness on early human embryonic stem cell differentiation. *J. Biol. Eng.*, **7**(1), 7.

41. Even-Ram, S., Artym, V., Yamada, K. M. (2006). Matrix control of stem cell fate. *Cell*, **126**(4), 645–647.

42. Fink, C., Ergun, S., Kralisch, D., Remmers, U., Weil, J., Eschenhagen, T. (2000). Chronic stretch of engineered heart tissue induces hypertrophy and functional improvement. *FASEB J.*, **14**(5), 669–679.

43. Forte, G., Pagliari, S., Ebara, M., Uto, K., Tam, J. K., Romanazzo, S., Escobedo-Lucea, C., Romano, E., Di Nardo, P., Traversa, E., Aoyagi, T. (2012). Substrate stiffness modulates gene expression and phenotype in neonatal cardiomyocytes in vitro. *Tissue Eng. Part A.*, **18**(17–18), 1837–1848.

44. Frey, M. T., Tsai, I. Y., Russell, T. P., Hanks, S. K., Wang, Y. L. (2006). Cellular responses to substrate topography: Role of myosin II and focal adhesion kinase. *Biophys. J.*, **90**(10), 3774–3782.

45. Galbraith, C. G., Yamada, K. M., Sheetz, M. P. (2002). The relationship between force and focal complex development. *J. Cell Biol.*, **159**(4), 695–705.

46. Ge, D., Liu, X., Li, L., Wu, J., Tu, Q., Shi, Y., Chen, H. (2009). Chemical and physical stimuli induce cardiomyocyte differentiation from stem cells. *Biochem. Biophys. Res. Commun.*, **381**(3), 317–321.

47. Georges, P. C., Janmey, P. A. (2005). Cell type-specific response to growth on soft materials. *J. Appl. Physiol.*, **98**(4), 1547–1553.

48. Ghazanfari, S., Shadpour, M. T., Shokrgozar, M. A. (2009). Effects of cyclic stretch on proliferation of mesenchymal stem cells and their differentiation to smooth muscle cells. *Biochem. Biophys. Res. Commun.*, **388**(3), 601–605.

49. Ghibaudo, M., Trichet, L., Le Digabel, J., Richert, A., Hersen, P., Ladoux, B. (2009). Substrate topography induces a crossover from 2D to 3D behavior in fibroblast migration. *Biophys. J.*, **97**(1), 357–368.

50. Ghibaudo, M. J., Di Meglio, M., Hersen, P., Ladoux, B. (2011). Mechanics of cell spreading within 3D-micropatterned environments. *Lab Chip.*, **11**(5), 805–812.

51. Gould, R. A., Chin, K., Santisakultarm, T. P., Dropkin, A., Richards, J. M., Schaffer, C. B., Butcher, J. T. (2012). Cyclic strain anisotropy regulates valvular interstitial cell phenotype and tissue remodeling in three-dimensional culture. *Acta Biomater.*, **8**(5), 1710–1719.

52. Gu, Y., Yu, J., Lum, L. G., Lee, R. J. (2007). Tissue engineering and stem cell therapy for myocardial repair. *Front. Biosci.*, **12**, 5157–5165.

53. Guan, J., Wang, F., Li, Z., Chen, J., Guo, X., Liao, J., Moldovan, N. I. (2011). The stimulation of the cardiac differentiation of mesenchymal stem cells in tissue constructs that mimic myocardium structure and biomechanics. *Biomaterials*, **32**(24), 5568–5580.

54. Guilak, F., Cohen, D. M., Estes, B. T., Gimble, J. M., Liedtke, W., Chen, C. S. (2009). Control of stem cell fate by physical interactions with the extracellular matrix. *Cell Stem Cell*, **5**(1), 17–26.

55. Guillemette, M. D., Cui, B., Roy, E., Gauvin, R., Giasson, C. J., Esch, M. B., Carrier, P., Deschambeault, A., Dumoulin, M., Toner, M., Germain, L., Veres, T., Auger, F. A. (2009). Surface topography induces 3D self-orientation of cells and extracellular matrix resulting in improved tissue function. *Integr. Biol.* (*Camb*)., **1**(2), 196–204.

56. Guo, X. M., Zhao, Y. S. Chang, H. X., Wang, C. Y., Zhang, X. A., Duan, C. M., Dong, L. Z., Jiang, H., Li, J., Song, Y., Yang, X. J. (2006). Creation of engineered cardiac tissue in vitro from mouse embryonic stem cells. *Circulation*, **113**(18), 2229–2237.

57. Gwak, S. J., Bhang, S. H., Kim, I. K., Kim, S. S., Cho, S. W., Jeon, O., Yoo, K. J., Putnam, A. J., Kim, B. S. (2008). The effect of cyclic strain on embryonic stem cell-derived cardiomyocytes. *Biomaterials*, **29**(7), 844–856.

58. Hinz, B., Dugina, V., Ballestrem, C., Wehrle-Haller, B., Chaponnier, C. (2003). Alpha-smooth muscle actin is crucial for focal adhesion maturation in myofibroblasts. *Mol. Biol. Cell*, **14**(6), 2508–2519.

59. Hinz, B. (2007). Formation and function of the myofibroblast during tissue repair. *J. Invest. Dermatol.*, **127**(3), 526–537.

60. Hinz, B. (2010). The myofibroblast: Paradigm for a mechanically active cell. *J. Biomechanics*, **43**(1), 146–155.

61. Holle, A. W., Engler, A. J. (2011). More than a feeling: Discovering, understanding, and influencing mechanosensing pathways. *Curr. Opin. Biotechnol.*, **22**(5), 648–654.

62. Huang, N. F., Yu, J., Sievers, R., Li, S., Lee, R. J. (2005). Injectable biopolymers enhance angiogenesis after myocardial infarction. *Tissue Eng.*, **11**(11/12), 1860–1866.

63. Ieda, M., Fu, J.-D., Delgado-Olguin, P., Vedantham, V., Bruneau, B. G., Srivastava, D. (2010). Direct reprogramming of fibroblasts into functional cardiomyocytes by defined factors. *Cell*, **142**(3), 375–386.

64. Ingber, D. E. (1997). Tensegrity: The architectural basis of cellular mechanotransduction. *Annu Rev. Physiol.*, **59**, 575–599.

65. Ingber, D. E. (2003). Mechanobiology and diseases of mechanotransduction. *Ann. Med.*, **35**(8), 564–577.

66. Ingber, D. E. (2003) Tensegrity II. How structural networks influence cellular information processing networks. *J. Cell Sci.*, **116**(Pt 8), 1397–1408.

67. Ingber, D. E. (2006). Cellular mechanotransduction: Putting all the pieces together again. *FASEB J.*, **20**(7), 811–827.

68. Jacot, J. G., Kita-Matsuo, H., Wei, K. A., Chen, H. S., Omens, J. H., Mercola, M., McCulloch, A. D. (2010). Cardiac myocyte force development during differentiation and maturation. *Ann. N. Y. Acad. Sci.*, **1188**, 121–127.

69. Jacot, J. G., Martin, J. C., Hunt, D. L. (2010). Mechanobiology of Cardiomyocyte Development. *J. Biomech.*, **43**(1), 93.

70. James, R., Toti, U. S., Laurencin, C. T., Kumbar, S. G. (2011). Electrospun nanofibrous scaffolds for engineering soft connective tissues. *Methods Mol. Biol.*, **726**, 243–258.

71. Jang, J. Y., Lee, S. W., Park, S. H., Shin, J. W., Mun, C., Kim, S. H., Kim, D. H., Shin, J. W. (2011). Combined effects of surface morphology and mechanical straining magnitudes on the differentiation of mesenchymal stem cells without using biochemical reagents. *J. Biomed. Biotechnol.*, **2011**, 860652.

72. Jawad, H., Lyon, A. R., Harding, S. E., Ali, N. N., Boccaccini, A. R. (2008). Myocardial tissue engineering. *Br. Med. Bull.*, **87**, 31–47.

73. Jiang, Y., Lu, S., Zeng, Y. (2011). Dermal fibroblast behaviour on micropatterned substrates with different pattern geometries. *J. Tissue Eng. Regen. Med.*, **5**(5), 402–409.

74. Johnston, P. V., Sasano, T., Mills, K., Evers, R., Lee, S.-T., Smit, R. R., Lardo, A. C., Lai, S., Steenbergen, C., Gerstenblith, G., Lange, R., Marbán, E. (2009). Engraftment, differentiation, and functional benefits of autologous cardiosphere-derived cells in porcine ischemic cariomyopathy. *Circulation*, **120**(12), 1075–1083.

75. Katsumi, A., Orr, A. W., Tzima, E., Schwartz, M. A. (2004). Integrins in mechanotransduction. *J. Biol. Chem.*, **279**(13), 12001–12004.

76. Kearney, E. M., Prendergast, P. J., Campbell, V. A. (2008). Mechanisms of strain-mediated mesenchymal stem cell apoptosis. *J. Biomech. Eng.*, **130**(6), 061004.

77. Keung, A. J., Kumar, S., Schaffer, D. V. (2010). Presentation counts: Microenvironmental regulation of stem cells by biophysical and material cues. *Annu. Rev. Cell. Dev. Biol.*, **26**, 533–556.

78. Kiang, J. D., Wen, J. H., Del Álamo, J. C., Engler, A. J. (2013). Dynamic and reversible surface topography influences cell morphology. *J. Biomed. Mater. Res. A.*, **101**(8), 2313–2321.

79. Kim, D. H., Han, K., Gupta, K., Kwon, K. W., Suh, K. Y., Levchenko, A. (2009). Mechanosensitivity of fibroblast cell shape and movement to anisotropic substratum topography gradients. *Biomaterials*, **30**(29), 5433–5444.

80. Krieg, M., Arboleda-Estudillo, Y., Puech, P. H., Käfer, J., Graner, F., Müller, D. J., Heisenberg, C. P. (2008). Tensile forces govern germ-layer organization in zebrafish. *Nat. Cell Biol.*, **10**(4), 429–436.

81. Krupnick, A. S., Kreisel, D., Engels, F. H., Szeto, W. Y., Plappert, T., Popma, S. H., Flake, A. W., Rosengard, B. R. (2002). A novel small animal model of left ventricular tissue engineering. *J. Heart Lung Transpl.*, **21**(2), 233–243.

82. Kshitiz, K., Park, J., Kim, P., Helen, W., Engler, A. J., Levchenko, A., Kim, D. H. (2012). Control of stem cell fate and function by engineering physical microenvironments. *Integr. Biol.* (*Camb*)., **4**(9), 1008–1018.

83. Kumar, A., Chaudhry, I., Reid, M. B., Boriek, A. M. (2002). Distinct signaling pathways are activated in response to mechanical stress applied axially and transversely to skeletal muscle fibers. *J. Biol. Chem.*, **277**(48), 46493–46503.

84. Kuo, C. H., Xian, J., Brenton, J. D., Franze, K., Sivaniah, E. (2012). Complex stiffness gradient substrates for studying mechanotactic cell migration. *Adv. Mater.*, **24**(45), 6059–6064.

85. Kuppan, P., Vasanthan, K. S., Sundaramurthi, D., Krishnan, U. M., Sethuraman, S. (2011). Development of poly(3-hydroxybutyrate-co-3-hydroxyvalerate) fibers for skin tissue engineering: Effects of topography, mechanical, and chemical stimuli. *Biomacromolecules*, **12**(9), 3156–3165.

86. Kurpinski, K., Chu, J., Hashi, C., Li, S. (2006). Anisotropic mechanosensing by mesenchymal stem cells. *Proc. Natl. Acad. Sci. U. S. A.*, **103**(44), 16095–16100.

87. Lele, T. P., Thodeti, C. K., Ingber, D. E. (2006). Force meets chemistry: Analysis of mechanochemical conversion in focal adhesions using fluorescence recovery after photobleaching. *J. Cell Biochem.*, **97**(6), 1175–1183.

88. Li, D., Zhou, J., Chowdhury, F., Cheng, J., Wang, N., Wang, F. (2011). Role of mechanical factors in fate decisions of stem cells. *Regen. Med.*, **6**(2), 229–240.

89. Li, Z., Gong, Y., Sun, S., Du, Y., Lü, D., Liu, X., Long, M. (2013). Differential regulation of stiffness, topography, and dimension of substrates in rat mesenchymal stem cells. *Biomaterials*, **34**(31), 7616–7625.

90. Lin, Y. H., Li, J., Swanson, E. R., Russell, B. (2013). CapZ and actin capping dynamics increase in myocytes after a bout of exercise and abates in hours after stimulation ends. *J. Appl. Physiol.*, **114**(11), 1603–1609.

91. MacKenna, D., Summerour, S. R., Villarreal, F. J. (2000). Role of mechanical factors in modulating cardiac fibroblast function and extracellular matrix synthesis. *Cardiovasc. Res.*, **46**(2), 257–263.

92. Mammoto, A., Ingber, D. E. (2009). Cytoskeletal control of growth and cell fate switching. *Curr. Opin. Cell Biol.*, **21**(6), 864–870.

93. Martinac, B. (2004). Mechanosensitive ion channels: Molecules of mechanotransduction. *J. Cell Sci.*, **117**(Pt 12), 2449–2460.

94. Maul, T. M., Chew, D. W., Nieponice, A., Vorp, D. A. (2011). Mechanical stimuli differentially control stem cell behavior: Morphology, proliferation, and differentiation. *Biomech. Model Mechanobiol.*, **10**(6), 939–953.

95. Motlagh, D., Hartman, T. J., Desai, T. A., Russell, B. (2003). Microfabricated grooves recapitulate neonatal myocyte connexin43 and N-cadherin expression and localization. *J. Biomed. Mater. Res. Part A.*, **67**(1), 148–157.

96. Motlagh, D., Senyo, S. E., Desai, T. A., Russell, B. (2003). Microtexture substrata alter gene expression, protein localization and the shape of cardiac myocytes. *Biomaterials*, **24**, 2463–2476.

97. Myrna, K. E., Mendonsa, R., Russell, P., Pot, S. A., Liliensiek, S. J., Jester, J. V., Nealey, P. F., Brown, D., Murphy, C. J. (2012). Substratum topography modulates corneal fibroblast to myofibroblast transformation. *Invest. Ophthalmol. Vis. Sci.*, **53**(2), 811–816.

98. Nieponice, A., Maul, T. M., Cumer, J. M., Soletti, L., Vorp, D. A. (2007). Mechanical stimulation induces morphological and phenotypic changes in bone marrow-derived progenitor cells within a three-dimensional fibrin matrix. *J. Biomed. Mater. Res. A.*, **81**(3), 523–530.

99. Norman, J. J., Desai, T. A. (2005). Control of cellular organization in three dimensions using a microfabricated polydimethylsiloxane-collagen composite tissue scaffold. *Tissue Eng.*, **11**(3–4), 378–386.

100. Norman, J. J., Desai, T. A. (2006). Methods for fabrication of nanoscale topography for tissue engineering scaffolds. *Ann. Biomed. Eng.*, **34**(1), 89–101.

101. Norman, J. J., Collins, J. M., Sharma, S., Russell, B., Desai, T. A. (2008). Microstructures in 3D biological gels affect cell proliferation. *Tissue Eng. Part A.*, **14**(3), 379–390.

102. O'Cearbhaill, E. D., Punchard, M. A., Murphy, M., Barry, F. P., Mchugh, P. E., Barron, V. (2008). Response of mesenchymal stem cells to the biomechanical environment of the endothelium on a flexible tubular silicone substrate. *Biomaterials*, **29**(11), 1610–1619.

103. Olivares-Navarrete, R., Hyzy, S. L., Hutton, D. L., Erdman, C. P., Wieland, M., Boyan, B. D., Schwartz, Z. (2010). Direct and indirect effects of microstructured titanium substrates on the induction of mesenchymal stem cell differentiation towards the osteoblast lineage. *Biomaterials*, **31**(10), 2728–2735.

104. Ott, H. C., Matthiesen, T. S., Goh, S. K. Black, L. D., Kren, S. M., Netoff, T. I., Taylor, D. A. (2008). Perfusion-decellularized matrix: Using nature's platform to engineer a bioartificial heart. *Nat. Med.*, **14**(2), 213–221.

105. Pan, J., Fukuda, K., Saito, M., Matsuzaki, J., Kodama, H., Sano, M., Takahashi, T., Kato, T., Ogawa, S. (1999). Mechanical stretch activates the JAK/STAT pathway in rat cardiomyocytes. *Circ. Res.*, **84**, 1127–1136.

106. Park, J. S., Chu, J. S., Cheng, C., Chen, F., Chen, D., Li, S. (2004). Differential effects of equiaxial and uniaxial strain on mesenchymal stem cells. *Biotechnol. Bioeng.*, **88**(3), 359–368.

107. Patel, A. A., Thakar, R. G., Chown, M., Ayala, P., Desai, T. A., Kumar, S. (2010). Biophysical mechanisms of single-cell interactions with microtopographical cues. *Biomed. Microdevices.*, **12**(2), 287–296.

108. Patel, A. A., Desai, T. A., Kumar, S. (2011). Microtopographical assembly of cardiomyocytes. *Integr. Biol. (Camb).*, **3**(10), 1011–1019.

109. Pot, S. A., Liliensiek, S. J., Myrna, K. E., Bentley, E., Jester, J. V., Nealey, P. F., Murphy, C. J. (2010). Nanoscale topography-induced modulation of fundamental cell behaviors of rabbit corneal keratocytes, fibroblasts, and myofibroblasts. *Invest. Ophthalmol. Vis. Sci.*, **51**(3), 1373–1381.

110. Puetzer, J., Williams, J., Gillies, A., Bernacki, S., Loboa, E. G. (2013). The effects of cyclic hydrostatic pressure on chondrogenesis and viability of human adipose- and bone marrow-derived mesenchymal stem cells in three-dimensional agarose constructs. *Tissue Eng. Part A.*, **19**(1–2), 299–306.

111. Qi, L., Li, N., Huang, R., Song, Q., Wang, L., Zhang, Q, Su, R., Kong, T., Tang, M., Cheng, G. (2013). The effects of topographical patterns and sizes on neural stem cell behavior. *PLoS ONE.*, **8**(3), e59022.

112. Rehfeldt, F., Engler, A. J., Eckhardt, A., Ahmed, F., Discher, D. E. (2007). Cell responses to the mechanochemical microenvironement: Implications for regenerative medicine and drug delivery. *Adv. Drug Deliv. Rev.*, **59**(13), 1329–1339.

113. Riveline, D., Zamir, E., Balaban, N. Q., Schwarz, U. S., Ishizaki, T., Narumiya, S., Kam, Z., Geiger, B., Bershadsky, A. D. (2001). Focal contacts as mechanosensors: Externally applied local mechanical force induces growth of focal contacts by an mDia1-dependent and ROCK independent mechanism. *J. Cell Biol.*, **153**(6), 1175–1186.

114. Rodriguez, A. G., Han, S. J., Regnier, M., Sniadecki, N. J. (2011). Substrate stiffness increases twitch power of neonatal cardiomyocytes in correlation with changes in myofibril structure and intracellular calcium. *Biophys. J.*, **101**(10), 2455–2464.

115. Rorth, P. (2011). Whence directionality: Guidance mechanisms in solitary and collective cell migration. *Dev. Cell*, **20**(1), 9–18.

116. Ruiz, S. A., Chen, C. S. (2008). Emergence of patterned stem cell differentiation within multicellular structures. *Stem Cells*, **26**(11), 2921–2927.

117. Russell, B., Motlagh, D., Ashley, W. W. (2000). Form follows function: How muscle shape is regulated by work. *J. Appl. Physiol.*, **88**(3), 1127–1132.

118. Russell, B., Curtis, M. W., Koshman, Y. E., Samarel, A. M. (2010). Mechanical stress-induced sarcomere assembly for cardiac muscle growth in length and width. *J. Mol. Cell Cardiol.*, **48**(5), 817–823.

119. Saha, K., Keung, A. J., Irwin, E. F., Li, Y., Little, L., Schaffer, D. V., Healy, K. E. (2008). Substrate modulus directs neural stem cell behavior. *Biophys. J.*, **95**(9), 4426–4438.

120. Saha, S., Ji, L., de Pablo, J. J., Palecek, S. P. (2006). Inhibition of human embryonic stem cell differentiation by mechanical strain. *J. Cell Physiol.*, **206**(1), 126–137.

121. Samarel, A. M. (2005). Costameres, focal adhesions, and cardiomyoctye mechanotransduction. *Am. J. Physiol. Heart Circ. Physiol.*, **289**, H2291–H2301.

122. Sargent, C. Y., Berguig, G. Y., Mcdevitt, T. C. (2009). Cardiomyogenic differentiation of embryoid bodies is promoted by rotary orbital suspension culture. *Tissue Eng. Part A.*, **15**(2), 331–342.

123. Sarraf, C. E., Otto, W. R., Eastwood, M. (2011). In vitro mesenchymal stem cell differentiation after mechanical stimulation. *Cell Prolif.*, **44**(1), 99–108.

124. Sawada, Y., Sheetz, M. P. (2002). Force transduction by Triton cytoskeletons. *J. Cell Biol.*, **156**(4), 609–615.

125. Schmelter, M., Ateghang, B., Helmig, S., Wartenberg, M., Sauer, H. (2006). Embryonic stem cells utilize reactive oxygen species as transducers of mechanical strain-induced cardiovascular differentiation. *FASEB J.*, **20**(8), 1182–1184.

126. Senyo, S. E., Koshman, Y. E., Russell, B. (2007). Stimulus interval, rate and direction differentially regulate phosphorylation for mechanotransduction in neonatal cardiac myocytes. *FEBS Lett.*, **581**(22), 4241–4247.

127. Seo, C. H., Furukawa, K., Montagne, K., Jeong, H., Ushida, T. (2011). The effect of substrate microtopography on focal adhesion maturation and actin organization via the RhoA/ROCK pathway. *Biomaterials*, **32**(36), 9568–9575.

128. Shah, U., Bien, H., Entcheva, E. (2010). Microtopographical effects of natural scaffolding on cardiomyocyte function and arrhythmogenesis. *Acta Biomater.*, **6**(8), 3029–3034.

129. Shimizu, N. (2008). Cyclic strain induces mouse embryonic stem cell differentiation into vascular smooth muscle cells by activating PDGF receptor beta. *J. Appl. Physiol.*, **104**(3), 766–772.

130. Spradling, A., Drummond-Barbosa, D., Kai, T. (2001). Stem cells find their niche. *Nature*, **414**(6859), 98–104.

131. Street, S. F. (1983). Lateral transmission of tension in frog myofibers: A myofibrillar network and transverse cytoskeletal connections are possible transmitters. *J. Cell Physiol.*, **114**(3), 346–364.

132. Sun, Y., Zhang, J. Q., Zhang, J., Lamparter, S. (2000). Cardiac remodeling by fibrous tissue after infarction in rats. *J. Lab Clin. Med.*, **135**(4), 316–323.

133. Sun, Y. (2009). Myocardial repair/remodelling following infarction: Roles of local factors. *Cardiovasc. Res.*, **81**(3), 482–490.

134. Tan, J. L., Tien, J., Pirone, D. M., Gray, D. S., Bhadriraju, K., Chen, C. S. (2003). Cells lying on a bed of microneedles: An approach to isolate mechanical force. *Proc. Natl. Acad. Sci. U. S. A.*, **100**(4), 1484–1489.

135. Tee, R., Lokmic, Z., Morrison, W. A., Dilley, R. J. (2010). Strategies in cardiac tissue engineering. *ANZ J. Surg.*, **80**(10), 683–693.

136. Teixeira, A. I., Abrams, G. A., Bertics, P. J., Murphy, C. J., Nealey, P. F. (2003). Epithelial contact guidance on well-defined micro- and nanostructured substrates. *J. Cell Sci.*, **116**(Pt 10), 1881–1892.

137. Thakar, R. G., Chown, M. G., Patel, A., Peng, L., Kumar, S., Desai, T. A. (2008). Contractility-dependent modulation of cell proliferation and adhesion by microscale topographical cues. *Small*, **4**(9), 1416–1424.

138. Thorburn, J., Carlson, M., Mansour, S. J., Chien, K. R., Ahn, N. G., Thorburn, A. (1995). Inhibition of a signaling pathway in cardiac muscle cells by active mitogen-activated protein kinase kinase. *Mol. Biol. Cell.*, **6**(11), 1479–1490.

139. Torsoni, A. S., Constancio, S. S., Nadruz, W. Jr., Hanks, S. K., Franchini, K. G. (2003). Focal adhesion kinase is activated and mediates the early hypertrophic response to stretch in cardiac myocytes. *Circ. Res.*, **93**(2), 140–147.

140. Truong, Y. B., Glattauer, V., Lang, G., Hands, K., Kyratzis, I. L., Werkmeister, J. A., Ramshaw, J. A. (2010). A comparison of the effects of fibre alignment of smooth and textured fibres in electrospun membranes on fibroblast cell adhesion. *Biomed. Mater.*, **5**(2), 25005.

141. Urbanek, K., Quaini, F., Tasca, G., Torella, D., Castaldo, C., Nadal-Ginard, B., Leri, A., Kajstura, J., Quaini, E., Anversa, P. (2003). Intense myocyte formation from cardiac stem cells in human cardiac hypertrophy. *Proc. Natl. Acad. Sci. USA.*, **100**, 10440–10445.

142. Varner, V. D., Voronov, D. A., Taber, L. A. (2010). Mechanics of head fold formation: Investigating tissue-level forces during early development. *Development*, **137**(22), 3801–3811.

143. Vogel, V., Sheetz, M. (2006). Local force and geometry sensing regulate cell functions. *Nat. Rev. Mol. Cell Biol.*, **7**, 265–275.

144. Wagner, S., Storbeck, C. J., Roovers, K., Chaar, Z. Y., Kolodziej, P., McKay, M., Sabourin, L. A. (2008). FAK/src-family dependent activation of the Ste20-like kinase SLK is required for microtubule-dependent focal adhesion turnover and cell migration. *PLoS One*, **3**(4), e1868.

145. Wang, Y. P., Fuchs, F. (1994). Length, force, and Ca(2+)-troponin C affinity in cardiac and slow skeletal muscle. *Am. J. Physiol.*, **266**(4 Pt 1), C1077–C1082.

146. Watt, F. M., Hogan, B. L. (2000). Out of Eden: Stem cells and their niches. *Science*, **287**(5457), 1427–1430.

147. Weber, K. T. (2004). Fibrosis in hypertensive heart disease: Focus on cardiac fibroblasts. *J. Hypertens.*, **22**(1), 47–50.

148. Weyts, F. A., Bosmans, B., Niesing, R., van Leeuwen, J. P., Weinans, H. (2003). Mechanical control of human osteoblast apoptosis and proliferation in relation to differentiation. *Calcif. Tissue Int.*, **72**(4), 505–512.

149. Winer, J. P., Janmey, P. A., Mccormick, M. E., Funaki, M. (2009). Bone marrow-derived human mesenchymal stem cells become quiescent on soft substrates but remain responsive to chemical or mechanical stimuli. *Tissue Eng. Part A*, **15**(1), 147–154.

150. Wozniak, M. A., Chen, C. S. (2009). Mechanotransduction in development: A growing role for contractility. *Nat. Rev. Mol. Cell Biol.*, **10**(1), 34–43.

151. Xiong, H. A., Rabie, B., Hagg, U. (2005). Mechanical strain leads to condylar growth in adult rats. *Front Biosci.*, **10**, 67–73.

152. Xu, B., Song, G., Ju, Y., Li, X., Song, Y., Watanabe, S. (2012). RhoA/ROCK, cytoskeletal dynamics and focal adhesion kinase are required for mechanical stretch induced tenogenic differentiation of human mesenchymal stem cells. *J. Cell Physiol.*, **227**(6), 2722–2729.

153. Yeung, T., Georges, P. C., Flanagan, L. A., Marg, B., Ortiz, M., Funaki, M., Zahir, N., Ming, W., Weaver, V., Janmey, P. A. (2005). Effects of substrate stiffness on cell morphology, cytoskeletal structure, and adhesion. *Cell Motil. Cytoskeleton.*, **60**(1), 24–34.

154. Yu, J. G., Russell, B. (2005). Cardiomyocyte remodeling and sarcomere addition after uniaxial static strain in vitro. *J. Histochem. Cytochem.*, **53**(7), 839–844.

## Chapter 2

# How Microbes Sense and Respond to Force: A Single-Molecule View

**Peter N. Lipke,[a] Jürgen J. Heinisch,[b] and Yves F. Dufrêne[c]**

[a]*Department of Biology, Brooklyn College of the City University of New York, New York, NY 11210, USA*
[b]*Fachbereich Biologie/Chemie, AG Genetik, Universität Osnabrück, Barbarastr. 11, D-49076 Osnabrück, Germany*
[c]*Institute of Condensed Matter and Nanosciences, Université catholique de Louvain, Croix du Sud 2/18, B-1348 Louvain-la-Neuve, Belgium*

Yves.Dufrene@uclouvain.be

## 2.1 Introduction

Living cells constantly sense and respond to external forces. Signal transduction pathways detect environmental changes of either physical or chemical nature at the cell surface and trigger the appropriate intracellular responses. These pathways frequently alter the transcriptional program in the nucleus and consequently the cells proteome. Most of these signal transduction pathways start with sensor proteins, which span the plasma membrane and either bind specific ligands or detect mechanical forces and other changes in the physical environment. Mechanosensing and mechanotransduction, which allow the cells to convert mechanical

*Cells, Forces, and the Microenvironment*
Edited by Charles M. Cuerrier and Andrew E. Pelling

ISBN 978-981-4613-36-1 (Hardcover), 978-981-4613-37-8 (eBook)
www.panstanford.com

forces into biochemical signals, play key roles in regulating processes such as cell adhesion, cell growth, differentiation, cell shape, and cell death.

Mechanical forces can alter protein conformation and biological activity by various mechanisms: (i) force-induced exposure of otherwise cryptic peptide sequences (such as in the cell-adhesion proteins fibronectin and integrin [1–4]), (ii) opening of mechanosensitive ion channels (such as the bacterial mechanosensitive K+ channel, MscL; [5]), (iii) strengthening of receptor–ligand interactions by tensile mechanical force (catch bonds such as those occurring in the *E. coli* fimbrial adhesive protein FimH [6,7]), and (iv) binding of cytoskeletal proteins (such as vinculin) to mechanically stretched cytoplasmic proteins (such as talin rod molecules) [8].

Several methods may be used to probe the interactions and mechanical properties of single molecules. Flow chamber experiments consist in using laminar flow to induce intermolecular forces between coated cell walls and spheres [9]. Microneedles are bendable glass coated with biomolecules that are brought in contact with a functionalized surface, while measuring the force generated on the microneedle by beam deflection. Biomembrane force probes (BFP) measure single-molecule forces using a force transducer made of a phospholipid vesicle maintained by a glass micropipette. In optical and magnetic tweezer experiments, force probes are microscopic beads manipulated by an external field (photon or magnetic fields). Atomic force microscopy (AFM) is a versatile technique for measuring biomolecular forces over a wide range of force strengths and length scales that are biologically relevant, going from weak intermolecular interactions to strong covalent bonds [10–16]. Currently, AFM is the only method capable to both localize and force-probe single molecules on live cells, while imaging the cell surface structure (Fig. 2.1).

Although single-molecule techniques have been increasingly used to probe the molecular basis of mechanosensing and mechanotransduction processes, applying these techniques to live cells remains very challenging. In this chapter, we discuss how the key capabilities of AFM have been applied to gain insight into the interactions, nanomechanics and supramolecular organization of their individual surface-associated proteins.

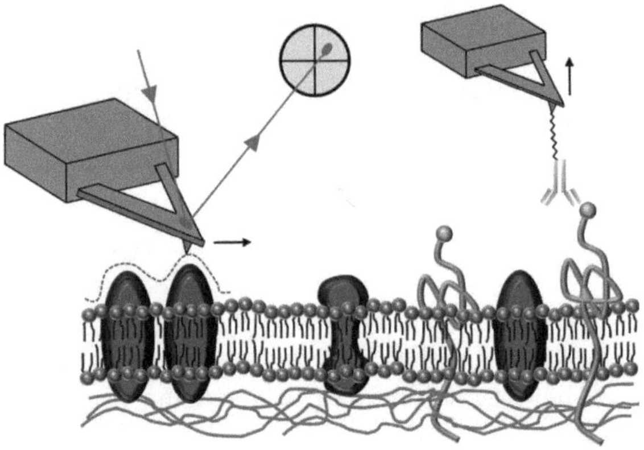

**Figure 2.1** Atomic force microscopy of the cell surface. In AFM imaging, a sharp tip scans the cell surface with nanometer resolution (left). In single-molecule force spectroscopy (SMFS), the small interaction force between the tip and cell surface molecules is measured while the distance between the tip and cell is varied (right). The right example shows a tip labelled with a ligand to detect, localize and manipulate individual cell surface proteins.

## 2.2 Atomic Force Microscopy of Live Cells

### 2.2.1 Live-Cell Imaging

AFM analysis of single cells requires their firm attachment onto a substrate. A straightforward approach is to exploit the ability of animal cells to spread and adhere to solid surfaces (Figs. 2.2a,b). Unlike animal cells, microbes have a well-defined shape and usually have no tendency to deform and spread on surfaces under experimental conditions. Consequently, the contact area between a cell and a support is small, often leading to cell detachment by the scanning tip. Therefore, several approaches have been developed to promote microbial cell attachment [17] (Figs. 2.2c,d). A first approach is to attach negatively charged cells onto substrates (e.g., glass) functionalized with positively charged macromolecules, such as poly-L-lysine and polyethylenimine. This method has been successfully applied to lactic bacteria, Gram-negative bacteria and diatoms, providing novel information on their nanoscale surface properties. Similarly, hydrophobic cells may be strongly adhered to hydrophobic substrates (Fig. 2.2d). Cells can also be mechanically immobilized either onto gelatin-coated supports or in porous polymer membranes (Fig. 2.2c). In the latter approach, a concentrated cell suspension is gently sucked through a porous membrane with pore sizes similar to the size of the cells to be investigated. This procedure offers two advantages: It is fairly simple and does not involve a macromolecular support, thus preventing the risk of cell surface or tip contamination. Using these methods, microbial cells can be routinely imaged in buffer or growth medium by scanning the AFM tip over the sample surface and measuring the interaction force between the tip and the surface (Fig. 2.1, left).

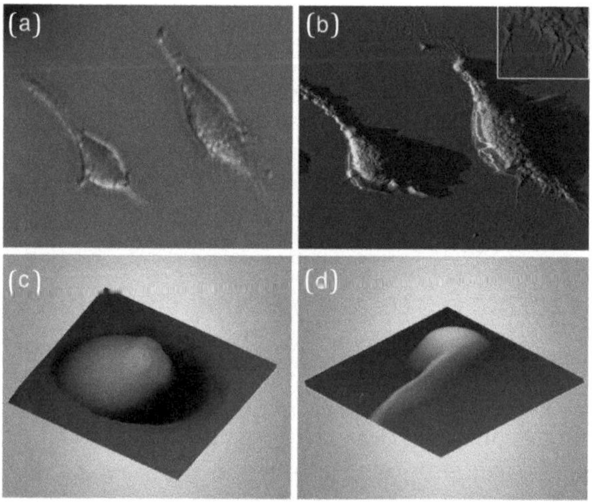


**Figure 2.2** Live-cell imaging. (a, b) Combined optical (a) and AFM (b) images (width: 100 μm) of gently fixed macrophages spread on glass. The inset in (b) shows a higher magnification of the cell edge. (c, d) Three-dimensional AFM images of yeast (8 μm × 8 μm) and hyphae (15 μm × 15 μm) from the fungal pathogen *Candida albicans*. The yeast cell shows cell wall heterogeneities in the form of circular bud scars left after detachment of the daughter cell, while the germinating cell clearly shows a germ tube.

## 2.2.2 Single-Molecule Detection and Manipulation

Single-molecule imaging by AFM is generally performed by single-molecule force spectroscopy (SMFS) [14], in which the cantilever deflection is recorded as a function of the vertical displacement of the scanner, i.e., as the sample is pushed toward the tip and retracted (Fig. 2.1, right). After appropriate corrections, this yields a so-called "force–distance curve." These curves can be recorded at multiple locations to yield spatially resolved force maps. The characteristic adhesion (or unbinding) force between tip and sample observed during retraction is then used to detect and localize specific molecules (Fig. 2.1, right).

Molecular recognition experiments imply functionalizing the AFM tip with appropriate ligands [14]. The forces that immobilize the molecules have to be stronger than the intermolecular force to be measured, and the attached biomolecules should have enough mobility to interact with complementary molecules. Non-specific adhesion forces should be minimized and the biomolecules should be linked at a low density to ensure single-molecule detection. A common method to covalently anchor proteins on tips is to use a polyethylene glycol (PEG) cross-linker, which provides motional freedom and prevents denaturation. Tips are first modified with amino groups, further reacted with PEG linkers carrying benzaldehyde functions that are then directly attach to proteins through their lysine residues. Biomolecules may also be attached via self-assembled monolayers of alkanethiols on gold tips.

Notably, SMFS can also subject single molecules to force in order to study their mechanical properties [18]. In the past years, such single-molecule manipulations have increased our understanding of the mechanical behaviour of a great variety of proteins [19,20]. Most of these experiments have been conducted in vitro, meaning the molecules were removed from the native cellular context. Here, we discuss recent studies in which SMFS has been used to investigate the force response and the distribution of single proteins on the surface of living cells.

## 2.3 Protein Mechanics

### 2.3.1 Membrane Sensors

Mechanosensors on cell surfaces convert external mechanical forces into intracellular biochemical signals. In yeast cell wall integrity (CWI) signaling, five proteins spanning the plasma membrane (Wsc1, Wsc2, Wsc3, Mid2, and Mtl1) are believed to detect mechanical forces acting on the cell wall or the plasma membrane, and to trigger an intracellular signaling cascade, with a central mitogen-activated protein kinase (MAPK) module. This ultimately leads to the expression of genes whose products ensure cell wall remodeling (i.e., they encode either cell wall biosynthetic enzymes or cell wall proteins) [21]. The sensors share a similar structure: They all contain an amino-terminal signal sequence, which guides them through the secretory pathway and is cleaved off before their insertion into the plasma membrane; this is followed by a large, highly mannosylated, external serine/threonine-rich region (STR); all sensors also possess a single predicted transmembrane domain (TMD) and a comparatively short cytoplasmic tail. To date, the exact biophysical mechanisms by which yeast sensors feel and respond to physical forces are poorly understood.

By integrating SMFS with genetic manipulations, the mechanical behavior of single Wsc1 sensors was unraveled in living *Saccharomyces cerevisiae* cells [22]. Genetic manipulations were employed to solve a major experimental problem: In essence, AFM is a surface technique; so how can it probe sensors that are embedded within the cell wall? A simple calculation indeed reveals that the native Wsc1 sensor supplied with a His-tag should extend no more than 86 nm above the plasma membrane. As the cell wall is approximately 100–110 nm thick, as determined by transmission electron microscopy, native sensors cannot reach the outermost cell surface. The construction of fusion proteins with >100 additional amino acids originating from a His-tagged Mid2-STR sequence, circumvented this problem and results in elongated sensors with external lengths of up to 153 nm. This indeed rendered the sensors accessible to probing by SMFS (Fig. 2.3a).

For SMFS experiments, *S. cerevisiae* cells expressing such His-tagged elongated Wsc1 proteins were immobilized on a polycarbonate membrane, and single sensors were detected by scanning the cell surface with a $Ni^{2+}$-nitrilotriacetate (NTA)-tip (Fig. 2.3a). Adhesion force data showed individual molecules and confirmed their cell surface penetration. In contrast, His-tagged Wsc1 sensors lacking the elongation could only be detected in the area of the bud scar. Upon pulling on single sensors, force-extension curves (Fig. 2.3a) were obtained in which the molecule was stretched at nearly constant force. This presumably reflects the straightening of the extracellular polypeptide chain and/or the stretching of the plasma membrane. The constant force region was followed by a linear region, where force was directly proportional to extension, thus characteristic of a Hookean spring with a spring constant of $4.6 \pm 0.4$ pN $nm^{-1}$. For other proteins tested so far, such a nanospring behavior is very unusual in nature and stands in sharp contrast to the properties of modular proteins. Generally, stretching modular proteins leads to non-linear force peaks that are well described by the worm-like-chain model, where each peak reflects the force-induced unfolding of secondary structures [18–20].

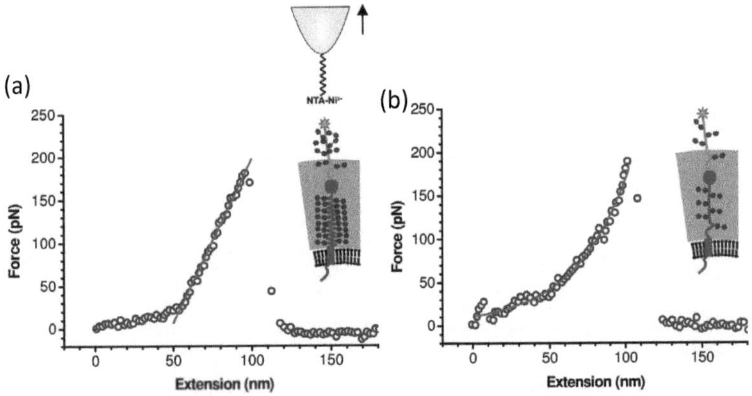


**Figure 2.3** AFM shows that yeast sensors behave like nanosprings. (a) Representative force curve obtained upon stretching a single Wsc1 molecule. As shown in the cartoon, *S. cerevisiae* cells expressing Wsc1 sensors with an extended His-tag were probed using $Ni^{2+}$-NTA-tips. Clearly visible are two extension regimes, reflecting elongation at nearly zero force, followed by a Hookean spring behaviour (red line). (b) Representative force curve obtained upon stretching single Wsc1 sensors in a *pmt4* deletion mutant in which mannosylation of the extracellular Ser/Thr-rich region is substantially reduced. The force increases are no longer linear, but are well-described with an exponential fit (red curve).

Hyper-mannosylation of the extracellular Ser/Thr-region of Wsc1 is believed to confer a stiff, extended conformation to this part of the sensor. To support this view, two different mutants altered in the extracellular Ser/Thr-rich region were analyzed (i.e., a genomic *pmt4* deletion, which causes reduced mannosylation of secreted proteins due to a defect in the encoded mannosyltransferase, and a strain producing a Wsc1 derivative with an insertion of a stretch of non-glycosylated glycines within the STR sequence) [22]. The mechanical properties of these sensor variants differed dramatically from those of the wild type and no longer exhibited a linear Hookean spring behavior (Fig. 2.3b). This demonstrates that glycosylation of the sensor adds to the stiffness of the extracellular Ser/Thr-rich region and is required for its spring properties. Importantly, non-elongated, His-tagged sensors detected in the bud scar display Hookean spring behavior, ruling out an artificial effect of the Mid2-STR region used for elongation on the spring behaviour of the hybrid, fully functional sensor [22].

Does the sensor stiffness respond to CWI stress? Lowering the salt concentration or increasing the incubation temperature of the yeast cells resulted in a substantial reduction of the sensor spring constant, indicating that the mechanical properties of Wsc1 are influenced by cell surface stress [22]. The reasons for such sensor softening may be direct or indirect. Indeed, stressing by hypoosmotic shock or heat shock may alter the elasticity and tension of the cell wall and/or plasma membrane, which would alter the apparent protein stiffness. On the other hand, temperature and turgor pressure could also directly affect the mechanical properties of the protein.

In summary, the single-molecule experiments described above open new avenues for investigating how proteins respond to forces in living cells—primarily those of yeast and bacteria, but also of mammals—and how mechanosensing events proceed in vivo. This will surely provide new insights into what has been suggested to be one of the most ancient sensory mechanisms in evolution.

### 2.3.2 Cell Adhesion Proteins

The force-induced deformation of cellular proteins also plays a major role in mediating cell adhesion. In animal cells, force-induced conformational changes in cell adhesion proteins such as integrins are known to increase the protein binding strength [1,2]. In microbes, the fimbrial adhesive protein FimH mediates weak adhesion at low flow, but strong adhesion at high flow through so-called "catch bonds," i.e., receptor–ligand bonds that are strengthened by tensile mechanical force. Another example is the force-induced unfolding of Als cell adhesion proteins from *C. albicans* [23]. Of the eight different Als proteins, Als5p is one of the most extensively characterized. Als proteins possess four functional regions, an immunoglobulin (Ig)-like region near the N-terminus that initiates cell adhesion, a threonine-rich region (T) containing a functional amyloid sequence, a tandem repeat (TR) region that participates in cell–cell aggregation, and a stalk region projecting the molecule away from the cell surface. Microscopic assays using *S. cerevisiae* surface display models have revealed that Als-mediated adherence involves two steps: initial adhesion via the specific binding of the Ig region to peptide ligands, followed by cell–cell aggregation mediated by the T and TR regions. Until recently, the molecular mechanisms underlying the mechanical strength of Als proteins were mysterious.

SMFS was recently used to study the adhesive and mechanical properties of Als5p proteins, both on isolated molecules and on cell surfaces [24]. Soluble Als fragments were attached on gold surfaces and picked up by their terminal Ig domain using an AFM tip. Force-extension curves showed saw-tooth patterns with well-defined force peaks, each peak corresponding to the force-induced unfolding of an individual TR domain (Fig. 2.4a). Urea altered the shape of the unfolding peaks, reflecting a loss of mechanical stability of the domains due to hydrogen bond disruption. Single Als proteins could also be unfolded directly on live cells. Force curves obtained on cells with proteins having six repeats displayed saw-tooth patterns similar to those found on isolated proteins, while cells not producing any repeats were unable to bind the AFM tip. The unfolding probability increased with the number of repeats and was correlated with the level of cell–cell adhesion, indicating these modular domains might play a role in fungal adhesion. Presumably, the force-induced unfolding of Als proteins leads to extended conformations in which hydrophobic groups are freshly exposed, thus favoring hydrophobic interactions between opposing cells.

Remarkably, stretching Als proteins through their amyloid sequence yielded characteristic force signatures corresponding to the mechanical unzipping of β-sheet interactions formed between surface-arrayed Als proteins (Fig. 2.4b) [25]. The unzipping probability increased with contact time, reflecting the time necessary for optimal inter β-strand associations. This shows that amyloid interactions provide cohesive strength to a major adhesion protein from a microbial pathogen, thereby strengthening cell adhesion.

Such functional amyloids may represent a generic mechanism for providing mechanical strength to cell adhesion proteins.

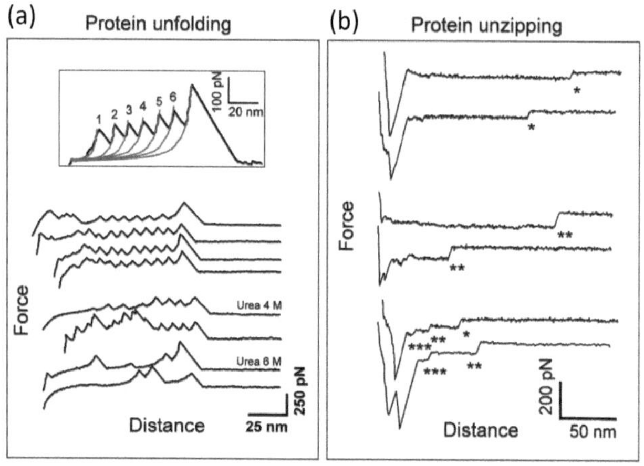


**Figure 2.4** Unfolding and unzipping of single cell adhesion proteins. (a) Force-extension curves obtained by stretching single Als molecules showed periodic features reflecting the sequential unfolding of multiple 36-amino acid tandem repeat (TR) domains (upper traces). Force peaks were well-described by the worm-like-chain model (inset; red line). Addition of urea dramatically altered the unfolding peaks due to hydrogen bond disruption (lower traces). (b) Representative force curves recorded in buffer between an AFM tip bearing an amyloid peptide and surface-displayed Als amyloids. Stretching Als proteins through their amyloid sequence yielded characteristic force plateaus corresponding to the mechanical unzipping of β-sheet interactions between surface-arrayed proteins. The number of force plateaus superimposed is highlighted by a corresponding number of asterisks (*). Reprinted with permission from Refs [24,25].

Accordingly, single-molecule AFM represents a valuable platform for understanding the mechanical properties of surface receptors such as sensors and adhesins and might help us to elucidate their potential implication in diseases.

## 2.4 Protein Clustering

A hot topic in current cell biology is to understand how cell surface proteins form nanodomains [1]. The small size and dynamic nature of these clusters make their direct visualization in live cells very challenging. Single-molecule imaging using spatially resolved SMFS has enabled researchers to map the distribution of single proteins on cell surfaces, thereby providing key insights into their dynamic clustering. Here below, we explain how the technique has enabled the discovery of functional protein nanocomplexes that form under stress, thereby activating cell signaling and cell adhesion.

### 2.4.1 Membrane Sensors

The distribution of single Wsc1 sensors was mapped on living *S. cerevisiae* cells by spatially resolved SMFS [26]. Molecular recognition images (1 μm × 1 μm) were recorded on cells expressing the above-mentioned His-tagged elongated sensors using $Ni^{2+}$-NTA-tips. Many sensor molecules appeared to cluster in areas of approximately 0.04 $\mu m^2$ and an equivalent diameter of 230 nm. These microdomains are clearly larger than the putative size of rafts in higher eukaryotes.

The number of wild-type Wsc1 sensors as well as the size of their clusters increased when cells were stressed by heat or low osmolarity, both conditions that activate the cell wall integrity pathway. Sensor clustering thus seems to be a stress-responsive process, which is intimately connected to signaling. In combination with the mutant phenotypes described below, we therefore proposed that clustering is a means developed by yeast to locally concentrate Wsc1 sensors and their interacting CWI downstream signaling components in so-called "sensosomes." This may be a means of enhancing the signal strength and/or providing a threshold to trigger the corresponding cellular response.

Activation of the downstream signaling cascade is abolished in mutants with cysteine to alanine exchanges in the cysteine-rich domain (CRD) of Wsc1, which produced non-functional sensors [26]. Although such mutant proteins displayed adhesion frequencies, adhesion values and spring behaviors similar to those of the wild-type protein, they were evenly distributed over the yeast cell surface, rather, i.e., they did not cluster in microdomains. It was thus proposed that in yeast, like in bacterial chemotaxis or for instance the immune system of higher eukaryotes, the function of sensors is coupled to their localized enrichment in membrane patches.

### 2.4.2 Cell Adhesion Proteins

By combining single-molecule imaging and manipulation using SMFS, we showed that mechanical force can trigger the formation and propagation of cell adhesion nanodomains in living yeast cells [27]. The surface distribution of Als5 proteins from *C. albicans* was probed on live cells using AFM with tips bearing antibodies directed specifically toward the protein. Pulling on single proteins was shown to induce the formation of "nanoadhesomes," i.e., adhesion domains of 100–500 nm size. The force-induced nanodomains were found to propagate over the entire cell surface. Control experiments demonstrated that Als5p nanodomains result from protein redistribution triggered by force-induced conformational changes in the initially probed proteins, rather than from non-specific cell wall perturbations. Als5p remodeling was independent of cellular metabolic activity since heat-killed cells show the same behavior as live cells.

Single-site mutation in the conserved amyloid-forming sequence of the protein revealed that amyloid interactions represent the driving force underlying Als5p clustering [28]. Hence, the strength of yeast adhesion results from the force-activated amyloid-like clustering of hundreds of proteins on the cell surface to form arrays of ordered multimeric binding sites [23]. These amyloid clusters explain why adhesins often show weak binding to specific ligands, yet mediate remarkably strong adherence. These results highlight the role that amyloids can play in microbial cell adhesion, both in clustering the adhesins to increase binding avidity, and in the potential to form stable amyloid-type bonds between cells. Similar amyloid-forming sequences are found in most fungal adhesins. In several gene families, they have been demonstrated to be important for activation of cellular aggregation and subsequent biofilm formation [28]. These activities are similar to those of bacterial amyloid-forming surface proteins as well. In gram-negative bacteria, amyloid interactions stabilize biofilm-mediating appendages called curli, which interact through the hydrophobic effect in biofilms [29]. Surface adhesins in gram-positives are interacting through amyloid interactions to attach cells to biofilm matrices, and form part of the matrix itself [30]. Thus, AFM has highlighted the role that functional amyloids can play in cell adhesion, a role that we expect will greatly expand as more cell adhesion systems are examined. As functional amyloids may be a general mechanism for activating the adhesion of pathogens, these analyses offer exciting prospects in therapeutics for developing new antimicrobial strategies.

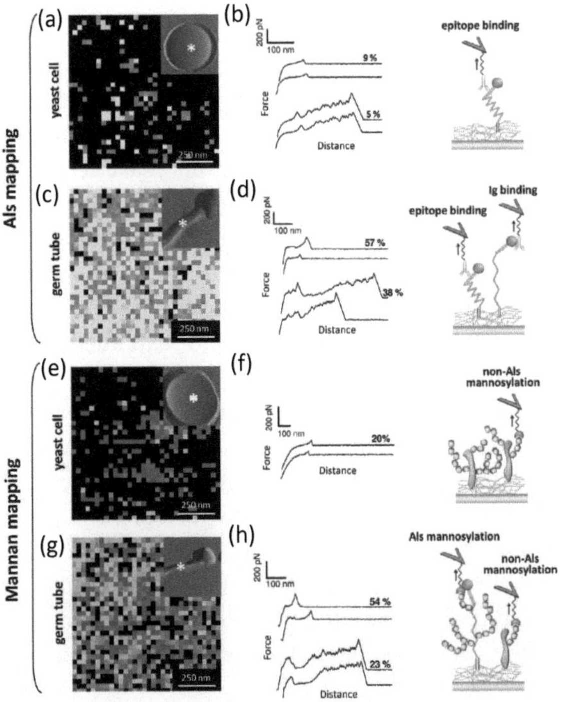


**Figure 2.5** Single-molecule imaging and functional analysis of Als adhesins and mannans during *Candida albicans* morphogenesis. (a, c) Adhesion force maps (1 μm × 1 μm, color scale: 300 pN) and representative force curves (b, d) recorded with an anti-Als3 tip on a *C. albicans* yeast cell (a, b) and on a *C. albicans* germ tube (c, d). (e–h) Same data obtained with a lectin-tip. The insets are topographic images of the cells in which the * symbol indicate where the force maps were recorded. The cartoons (right) emphasize the major changes in Als3 and mannan biophysical properties: while non-germinating cells showed essentially single weak adhesion peaks reflecting Als epitope recognition (b) or non-Als mannose recognition (f), germ tubes featured saw-tooth patterns with multiple force peaks documenting unfolding of entire Als proteins (d, h). The major increase in the distribution and extension of Als3 proteins and their associated mannans is critical for cell adhesion, hence for pathogenesis and immune response. Reprinted with permission from Ref. [31].

More recently, we demonstrated that cellular morphogenesis (yeast-to-hyphae transition) in *C. albicans* leads to major remodeling of cell adhesion proteins and their associated glycoconjugates (Fig. 2.5), a behavior that is critical for biofilm formation and immune responses [31]. AFM tips functionalized with specific antibodies revealed that the distribution and biophysical properties of Als adhesins dramatically changed upon germination, their exposure, adhesion, unfolding and extension being much higher on the germ tubes (Figs. 2.5c,d) than on yeast cells (Figs. 2.5a,b). The use of hydrophobic tips showed that these molecular changes were accompanied by a major increase in cell surface hydrophobicity. These data support a model in which the large amounts of Als tandem-repeat domains on germ tubes, and their force-induced unfolding, lead to extended conformations favoring hydrophobic interactions, thus strengthening adhesion. In parallel, specific lectin probes demonstrated that formation of hyphae is also accompanied by an increase in the exposure and extension of mannan glycoconjugates (Figs. 2.5e–h), a finding likely to be important for the interaction with epithelium and immune cells. In summary, single-molecule AFM provides a means to simultaneously image and manipulate the individual components of microbial pathogens, thereby contributing to increase our understanding of their biofilm and immune interactions.

## 2.5 Conclusions and Future Prospects

Microbes use a variety of sensing proteins to probe their environment. However, the way in which these mechanosensors respond to mechanical stimuli, and assemble into large complexes remains poorly understood at the molecular level. This Chapter demonstrates that AFM techniques, combined with genetic manipulations, have revolutionized the way in which microbiologists analyze cell surface proteins to molecular resolution. Such an integrated platform can be applied for measuring the mechanical properties of single proteins, and for imaging their localization, thus unraveling their molecular elasticity and revealing whether they are isolated or form clusters. In the future, these measurements may contribute to trigger a paradigm shift in cell surface biology, so that new questions can be addressed.

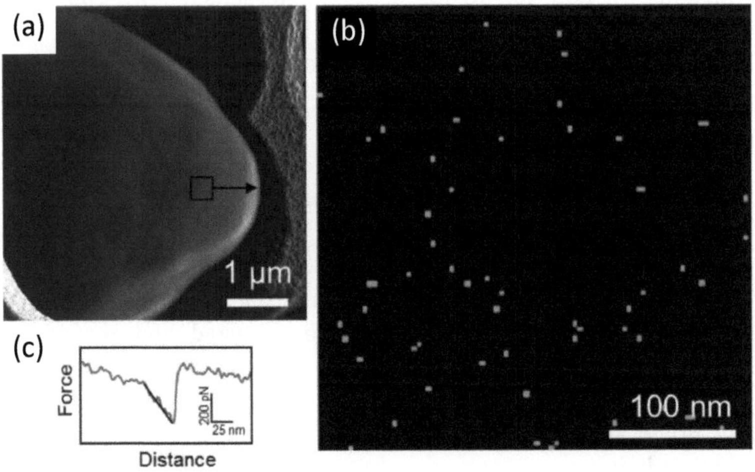


**Figure 2.6** Toward fast single protein detection and manipulation using peak force tapping technology. (a) AFM topography of a *Saccharomyces cerevisiae* cell expressing His-tagged sensors, recorded in acetate buffer using tips functionalized with $Ni^{2+}$-NTA groups. (b) Adhesion force map recorded simultaneously as the topography in the small square region shown in (a). Each green pixel documents the detection of single sensors to a resolution of 4 nm. (c) Pulling on individual sensors reveal adhesion events with linear shapes, thus indicating that they behave as nanosprings. Reprinted with permission from Ref. [32].

A critical challenge for future single protein manipulation and imaging by AFM is to increase the speed of the measurements so that dynamic cellular processes can be studied. Until now, the low speed and poor spatial resolution of spatially resolved SMFS (force-volume imaging) have severely hampered its widespread use in cell biology. We recently used peak force tapping (PFT) with chemically functionalized tips for probing the localization and interactions of biological sites on living cells at high speed and high resolution [32]. The distribution of single His-tagged cell surface sensors was mapped to molecular resolution by imaging the cell surface with tips bearing $Ni^{2+}$-NTA groups (Fig. 2.6b). AFM force data were acquired at an imaging speed

(8 min/frame, at 512 × 512 pixels) similar to that of conventional topographic imaging, thus much better than force-volume imaging. In addition to localizing single sensors, we could also measure their mechanical properties by analyzing individual force curves (Figs. 2.6c). Most adhesion peaks showed a linear profile characteristic of a Hookean spring. This indicates that sensor proteins have unusual nanospring properties and can resist high mechanical forces without undergoing secondary structure unfolding, which makes them ideally suited to function as mechanosensors. Accordingly, PFT is able to achieve single-molecule imaging and manipulation at (near) molecular lateral resolution and fairly high speed, offering new possibilities to understand how single proteins are localized and respond to force in living cells. Owing to its key capabilities, this novel biochemically sensitive imaging technique is a powerful complement to other advanced AFM modes for quantitative, high-resolution bioimaging.

## Acknowledgments

Work in the team of Y. F. D. was supported by the National Foundation for Scientific Research (FNRS), the Foundation for Training in Industrial and Agricultural Research (FRIA), the Université catholique de Louvain (Fonds Spéciaux de Recherche), the Federal Office for Scientific, Technical and Cultural Affairs (Interuniversity Poles of Attraction Programme), and the Research Department of the Communauté française de Belgique (Concerted Research Action). Y. F. D. is a Senior Research Associate of the FNRS. Work in P. N. Lipke's lab was supported by NIH grants SC1 GM083756 and R01 GM098616. Work on CWI sensors in the lab of J. J. H. is supported by a grant from the Deutsche Forschungsgemeinschaft in the frame of the SFB944.

## References

1. Vogel, V., Sheetz, M. (2006). Local force and geometry sensing regulate cell functions, *Nat. Rev. Mol. Cell Biol.*, **7**, 265–275.
2. Brown, A. E. X., Discher, D. E. (2009). Conformational changes and signaling in cell and matrix physics, *Curr. Biol.*, **19**, R781–R789.

3. Schwartz, M. A. (2009). The force is with us, *Science*, **323**, 588–589.
4. Friedland, J. C., Lee, M. H., Boettiger, D. (2009). Mechanically activated integrin switch controls α5β1 function, *Science*, **323**, 642–644.
5. Li, Y., Wray, R., Eaton, C., Blount, P. (2009). An open-pore structure of the mechanosensitive channel MscL derived by determining transmembrane domain interactions upon gating, *FASEB J.*, **23**, 2197–2204.
6. Sokurenko, E. V., Vogel, V., Thomas, W. E. (2008). Catch-bond mechanism of force-enhanced adhesion: Counterintuitive, elusive, but ... widespread? *Cell Host Microbe*, **4**, 314–323.
7. Yakovenko, O., Sharma, S., Forero, M., Tchesnokova, V., Aprikian, P., Kidd, B., Mach, A., Vogel, V., Sokurenko, E., Thomas, W. E. (2008). FimH forms catch bonds that are enhanced by mechanical force due to allosteric regulation, *J. Biol. Chem.*, **283**, 11596–11605.
8. del Rio, A., Perez-Jimenez, R., Liu, R., Roca-Cusachs, P., Fernandez, J. M., Sheetz, M. P. (2009). Stretching single talin rod molecules activates vinculin binding, *Science*, **323**, 638–641.
9. Pierres, A., et al. (2002). Dissecting streptavidin-biotin interaction with a Laminar flow chamber, *Biophys. J.*, **82**, 3214–3223.
10. Evans, E. A., Calderwood, D. A. (2007). Forces and bond dynamics in cell adhesion, *Science*, **316**, 1148–1153.
11. Bustamante, C., Macosko, J. C., Wuite, G. J. L. (2000). Grabbing the cat by the tail: Manipulating molecules one by one, *Nat. Rev. Mol. Cell Biol.*, **1**, 130–136.
12. Sotomayor, M., Schulten, K. (2007). Single-molecule experiments in vitro and in silico, *Science*, **316**, 1144–1148.
13. Muller, D. J., Dufrêne, Y. F. (2008). Atomic force microscopy as a multifunctional molecular toolbox in nanobiotechnology, *Nat. Nanotechnol.*, **3**, 261–269.
14. Hinterdorfer, P., Dufrêne, Y. F. (2006). Detection and localization of single molecular recognition events using atomic force microscopy, *Nat. Methods*, **3**, 347–355.
15. Dufrêne, Y. F. (2008). Towards nanomicrobiology using atomic force microscopy, *Nat. Rev. Microbiol.*, **6**, 674–680.
16. Neuman, K. C., Nagy, A. (2008). Single-molecule force spectroscopy: Optical tweezers, magnetic tweezers and atomic force microscopy, *Nat. Methods*, **5**, 491–505.
17. Dufrêne, Y. F. (2008). Atomic force microscopy and chemical force microscopy of microbial cells, *Nat. Protocols*, **3**, 1132–1138.

18. Rief, M., Gautel, M., Oesterhelt, F., Fernandez, J. M., Gaub, H. E. (1997). Reversible unfolding of individual titin immunoglobulin domains by AFM, *Science*, **276**, 1109–1112.

19. Li, H., Cao, Y. (2010). Protein mechanics: From single molecules to functional biomaterials. *Acc. Chem. Res.*, **43**, 1331–1341.

20. Puchner, E. M., Gaub, H. E. (2009). Force and function: Probing proteins with AFM-based force spectroscopy, *Curr. Opin. Struct. Biol.*, **19**, 605–614.

21. Levin, D. E. (2005). Cell wall integrity signaling in Saccharomyces cerevisiae, *Microbiol. Mol. Biol. Rev.*, **69**, 262–291.

22. Dupres, V., Alsteens, D., Wilk, S., Hansen, B., Heinisch, J. J., Dufrêne, Y. F. (2009). The yeast Wsc1 cell surface sensor behaves like a nanospring in vivo, *Nat. Chem. Biol.*, **5**, 857–862.

23. Lipke, P. N., Garcia, M. C., Alsteens, D., Ramsook, C. B., Klotz, S. A. and Dufrêne, Y. F. (2012). Strengthening relationships: Amyloids create adhesion nanodomains in yeasts, *Trends Microbiol.*, **20**, 59–65.

24. Alsteens, D., Dupres, V., Klotz, S. A., Gaur, N. K., Lipke, P. N., Dufrêne, Y. F. (2009). Unfolding individual Als5p adhesion proteins on live cells, *ACS Nano*, **3**, 1677–1682.

25. Alsteens, D., Ramsook, C. B., Lipke, P. N., Dufrêne, Y. F. (2012). Unzipping a functional microbial amyloid, *ACS Nano*, **6**, 7703–7711.

26. Heinisch, J. J., Dupres, V., Wilk, S., Jendretzki, A., Dufrêne, Y. F. (2010). Single-molecule atomic force microscopy reveals clustering of the yeast plasma-membrane sensor Wsc1, *PLoS ONE*, **5**, e11104.

27. Alsteens, D., Garcia, M. C., Lipke, P. N., Dufrêne, Y. F. (2010). Force-induced formation and propagation of adhesion nanodomains in living fungal cells, *Proc. Natl. Acad. Sci. U. S. A.*, **107**, 20744–20749.

28. Garcia, M. C., Lee, J. T., Ramsook, C. B., Alsteens, D., Dufrêne, Y. F., Lipke, P. N. (2011). A role for amyloid in cell aggregation and biofilm formation, *PLoS ONE*, **6**, e17632.

29. Blanco, L. P., Evans, M. L., Smith, D. R., Badtke, M. P., Chapman, M. R. (2012). Diversity, biogenesis and function of microbial amyloids, *Trends Microbiol.*, **20**, 66–73.

30. Romero, D., Aguilar, C., Losick, R. and Kolter, R. (2010). Amyloid fibers provide structural integrity to Bacillus subtilis biofilms, *Proc. Natl. Acad. Sci. U. S. A.*, **107**, 2230–2234.

31. Beaussart, A., Alsteens, D., El-Kirat-Chatel, S., Lipke, P. N., Kucharíková, S., Van Dijck, P., Dufrêne, Y. F. (2012). Single-molecule imaging and

functional analysis of Als adhesins and mannans during Candida albicans morphogenesis, *ACS Nano*, **12**, 10950–10964.

32. Alsteens, D., Dupres, V., Yunus, S., Latgé, J. P., Heinisch, J. J., Dufrêne, Y. F. (2012). High-resolution imaging of chemical and biological sites on living cells using peak force tapping atomic force microscopy, *Langmuir*, **28**, 16738–16744.

# Chapter 3

# Modulating Cell Adhesion by Non-Covalent Ligand Attachment

**Andreas Müller, Christina Müller, and Tilo Pompe**

*Institute of Biochemistry, Universität Leipzig,*
*Johannisallee 21-23 04103 Leipzig, Germany*

tilo.pompe@uni-leipzig.de

## 3.1 Introduction

Cell adhesion to in vivo extracellular matrices (ECM), implant materials, and in vitro cell culture scaffolds has a central role in regulating cellular functions. A lot of cellular processes are tightly regulated by strong interactions of cells with their microenvironment via adhesion ligands including proliferation, differentiation and apoptosis [9]. In the past decade, it was clearly shown that mechanical properties of the ECM in conjunction with their intracellular counterparts are fundamental to control and explain cellular behavior. The ECM, composed of fibrillar protein networks with glycosaminoglycan or mineral modifications, stay in direct mechanical contact with intracellular structures, like the actin network and stress fibers, microtubules and intermediate filaments, via ligand–receptor pairs at the cell membrane. This

*Cells, Forces, and the Microenvironment*
Edited by Charles M. Cuerrier and Andrew E. Pelling

ISBN 978-981-4613-36-1 (Hardcover), 978-981-4613-37-8 (eBook)
www.panstanford.com

mechanical balance depends on the full range of viscoelastic properties arising from the biopolymer characteristics of the involved components as well as their dynamical changes and the active probing by the cells.

Therefore, mechanical control mechanisms of cell behavior are heavily addressed in many areas of life science ranging from basic biophysical and biochemical questions to applications in biomedicine for therapeutic strategies using biomaterials scaffolds [4,8,9,20]. While the impact of elastic properties is widely discussed in literature, the cellular reaction to viscous properties of the ECM and dissipative contributions of the environment are barely studied.

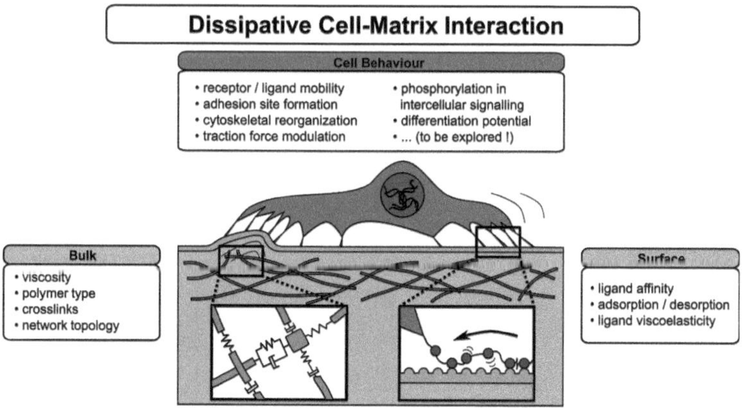


**Figure 3.1** Scheme of dissipative phenomena at the cell-materials interface during cell adhesion. Reprinted with permission from Müller et al. [24]. Copyright by the Royal Chemical Society.

We advocate that current research on mechanotransduction in cell signaling and the general impact of mechanical materials properties on cellular functions needs to address viscous contributions. In vivo, dynamic changes in ECM composition, non-linear elastic or fluid-like behavior of tissue are ubiquitous due to the complex mechanics of the respective constituents. Adhesion-enhancing ligands frequently bind in vivo as well as in vitro via non-covalent binding mechanisms as they are dynamically adsorbed from surrounding fluids and are actively secreted by cells to their ECM. In this process, they are constantly reorganized in the matrices and at materials interfaces. Hence, it can be assumed that cellular processes and the interaction of cells with the surrounding ECM are influenced by dissipative phenomena, including loss in energy by molecular friction and variation of the temporal responses of the involved matrix and cell components due to a broad range of damping processes, as summarized in Fig. 3.1. However, these considerations are in contrast to the frequently used simplified systems, which exhibit linear elastic characteristics and covalently attached adhesion ligands.

This chapter deals with these issues and provides examples of experimental indications that non-covalent ligand–substrate interaction can be handled as a dissipative process and leads to a control of cellular response in cell adhesion.

## 3.2 Viscous Properties of Cell Culture Substrates

Before we address the impact of non-covalent ligand attachment on cell adhesion phenomena, we briefly summarize other dissipative processes within cell adhesion substrates and at the cell–material interface, which are also in place if covalent attachment schemes are applied.

### 3.2.1 Viscous Bulk Properties

Dissipative bulk effects are commonly elicited by viscous flow within bulk biomaterials. This process mostly arises due to the inherent viscoelastic behavior of their biopolymeric constituents. As an example, collagen networks composed of different types of collagen show different viscoelastic properties [27,39]. However, even synthetic polymer materials, frequently used as ideally elastic substrates, can exhibit viscous behavior under certain conditions [7,25]. By the choice of monomer and cross-linker concentration, polyacrylamide hydrogels with a viscoelastic response (high loss modulus $G''$) can be fabricated as demonstrated by Cameron et al. [7] with significant impact on the differentiation of human mesenchymal stem cells. Similarly, Murrell et al. [25] varied the composition of PDMS (polydimethylsiloxane) substrates to obtain substrates with a more elastic (storage modulus $G'$) or viscous ($G''$) behavior, thereby achieving different degrees of coupling between the movement of epithelial cell sheets and substrate deformation. Such viscous contributions can be explained by the mechanics of polymers and proteins as well as their linkage inside the networks, which may include both chemical as well as physical mechanisms [5,22,44].

### 3.2.2 Viscous Processes, Including (Covalently Attached) Ligands

The interfacial layer between the cell culture substrates and the cell membrane also has to be treated as a dynamic system giving rise to viscous processes—even with covalently attached ligands. This may again be explained by the polymeric nature of the involved large adhesion proteins, such as fibronectin or collagen. Ligand attachment may differ by the density of covalent bonds, thereby affecting the flexibility of the viscoelastic biopolymeric ligands. This issue arises for PDMS and polyacrylamide substrates, which have different mesh sizes at comparable stiffness, leading to higher or lower flexibility of the bound ligands (i.e., collagen I) and subsequent different behavior of adherent cells [40].

## 3.3 Non-Covalent Substrate-Ligand Interactions

While the above-mentioned facts are inherent to all cell–material interactions, the frequent occurrence of a non-covalent attachment of adhesion ligands on materials interfaces leads to additional dissipative interactions.

### 3.3.1 Dissipation Processes during Ligand Sorption

Intermolecular forces between protein ligands and materials interfaces drive the adsorption and desorption of adhesion ligands at cell culture scaffolds. This is a common process by which cell culture materials are frequently modified prior to use, in order to enhance spreading and adhesion of cells. This issue is even more important in the dynamic modification of interfaces by adsorption, desorption, and exchange during prolonged cell culture by other proteins present in cell culture media, body fluids, and secretions from adherent cells. In general, adsorption and desorption phenomena of proteins on materials interfaces are widely investigated [34]; however, temporal changes and dissipative processes are often ignored.

Besides polar interactions, hydrogen bonds and dispersion forces, hydrophobic interactions are regarded to be one of the most important factors contributing to the interaction of protein ligands and materials interfaces. Proteins' hydrophobic moieties will assemble at a hydrophobic interface, leading to a conformational change of the protein. Proteins may transform from a soluble to a reversibly bound and even to an irreversibly bound state [21,42]. The binding and denaturation of proteins lead to slow dissipation processes and long equilibration times at the interface [6].

Such interactions were modulated between the adhesion protein fibronectin (FN) and material surfaces by coatings of maleic anhydride copolymers [35,36]. Therein, a variation of the polarity and hydrophobicity of the coatings controlled the affinity of FN to the substrate interfaces. FN was shown to bind more strongly to hydrophobic interfaces and to adopt different conformations in dependence on the substrate properties. Furthermore, the exchange dynamics of FN and serum albumin were demonstrated to depend on hydrophobicity and polarity of the interface. As shown in Fig. 3.2, the analysis of the time dependence of FN–albumin

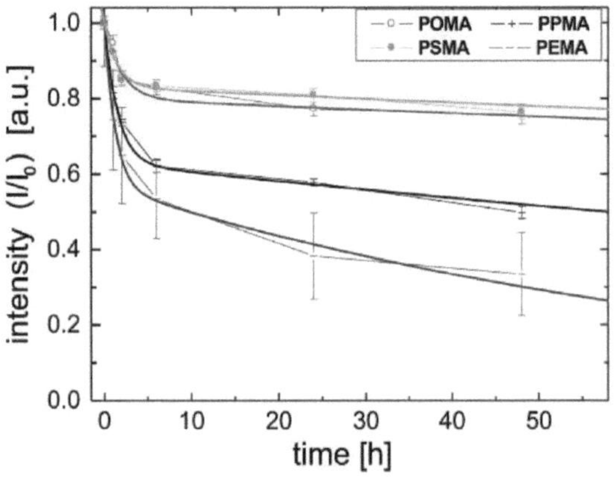


**Figure 3.2** FN exchange by serum albumin at various maleic anhydride copolymer surfaces (POMA–poly(octadecene-*alt*-maleic anhydride), PSMA–poly(styrene-*alt*-maleic anhydride), PPMA–poly(propene-*alt*-maleic anhydride), PEMA–poly(ethene-*alt*-maleic anhydride)) investigated by fluorescence microscopy. The normalized fluorescence intensity of surface-adsorbed rhodamine-labeled FN is shown over a time range of 48 h. For details see [35]. Reprinted with permission from Pompe et al. [33]. Copyright 2011, Elsevier.

exchange yielded a substrate-dependent double-exponential decay, which was explained by the presence of two different FN species—one weakly bound and a more strongly bound and denatured one. Herein a correlation of FN affinity to the polymer substrates is observed with strong interaction on hydrophobic surfaces and weak interaction on more polar surfaces.

As most adhesion ligands are large biopolymers, cooperative binding mechanisms can explain the complex sorption processes. By the Vroman effect, smaller and fast-diffusing proteins adsorb at an interface first but are displaced by bigger proteins later, because of the latter having greater affinity for the interface [43]. Opposite effects can also be observed as large proteins can be displaced in a concentration-dependent manner by small proteins from substrate interfaces due to the subsequent disruption of the multi-bonds of the large proteins to the substrate interface [21,35].

### 3.3.2 Ligand Mobility

Sorption processes do not always have to result in proteins being completely exchanged at the interface; instead, partial desorption or adsorption of ligands can also occur, resulting in movement across the interface, be it in a diffusive fashion or directed by a cell. Along with motion across an interface, one has to consider frictional processes, which on a molecular level may be modeled as successive binding and unbinding events [33]. Such molecular friction of ligands due to thermal breaking of multiple non-covalent bonds under external load will give rise to dissipation, as it was proposed for FN reorganization by adherent cells (Figs. 3.3a,b). Although the naturally occurring FN fibril formation is associated with cell-mediated reorganization of FN at interfaces, molecular friction is oftentimes overlooked in this context. However, the above study showed, that in such a case the drift velocity, $v_d$, of FN ligands along the substrate surface can be approximated as

$$v_d = \mu F_f, \tag{3.1}$$

where the ligand mobility $\mu$ has been introduced with

$$\mu = \delta^2 \omega_0 / k_B T \cdot \exp(-\Delta G_{FN} / k_B T). \tag{3.2}$$

Therein the laterally acting force $F_f$ changes the probability for the ligand to cross the energy barrier $\Delta G_{FN}$ between adjacent binding sites by thermal activation due to a thermal energy $k_B T$. Further parameters are $\delta$, the distance between intracellular binding sites, and $\omega_0$, a characteristic vibrational frequency.

(a)

(a)

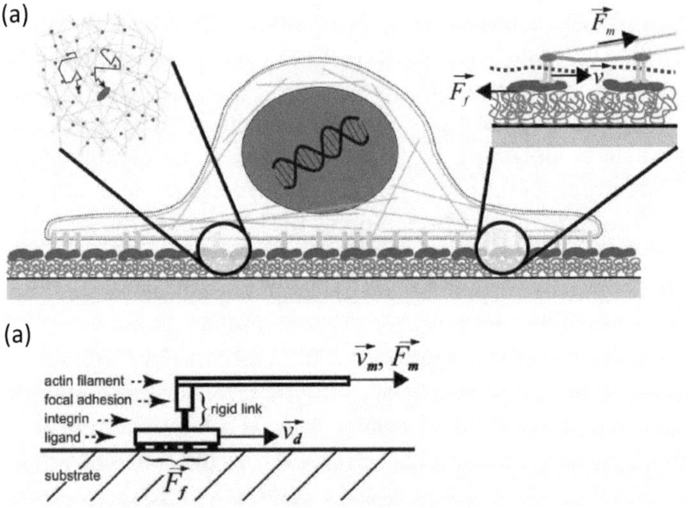


**Figure 3.3** Scheme of molecular friction processes during cell adhesion on non-covalently bound FN ligands. (a) Schemes of slow, random, and fast, directed motion of integrin-FN complexes within the interlayer between the cell membrane and the substrate surface. Left inset: Random reorganization of FN ligands by a myosin-driven translocation along the actin cortex network. Right inset: Constant motion of integrin-FN complexes inside a focal adhesion with equilibrium between molecular friction at the ligand-polymer interface and myosin-motor forces transferred via an integrin-ligand complex, proteins inside the focal adhesion, and the cytoskeleton. (b) Scheme of the simplifying model for explaining the friction force at the ligand-substrate interface. The binding of integrin receptors to extracellular adhesion ligands like fibronectin couples the myosin-driven, force-generating actin stress fibers via a rigid link to the frictional movement of the ligand along the surface of the supporting material surface. This movement of the noncovalently anchored ligand generates a sliding friction force due to the thermally activated, stochastic breakage of the weak bonds between protein and substrate. Reprinted with permission from Pompe et al. [33]. Copyright 2011, Elsevier.

Within this model a correlation between ligand mobility and the affinity of the ligands to the substrate surface could be demonstrated [33] leading to a regulation of cell traction forces. On hydrophobic surfaces with a high FN affinity the diffusive mobility of FN was decreased leading to an increase in cell traction force. A similar correlation was observed for the corresponding cell receptors, namely integrins, in another study with a modulated mobility on materials interfaces with varying hydrophobicity and respective ligand affinity [46].

Computer simulations also offer insight into molecular friction between biopolymers and interfaces. In a simulation of lateral pulling experiments of a small peptide, hydrogen bonds were shown to be the source of frictional forces [37]. However, the generality of such results may be limited due to the simulations' time scales and speed regimes with pulling speeds far above physiological conditions. For instance, the finding from the computer simulation seems to contradict the above-mentioned influence of hydrophobicity on ligand mobility as the specific interactions between amino acids of the peptide and interfaces by hydrogen bonds inhibit mobility while peptides can move fairly freely across hydrophobic interfaces, as there are no local and specific interactions [37]. Nevertheless, simulations have to be used in future to learn more about mechanisms involved in ligand mobility and understanding ligand friction in cell adhesion, although the used models have to be adapted to other time and length scales.

## 3.4 Cell Response to Non-Covalently Attached Adhesion Ligands

### 3.4.1 Characteristics of Cell Adhesion Sites in Response to Non-Covalently Attached Ligands in 2D

The process of cell adhesion via different receptors, ligands and factors in a 2D environment has been the subject of numerous studies [10,12,29]. On the subject of dissipative cell–matrix interactions, there are only a few studies dealing with adhesion to well-defined amounts and kinds of preadsorbed adhesion ligands, i.e., before cells can generate any substantial ECM themselves.

Two of the most commonly used adhesion ligands in these studies are vitronectin and the above-mentioned FN, and their corresponding cell receptors, integrins $\alpha_v\beta_3$ and $\alpha_5\beta_1$, respectively.

The behavior of the integrin receptors coupled to non-covalently bound ligands (as opposed to covalently bound ligands) has been intensely investigated [14,17,26,41]. In one approach, the co-localization of integrins and FN during the remodeling of ECM by fibroblasts on glass substrates was investigated [26]. During this process, two types of adhesion clusters containing integrins were identified: focal adhesions in the cell periphery and fibrillar adhesions below the central cell region. Focal adhesions mainly contain $\alpha_v\beta_3$ integrins along with vinculin, paxillin, focal adhesion kinase (FAK), and phosphotyrosines [26,45]. These adhesion clusters are in close proximity to the substrate surface, endure strong tension as they are coupled to actin stress fibers and are important for cell migration [38]. On the other hand, fibrillar adhesions primarily consist of $\alpha_5\beta_1$ integrin and have a high tensin content [26]. They originate in focal adhesions by a process, which has been shown to depend on mobile FN ligands [17,45]. Thereby FN is organized into fibrils due to the translocation of FN-coupled $\alpha_5\beta_1$ integrins out of focal adhesions caused by actomyosin activity, leading to the formation of elongated adhesion sites and self-assembly of fibers, called FN fibrillogenesis [26]. Despite the close association between fibrillar adhesions and actin stress fibers, along which the $\alpha_5\beta_1$ integrin translocation occurs, no strong forces act on fibrillar adhesions.

### 3.4.2 Fibronectin Matrix Remodeling during Cell Adhesion

On materials surfaces the strength of interaction between FN and the underlying substrate is a main parameter controlling adhesion site formation and matrix remodeling. By covalent attachment of FN to interfaces, and thus immobilization, the appearance of FN fibrils and fibrillar adhesions is suppressed and accordingly, no actin stress fibers can be observed [17,28,41]. Additionally, materials are designed to hold certain degrees of ligand mobility, thus eliciting a differential cell response in terms of matrix remodeling, adhesion site arrangement and force generation by the cells.

One major contribution to the affinity of FN to interfaces is made by polar and hydrophobic interactions. Surface modification of glass substrates to alter the wettability of the substrate, for instance, changes in the chemical composition [14,41,46] or copolymer coatings with variable proportions of polar and apolar groups [29], are utilized to investigate the influence of ligand affinity on cell adhesion. For example, studies showed an impaired FN remodeling and fibrillar adhesion formation for strongly bound FN [28,41] and, accordingly, a fast moving $\alpha_5\beta_1$ integrin species associated with FN fibrillogenesis could not be detected [41,46].

Furthermore, electrostatic interactions also have an impact on ligand–substrate interactions, as, for instance, FN bears a negative charge under physiological buffer conditions. Self-assembled monolayers (SAM) bearing carboxylic acid groups provide a higher FN mobility along with enhanced translocation of $\alpha_5\beta_1$ integrins out of focal adhesions and FN fibril formation, contrary to stronger FN anchorage to amine groups [11,18].

### 3.4.3 Cell Force Balance on Gradually Mobile Ligands

The correlation between cell response and ligand–substrate interactions might indicate a global force regulation at adhesion sites throughout the whole cell, which is controlled by the strength of ligand attachment. Not only does the amount of fibril formation change with ligand mobility, but in addition the size and density of adhesion sites correspond to gradual differences in ligand–substrate interactions [30]. Thereby, the cell reacts to high FN mobility by formation of smaller, and thus weaker, adhesion sites. This is further reflected by differences in the microscopic fibril structure as well [23,30,31].

These findings are further corroborated by traction force measurements on polyacrylamide hydrogels with different kinds of maleic acid copolymer coatings to tune the hydrophobicity of the interface [32]. Figure 3.4a shows how hydrophobicity, and thus ligand affinity, influences the maximum traction stress $T_{max}$ that human umbilical cord vein cells exert on these substrates. This is in line with the above-mentioned hypothesis that cell receptor forces correlate with dissipative molecular friction caused by the non-covalent binding of the ligand to the substrate [33]. Figure 3.4b depicts a decrease of the strain energy $U$ of the cells with increasing elastic modulus, which can be attributed to the modulation of the traction force by the molecular ligand friction on the substrate surface independent of the elastic modulus of the substrate itself [32]. These different responses of the cell's force apparatus consolidate the idea of different biophysical mechanisms for the detection of ligand affinity and substrate stiffness.

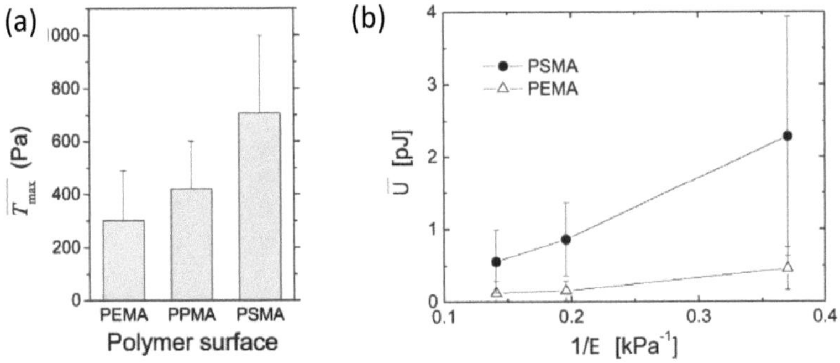


**Figure 3.4** Cell traction forces on substrates with graded ligand affinity and stiffness. (a) Maximum traction force $T_{max}$ on maleic acid copolymers with graded ligand affinity (PSMA–poly(styrene-*alt*-maleic anhydride > PPMA–poly(propene-*alt*-maleic anhydride) > PEMA–poly(ethene-*alt*-maleic anhydride)) (b) Strain energy U and elastic modulus E for PSMA and PEMA. Reprinted with permission from Pompe et al. [32,33]. Copyright 2009 and 2011, Elsevier.

### 3.4.4 Pattern Formation of Cell Adhesion Sites on Non-Covalently Attached Fibronectin

The adaptive scaling of adhesion sites in dependence on ligand affinity as proposed from the above-mentioned studies [30,31] can be paralleled by the patterns of adhesive contacts at the feet of animals like insects, spiders, and lizards. Instead of one single large contact face, fragmentation into a microscopic pattern of many tiny contact faces allows them to cling to walls and move across highly inclined planes due to optimized load transfer. The contact forces can be described by a micromechanical mechanism [1,2].

The pull-off force $F_c$ of an adhesive contact with the size $R$ (see Fig. 3.5) scales with $\alpha \cdot \gamma \cdot R$ according to Johnson, Kendall, and Roberts (JKR model) [16]. Herein, $\gamma$ is the interfacial energy and the factor $\alpha$ accounts for attractive forces outside the contact region [15]. An increase of the overall contact force $F_{c,total}$ can be attained by patterning of the single contact area into an array of numerous smaller contacts, assuming a distance of $z$ between contacts and an equal size reduction of the contacts with $R \propto z$. This yields a scaling of $\gamma/z$ for the contact force $F_{c,total}$.

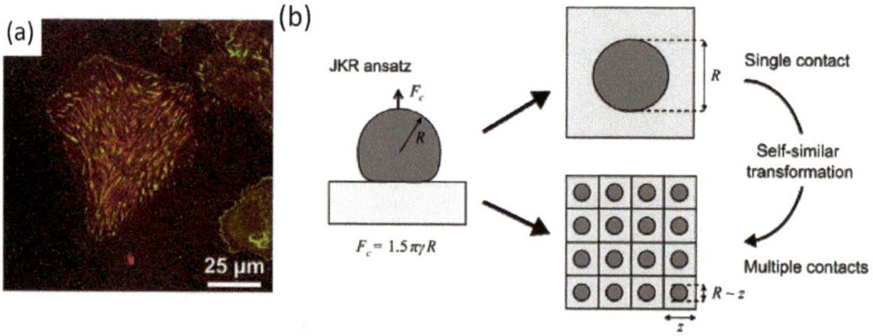


**Figure 3.5** Model for adaption of cell adhesion site pattern in response to substrate-ligand affinity. (a) The adhesion sites (red: FN fibrils, green: focal adhesions with phosphotyrosine staining) of an adherent endothelial cell 1 h after seeding. (b) Scheme of the model explaining the overall contact force after the micromechanical JKR model to scale with the density of the adhesive contacts. Reprinted with permission from Müller et al. [24]. Copyright by the Royal Chemical Society.

This model can be transferred to the above-mentioned case of focal adhesion patterns of FN fibrils (see Fig. 3.5) by approximating the focal adhesions as contact planes with radius $R$ and spacing $z$. The interfacial energy $\gamma$ comprises the affinity of the FN ligands to the substrate and, in terms of an activation energy $E$, can be related to the time constant $\tau$ of the FN displacement experiment from Section 3.3.1 by a Boltzmann factor $\tau \propto \exp(E/k_B T)$. This approach results in the scaling of

$$F_{c,total} \propto \gamma/z \propto \ln(\tau)/z. \quad (3.3)$$

For different maleic acid copolymers, time constants of FN displacement (see Fig. 3.2) were determined that correlated with the ligand affinity of the substrate—ranging from 526 h for the hydrophobic PSMA/POMA, to 189 h for PEMA and 83 h for PPMA ([35] and unpublished data, see Table 3.1). In addition, the spacing between the FN fibrils, and thus focal adhesions, showed substrate-dependence as well with PSMA/POMA 2.6 μm, PEMA 2.2 μm, PPMA 1.8 μm [30]. If these values are combined to $\ln(\tau)/z$, all the polymer surfaces yield similar numbers (Table 3.1), indicating that the assumption of a scaling focal adhesion pattern regulated by changes in the interfacial energy might be correct. In addition, the values of the maximum traction stress $T_{max}$ [33] (see Fig. 3.4) can be related to the energy of a single adhesive patch leading to a constant ratio

$$T_{max}/(\ln\tau \cdot z). \tag{3.4}$$

Here the same agreement between all polymer surfaces was found (see Table 3.1), being consistent with the above-stated fact that the maximum traction stress is in close relationship to the interfacial energy and thus the ligand affinity.

**Table 3.1** Comparison of the micromechanical model of pattern formation of adhesion sites with different experimental data sets

| Surface | POMA/PSMA | PPMA | PEMA |
|---|---|---|---|
| $\tau$ [h] | 526 | 189 | 83 |
| $T_{max}$ [Pa] | 707 | 419 | 301 |
| $z$ [μm] | 2.6 | 2.2 | 1.8 |
| $\ln(\tau)/z$ | 2.4 | 2.4 | 2.5 |
| $T_{max}/(\ln(\tau)\cdot z)$ | 43 | 36 | 38 |

As shown in Table 3.1, experimental data of FN displacement studies [35], measurements of focal adhesion density [30] and traction force measurements [33] support the suggested micromechanical model of pattern formation of cell adhesion sites. The time constant $\tau$ of FN displacement and the maximum traction force $T_{max}$ correlate with the ligand affinity of the substrate. Modeling focal adhesions as contact planes of spacing $z$ and radius $R$ yields similar numbers for the ratio $\ln(\tau)/z$ that is connected to the contact force $F_c$ (see Eq. (3.3)). The energy of the contact is proportional to the ratio $T_{max}/(\ln(\tau)\cdot z)$ and independent of the ligand affinity of the surface.

### 3.4.5 Influence of Ligand Attachment on Intracellular Signaling

Structural changes in the cell adhesion sites and ECM are closely related to alterations in the signal transduction pathways on the intracellular side—an interdependency called mechanotransduction. In this context, adhesion receptors and other mechanosensitive proteins, including their cytoplasmatic domains, are subjected to conformational changes and phosphorylation events when forces act on them through the connection to the ECM [3,13]. Focal adhesions are rich in phosphotyrosine [17] and hence they are key players for the evolution of signaling cascades during cell adhesion.

Several studies focus on phosphorylation levels of FAK in correlation to a non-covalent ligand attachment scheme as this protein is involved in cell spreading and migration. Tyrosine-397 and tyrosine-861 phosphorylation has been investigated on SAM with different end groups [19]. Thereby high levels of phosphorylation were found for SAM with $NH_2$-end groups compared to COOH, consistent with the assumption of conformational changes of FN during adsorption. In another study, different correlations of FAK tyrosine-397 and tyrosine-861 phosphorylation with substrate stiffness and ligand affinity were revealed (Fig. 3.6) [32]. Whereas

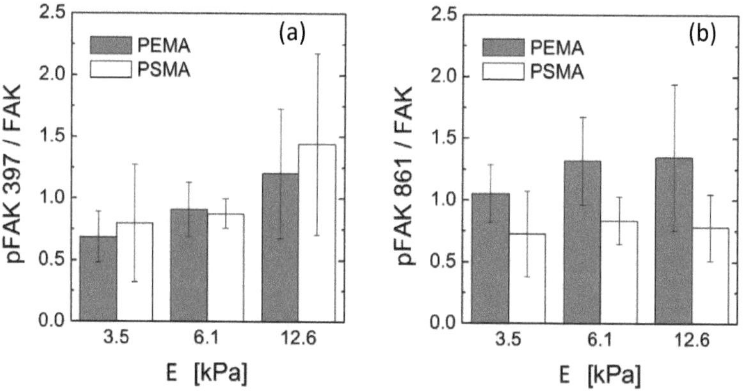


**Figure 3.6** FAK phosphorylation in dependence of substrate stiffness (elastic modulus *E*) and ligand affinity (PEMA–poly(ethene-*alt*-maleic anhydride) < PSMA–poly(styrene-*alt*-maleic anhydride)). (a) Phosphorylation on tyrosine-397. (b) Phosphorylation on tyrosine-861. Reprinted with permission from Pompe et al. [32]. Copyright 2009, Elsevier.

tyrosine-397 phosphorylation was augmented on stiffer substrates, no stiffness dependency could be observed for tyrosine-861. In contrast, levels of tyrosine-861 phosphorylation were significantly higher for weakly bound FN, though cell traction forces were lower on these substrates as well (see Fig. 3.4).

## 3.5 Conclusion

From the examples provided in this chapter, we believe that it is obvious that viscous and dissipative phenomena have to be addressed in studies on mechanotransduction in cell adhesion. The involved damping mechanisms in the biopolymer components will lead to different time regimes of the signaling components, both intracellular and extracellular. The impact of viscoelasticity of biomaterials and dynamic ligand–substrate interaction is currently underestimated in in vitro and in vivo studies, both on fundamental biophysical and biochemical levels as well as in biomedical applications. There is a need to consider non-covalent attachment schemes with a tuned affinity of adhesion ligands to cell culture scaffolds and to include viscoelastic characteristics in functional biomaterials systems. Similarly, methods and models for the analysis of viscoelastic properties of substrate materials and adhesion ligands as well as cellular responses on micrometer and molecular scale have to be developed and applied in order to reveal a comprehensive understanding of materials mechanical properties' effects on cell behavior. Based on that, we envision new criteria to arise for the design of materials scaffolds in biomedical applications.

### Acknowledgments

T. P., C. M., and A. M. acknowledge the financial support by the Deutsche Forschungsgemeinschaft (DFG PO 713/7-1 and Graduate School BuildMoNa).

### References

1. Arzt, E., Enders, S., and Gorb, S. (2002). Towards a micromechanical understanding of biological surface devices, *Z. Metallkd.*, **93**, 345–351.

2. Arzt, E., Gorb, S., and Spolenak, R. (2003). From micro to nano contacts in biological attachment devices, *Proc. Natl. Acad. Sci. U. S. A.*, **100**, 10603–10606.
3. Askari, J. A., Tynan, C. J., Webb, S. E., Martin-Fernandez, M. L., Ballestrem, C., and Humphries, M. J. (2010). Focal adhesions are sites of integrin extension, *J. Cell. Biol.*, **188**, 891–903.
4. Brown, A. E., and Discher, D. E. (2009). Conformational changes and signaling in cell and matrix physics, *Curr. Biol.*, **19**, R781–R789.
5. Buehler, M. J. (2006). Atomistic and continuum modeling of mechanical properties of collagen: Elasticity, fracture, and self-assembly, *J. Mater. Res.*, **21**, 1947–1961.
6. Calonder, C., Tie, Y., and Van Tassel, P. R. (2001). History dependence of protein adsorption kinetics, *Proc. Natl. Acad. Sci. U. S. A.*, **98**, 10664–10669.
7. Cameron, A. R., Frith, J. E., and Cooper-White, J. J. (2011). The influence of substrate creep on mesenchymal stem cell behaviour and phenotype, *Biomaterials*, **32**, 5979–5993.
8. Cranford, S. W., de Boer, J., van Blitterswijk, C., and Buehler, M. J. (2013). Materiomics: An-omics approach to biomaterials research, *Adv. Mater.*, **25**, 802–824.
9. Discher, D. E., Mooney, D. J., and Zandstra, P. W. (2009). Growth factors, matrices, and forces combine and control stem cells, *Science*, **324**, 1673–1677.
10. Faucheux, N., Schweiss, R., Lutzow, K., Werner, C., and Groth, T. (2004). Self-assembled monolayers with different terminating groups as model substrates for cell adhesion studies, *Biomaterials*, **25**, 2721–2730.
11. Faucheux, N., Tzoneva, R., Nagel, M. D., and Groth, T. (2006). The dependence of fibrillar adhesions in human fibroblasts on substratum chemistry, *Biomaterials*, **27**, 234–245.
12. Garcia, A. J., and Boettiger, D. (1999). Integrin-fibronectin interactions at the cell–material interface: Initial integrin binding and signaling, *Biomaterials*, **20**, 2427–2433.
13. Geiger, B., Spatz, J. P., and Bershadsky, A. D. (2009). Environmental sensing through focal adhesions, *Nat. Rev. Cell Biol.*, **10**, 21–33.
14. Groth, T., Altankov, G., Kostadinova, A., Krasteva, N., Albrecht, W., and Paul, D. (1999). Altered vitronectin receptor (alpha(v) integrin) function in fibroblasts adhering on hydrophobic glass, *J. Biomed. Mater. Res.*, **44**, 341–351.

15. Horn, R. G., Israelachvili, J. N., and Pribac, F. (1987). Measurement of the deformation and adhesion of solids in contact, *J. Colloid Interface Sci.*, **115**, 480–492.

16. Johnson, K. L., Kendall, K., and Roberts, A. D. (1971). Surface Energy and Contact of Elastic Solids, *Proc. R. Soc. A*, **324**, 301–313.

17. Katz, B. Z., Zamir, E., Bershadsky, A., Kam, Z., Yamada, K. M., and Geiger, B. (2000). Physical state of the extracellular matrix regulates the structure and molecular composition of cell–matrix adhesions, *Mol. Biol. Cell.*, **11**, 1047–1060.

18. Keselowsky, B. G., Collard, D. M., and Garcia, A. J. (2003). Surface chemistry modulates fibronectin conformation and directs integrin binding and specificity to control cell adhesion, *J. Biomed. Mater. Res. A*, **66**, 247–259.

19. Keselowsky, B. G., Collard, D. M., and Garcia, A. J. (2004). Surface chemistry modulates focal adhesion composition and signaling through changes in integrin binding, *Biomaterials*, **25**, 5947–5954.

20. Lecuit, T., Lenne, P. F., and Munro, E. (2011). Force generation, transmission, and integration during cell and tissue morphogenesis, *Annu. Rev. Cell. Dev. Biol.*, **27**, 157–184.

21. Lundstrom, I., and Elwing, H. (1990). Simple kinetic-models for protein exchange-reactions on solid-surfaces, *J. Colloid. Interfce Sci.*, **136**, 68–84.

22. MacKintosh, F. C., Käs, J., and Janmey, P. A. (1995). Elasticity of semiflexible biopolymer networks, *Phys. Rev. Lett.*, **75**, 4425–4428.

23. Müller, C., Stamov, D. R., Werner, C., and Pompe, T. (2012). Nanoscale characterization of cell receptors and binding sites on cell-derived extracellular matrices, *Ultramicroscopy*, **118**, 44–52.

24. Müller, C., Müller, A., and Pompe, T. (2013). Dissipative interactions in cell–matrix adhesion, *Soft Matter*, **9**, 6207–6216.

25. Murrell, M., Kamm, R., and Matsudaira, P. (2011). Substrate viscosity enhances correlation in epithelial sheet movement, *Biophys. J.*, **101**, 297–306.

26. Pankov, R., Cukierman, E., Katz, B. Z., Matsumoto, K., Lin, D. C., Lin, S., Hahn, C., and Yamada, K. M. (2000). Integrin dynamics and matrix assembly: tensin-dependent translocation of alpha(5)beta(1) integrins promotes early fibronectin fibrillogenesis, *J. Cell. Biol.*, **148**, 1075–1090.

27. Piechocka, I. K., van Oosten, A. S. G., Breuls, R. G. M., and Koenderink, G. H. (2011). Rheology of heterotypic collagen networks, *Biomacromolecules*, **12**, 2797–2805.

28. Pompe, T., Kobe, F., Salchert, K., Jorgensen, B., Oswald, J., and Werner, C. (2003). Fibronectin anchorage to polymer substrates controls the initial phase of endothelial cell adhesion, *J. Biomed. Mater. Res. A*, **67**, 647–657.

29. Pompe, T., Zschoche, S., Herold, N., Salchert, K., Gouzy, M. F., Sperling, C., and Werner, C. (2003). Maleic anhydride copolymers—a versatile platform for molecular biosurface engineering, *Biomacromolecules*, **4**, 1072–1079.

30. Pompe, T., Keller, K., Mitdank, C., and Werner, C. (2005). Fibronectin fibril pattern displays the force balance of cell–matrix adhesion, *Eur. Biophys. J.*, **34**, 1049–1056.

31. Pompe, T., Starruss, J., Bobeth, M., and Pompe, W. (2006). Modeling of pattern development during fibronectin nanofibril formation, *Biointerphases*, **1**, 93–97.

32. Pompe, T., Glorius, S., Bischoff, T., Uhlmann, I., Kaufmann, M., Brenner, S., and Werner, C. (2009). Dissecting the impact of matrix anchorage and elasticity in cell adhesion, *Biophys. J.*, **97**, 2154–2163.

33. Pompe, T., Kaufmann, M., Kasimir, M., Johne, S., Glorius, S., Renner, L., Bobeth, M., Pompe, W., and Werner, C. (2011). Friction-controlled traction force in cell adhesion, *Biophys. J.*, **101**, 1863–1870.

34. Rabe, M., Verdes, D., and Seeger, S. (2011). Understanding protein adsorption phenomena at solid surfaces, *Adv. Colloid. Interface Sci.*, **162**, 87–106.

35. Renner, L., Jorgensen, B., Markowski, M., Salchert, K., Werner, C., and Pompe, T. (2004). Control fibronectin displacement on polymer substrates to influence endothelial cell behaviour, *J. Mater. Sci. Mater. Med.*, **15**, 385–388.

36. Renner, L., Pompe, T., Salchert, K., and Werner, C. (2005). Fibronectin displacement at polymer surfaces, *Langmuir*, **21**, 4571–4577.

37. Serr, A., Horinek, D., and Netz, R. R. (2008). Polypeptide friction and adhesion on hydrophobic and hydrophilic surfaces: A molecular dynamics case study, *J. Am. Chem. Soc.*, **130**, 12408–12413.

38. Stricker, J., Aratyn-Schaus, Y., Oakes, P. W., and Gardel, M. L. (2011). Spatiotemporal constraints on the force-dependent growth of focal adhesions, *Biophys. J.*, **100**, 2883–2893.

39. Stuart, K., and Panitch, A. (2008). Characterization of gels composed of blends of collagen I, collagen III, and chondroitin sulfate, *Biomacromolecules*, **10**, 25–31.

40. Trappmann, B., Gautrot, J. E., Connelly, J. T., Strange, D. G., Li, Y., Oyen, M. L., Cohen Stuart, M. A., Boehm, H., Li, B., Vogel, V., Spatz, J. P., Watt, F. M., and Huck, W. T. (2012). Extracellular-matrix tethering regulates stem-cell fate, *Nat. Mater.*, **11**, 642–649.

41. Tzoneva, R., Groth, T., Altankov, G., and Paul, D. (2002). Remodeling of fibrinogen by endothelial cells in dependence on fibronectin matrix assembly. Effect of substratum wettability, *J. Mater. Sci. Mater. M.*, **13**, 1235–1244.

42. Van Tassel, P. R., Talbot, J., Tarjus, G., and Viot, P. (1996). Kinetics of irreversible adsorption with a particle conformatical change: A density expansion apporoach, *Phys. Rev. E*, **53**, 785–798.

43. Vroman, L., Adams, A. L., Fischer, G. C., and Munoz, P. C. (1980). Interaction of high molecular-weight kininogen, factor-Xii, and fibrinogen in plasma at interfaces, *Blood*, **55**, 156–159.

44. Wolff, L., and Kroy, K. (2012). Minimal model for the inelastic mechanics of biopolymer networks and cells, *Phys. Rev. E*, **86**, 040901.

45. Zamir, E., Katz, M., Posen, Y., Erez, N., Yamada, K. M., Katz, B. Z., Lin, S., Lin, D. C., Bershadsky, A., Kam, Z., and Geiger, B. (2000). Dynamics and segregation of cell–matrix adhesions in cultured fibroblasts, *Nat. Cell Biol.*, **2**, 191–196.

46. Zlatanov, I., Groth, T., Lendlein, A., and Altankov, G. (2005). Dynamics of beta 1-integrins in living fibroblasts: Effect of substratum wettability, *Biophys. J.*, **89**, 3555–3562.

## Chapter 4

# Traction Microscopy

**Ramaswamy Krishnan,[a] Dhananjay Tambe,[b] Robert L. Steward,[b] Corey C. Hardin,[c] and James P. Butler[b,d]**

[a]*Center for Vascular Biology Research,*
*Beth Israel Deaconess Medical Center, Boston, MA 02215, USA*
[b]*Department of Environmental Health,*
*Harvard School of Public Health, Boston, MA 02115, USA*
[c]*Division of Pulmonary and Critical Care Medicine,*
*Massachusetts General Hospital, Boston, MA 02114, USA*
[d]*Division of Sleep Medicine, Department of Medicine,*
*Harvard Medical School, Boston, MA 02115, USA*

jbutler@hsph.harvard.edu

Every adherent cell exerts contractile forces—called tractions—on its substrate. These forces can now be visualized on the submicron scale through the implementation of one of a variety of technologies collectively known as traction microscopy (TM). We provide here a historical review of TM and focus on its application to mechanosensing of the adherent cell, i.e., how the adherent cell senses and responds to mechanical properties of its microenvironment.

## 4.1 Introduction

Major advances in cell biology rely in large part on advances in imaging technologies. As an example, consider the now burgeoning

*Cells, Forces, and the Microenvironment*
Edited by Charles M. Cuerrier and Andrew E. Pelling

ISBN 978-981-4613-36-1 (Hardcover), 978-981-4613-37-8 (eBook)
www.panstanford.com

field of mechanobiology. Despite the widely held view that the single cell is a force-bearing member capable of both generating and responding to physical force, underlying forces were entirely invisible to observation. These hidden forces can now be visualized through the force imaging technology called TM. Using TM, forces have been implicated in cellular functions as diverse as migration, spreading, differentiation, and growth, and diseases as profound as asthma, acute lung injury, and cancer [1–13].

In this chapter, we will provide a historical review of TM with an emphasis on key technological advances. We will bring these advances to bear upon the topic of mechanosensing in the adherent cell, i.e., how the cell senses and responds to physical properties of its microenvironment. By physical properties, we refer to stiffness, thickness, mechanical stretch, and fluid flow, as well as mechanical interactions among neighboring cells. The roles of specific biological features such as soluble factors, biochemical constituents of the extracellular matrix, and downstream effectors of cell signaling have been elegantly described elsewhere [14–26]. Here we will focus specifically on the tractions themselves and describe their spatio-temporal regulation and evolution in the cell through the newly discovered physical principles of fluidization, plithotaxis, and kenotaxis.

## 4.2 What Are Tractions?

In the context of cells grown on substrates, traction is defined as the local force per unit area that is imposed on the microenvironment by an adherent cell (Fig. 4.1). It arises from physical interactions between the microenvironment, adhesion molecules, cytoskeletal filaments, and molecular motors. These interactions are powered by three cellular processes—the hydrolysis of adenosine triphosphate (ATP), the polymerization of actin and other cytoskeletal fibers, and the assembly/disassembly of cell–cell and cell–substrate adhesions. Direct visualization of these processes together with the traction forces they prescribe is now possible through TM. For example, using Fourier transform traction microscopy (see Section 4.3.3), Prager-Khoutorsky et al. [10] measured tractions during cellular polarization and have implicated protein tyrosine kinase molecules in regulating those forces (Fig. 4.2).

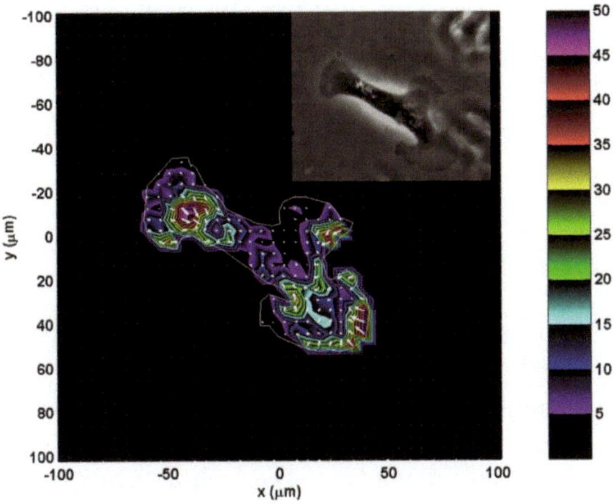


**Figure 4.1** Representative traction map and (inset) phase contrast for a rat pulmonary microvascular endothelial cell. The color map shows the traction magnitude in Pascal (Pa). Reprinted with permission from An et al. [100].

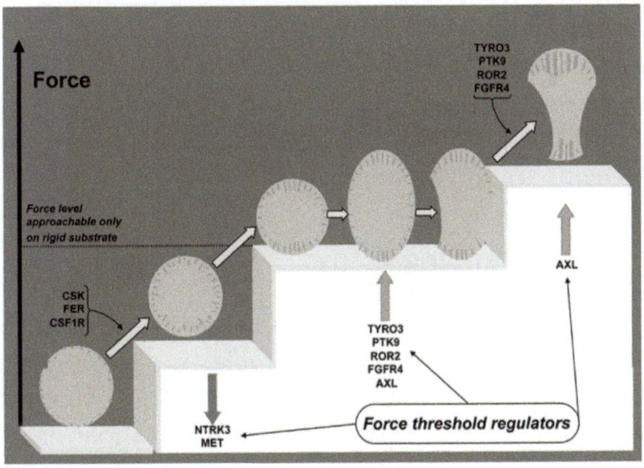


**Figure 4.2** Tractions guide cell polarization. For the cell to transition from a rounded to a polarized morphology, it is required to overcome threshold levels of traction (the cumulative magnitude of traction is plotted on the $Y$ axis). Such traction regulation is mediated by rigidity of the cellular substrate as well as protein tyrosine kinase dependent molecule checkpoints. Reprinted with permission from Prager-Khoutorsky et al. [10].

Similar structure–function relationships have been established during cell stretch, spreading, adhesion, migration, and barrier function [2,4–6,9,11,13,27–33]. To make such relationships visible, key technological advances in TM have been pivotal, as described next.

## 4.3 How Are Tractions Measured?

### 4.3.1 Wrinkling Assays

The very first assay to measure traction is credited to Harris and colleagues [34]. They cultured individual fibroblasts on the surface of an elastic substrate. As the cell crawls, it wrinkles the substrate. Based on the wrinkles, together with a priori knowledge of the elastic properties of the substrate, they calculated tractions. This approach has since been implemented in diverse cell types and physiological contexts [34–45]. While seminal in approach, wrinkling assays have four major limitations. First, the substrates are cumbersome to manufacture, and difficult to reproduce. Second, the substrates need to be extensively pre-calibrated on a well by well basis. Third, the substrates are limited to a narrow range of elastic stiffness. Finally, wrinkling is a strongly non-linear and often hysteretic phenomenon, similar to Euler buckling of beams; it follows that quantitative measurements of widely distributed traction fields are correspondingly difficult [46].

### 4.3.2 Beam-Bending Assays

The next embodiment of TM is credited to Sheetz and colleagues [47]. They cultured individual fibroblasts on the surface of a micromachined substrate. Embedded in the substrate were thousands of pads that rest on the ends of cantilever beams. As the cell craws over the pads, the beams bend laterally. The extent of bending multiplied by the stiffness of the cantilever beam is proportional to the local traction. Subcellular tractions were found to fluctuate strongly and dramatically in space and in time. While elegant in design, this technology was limited in spatial extent because measurements were confined to one spatial axis alone.

To overcome this limitation, a second generation of micropillar assays soon followed [48,49]. In these approaches to traction mapping, the isolated cell or the cell monolayer is cultured upon a bed of microneedles, whose sizes are sufficiently small to permit probes of subcellular changes. Each microneedle deflects independently and therefore works as an individual force sensor capable of measuring the in-plane traction field. External perturbations can be imposed as well [50]. Finally, recent innovations make this technique amenable to the measurements of intercellular forces [32,51].

While innovative in approach, bending assays are technically challenging to implement. Moreover, externally imposed perturbations are non-uniform since they are currently implemented through point loads applied via magnetic microneedles, and characterizing the imposed strain fields are correspondingly difficult.

### 4.3.3 Deformation Assays

The most commonly used embodiment of TM is the deformation assay. It was first described by Dembo and Wang [46]; they demonstrated that tractions could be mapped quantitatively from knowledge of the displacement field induced by the cell in a flexible substrate upon which it is adherent. The substrate is prepared with a layer of fluorescent beads (typically 0.2–0.5 μm in diameter) embedded within it. As the cell contracts, the embedded beads are displaced. By comparing positions of the displaced beads against a reference image of undisplaced beads (typically an image taken after detaching the cells from the substrate) one can compute the displacement field. From the displacement field, together with knowledge of the modulus of rigidity of the cellular substrate, the traction field can be computed through implementation of the inverse Boussinesq solution [52]. A key computational difficulty associated with this implementation is that displacements are not determined just by the local tractions in that neighborhood; this follows from the fact that point-like local tractions induce displacements over an extended spatial range. In a sense, the displacement field is a special kind of "smeared" version of the tractions. Reversing this process, i.e., recovering the tractions from the displacement field, involves "undoing" this smearing, a so-called ill-posed inverse problem. It is ill-posed in the sense that errors in the displacement field result in errors in the recovered tractions which increase with higher spatial frequencies. Due to this spatial coupling of tractions to displacements at a distance, inverse methods restricted to real space in recovering the tractions therefore necessarily involves inversion of very large matrices, and is computationally intensive.

The insight of Butler et al. [1] was to realize that this difficulty can be alleviated by solving the traction problem in Fourier space. Fourier transform traction cytometry (FTTC) starts with the well-known Boussinesq solution to the problem of computing displacements at a point $\vec{r}'$ resulting from a single point traction applied at a point $\vec{r}$. Due to translation invariance, this can only be a function of $(\vec{r}-\vec{r}')$. Denoting the Green's function or kernel mapping the traction to the displacement by $\vec{K}=\vec{K}(\vec{r}-\vec{r}')$ it follows that the full displacement field $\vec{u}(\vec{r})$ is obtained by convolution of the traction field with this kernel, i.e., by integrating over the traction distribution, $\vec{T}$:

$$\vec{u}(\vec{r})=\vec{K}\otimes\vec{T}=\int\vec{K}(\vec{r}-\vec{r}')T(\vec{r}')d\vec{r}',$$

where $T(\vec{r}')$ is the traction vector at $\vec{r}'$and $d\vec{r}'$ is the derivative with respect to $\vec{r}'$. By the convolution theorem, its Fourier transform is the product of the Fourier transforms of the individual functions. The significance of these manipulations is that the system is spatially uncoupled in Fourier space—whereas displacements at a point, $x$, $y$ may arise from tractions at many other points, displacements for each Fourier wave number $k$ are independent of those arising from other wave numbers. Mathematically, we have moved from a large non-diagonal matrix in $x, y$ to a two-dimensional matrix in $k$ space, which when coupled with the fast Fourier transform greatly reduces the difficulty of the computational task. This in turn implies not just a quantitative increase in computational efficiency but a qualitative advance insofar as multiple images taken over time can now be analyzed for probes of dynamic behavior of cells or cellular monolayer sheets either in baseline conditions or in response to specific perturbations, both pharmacologic as well as mechanical.

An additional attractive feature of the FTTC approach is that it yields expressions for low order moments of the tractions such as the net contractile moment (a measure of the net tendency of the cell to dilate or contract the substrate) and the total strain energy applied to the substrate, two convenient physical metrics to summarize the spatially complex forces.

Using this approach, tractions that are recovered from the displacement field are exact, i.e., when inverted the recovered tractions yield precisely the observed displacements. However, some of these displacements are still spurious—arising not from application of traction by a cell, but from intrinsic noise as well as small discrepancies in registration with the reference image. To solve this problem, all tractions outside of the cell boundary can be constrained to be zero. This approach is called constrained FTTC and requires an iterative solution as well as manual tracing of the cell boundary [1]. Here the iterations converge rapidly, but one must be careful about the user-defined boundaries of the cell. Even with DIC microscopy, some cellular processes may not be visible; treating these remote tractions as artifact is itself an error introduced by cell boundary tracing.

Recent advances to the FTTC method include traction reconstruction with point forces (TRPF) [33,53–55] and micropatterning of displacement grids [56]. While TRPF has improved the spatial resolution of TM to approximately 1 μm, micropatterning of displacement grids has simplified measurements since multiple cellular fields of views can now be incorporated without the need for cell removal. Finally, FTTC is now being implemented in a higher throughput format [57,58], and instead of adopting the Boussinesq solution, a 3D finite element method has been implemented to compute tractions from deformation of the substrate [59].

## 4.4 How Are Tractions Regulated?

### 4.4.1 Mechanical Stress Associated with Stiffness of the Extracellular Matrix

A powerful tactile set-point of traction is stiffness of the extracellular matrix (ECM) [16,60–67]. In principle, the stiffer the ECM, the greater the traction. However, this relationship is also critically influenced by cell spreading area [27–29] and cell confluence [68]. To what extent these factors might be intertwined remains an area of much recent investigation.

In the context of human health, the relationship between substrate stiffness and traction has profound importance. For example, in many serious diseases, including cancer, diabetes, hypertension, pulmonary fibrosis, and acute lung injury, the cellular substrate stiffens. In response to similar levels of substrate stiffening in vitro, the macrophage, fibroblast, epithelial cell or the endothelial cell enhances its traction [9,10,31,69]. Such traction enhancement can promote maladaptive changes; in the endothelial monolayer, for instance, increased substrate stiffness promotes paracellular gap formation [9,70].

### 4.4.2 Mechanical Stress Associated with Thickness of the Extracellular Matrix

For the cell that is adherent upon a flat elastic substrate, tractions are strongly influenced by the presence of a rigid foundation below the substrate. While this phenomenon is now well elucidated, there is limited consensus on the depth, $h_0$, to which the cells sense [22,59,71–76]. Indeed, it has been postulated that $h_0$ is proportional to focal adhesion size [71,74,77], cell size [22], and the ratio of cell size to substrate thickness [76]. Accordingly, estimates of $h_0$ have varied from a few microns to hundreds of microns [72,73]. These discrepancies may be reconciled by considering the limits for the validity of the Boussinesq solution, as described below [1,76].

A cell adherent on a substrate of finite depth has three physical length scales: depth of the substrate ($h$), size of the focal adhesion ($a$), and size of the cell ($L$). When $h >> a$ and $h >> L$ the effects of substrate depth are seen to be negligible. By contrast, when $h >> a$ but $h \approx L$, the presence of a rigid base—e.g., the plastic or glass coverslip in the laboratory preparation—does not affect the distribution of forces within the focal adhesion but does affect the distribution of forces within the cell. Finally, when $h \approx L$ and $h << L$, the presence of a rigid base affects the distribution of forces both within the focal adhesion and within the cell. While now well established in the laboratory preparation, it remains to be seen how these effects might impact cellular responses in vivo especially in the vicinity of a tumor or plaque that is characterized by stiffening and thickening of the cellular ECM.

### 4.4.3 Mechanical Stress Associated with 3-D Deformation: Out-of-Plane Tractions

When the cells adhere on a flat substrate, they exert tractions which not only have in-plane components ($T_xT_y$) but may also have an out-of-plane component ($T_z$). Until recently, however, $T_z$ was assumed to be negligible. Using a sophisticated confocal microscope together with intricate imaging and numerical analysis tools, it has recently been shown that a single, isolated, and migrating cell can exert tractions where $T_z$ is comparable to $T_x$ or $T_y$ [78,79]. A consensus model arising from these studies is that the periphery of the cell experiences an out-of-plane moment [80]. A similar claim is also made for an advancing sheet of cells [81].

For accuracy of tractions, is it necessary to perform sophisticated three-dimensional measurements? Or is it sufficient to perform straightforward two-dimensional measurement? The answer to these questions, pivots on the ratio of magnitude of cell height ($Z$) to magnitude of cell size ($L$) [12]. Geometrically, the ratio $Z/L$ is expected to be proportional to the ratio ($T_z/T_x$ or $T_z/T_y$) [12]. For single, isolated, and migrating cells, such as dictyostelium, $Z/L$ can be rather high [79]. By contrast, for an advancing front of a cell sheet, $Z/L$ is quite low [2,82]. Although both of these limits and associated situations are well studied, the value of $Z/L$ where $T_z$ begins to become important remains open [12,79].

### 4.4.4 Mechanical Stress Associated with Stretch of the Extracellular Matrix

With every breath, heartbeat or peristaltic movement diverse cells in the body are subjected to appreciable mechanical stretch. In response to comparable stretch in vitro, the cell is seen to modulate its traction in one of two distinct ways: (i) traction enhancement via reinforcement [83–85] or (ii) traction ablation via fluidization [4,5,86–89]. Reinforcement responses are mediated by signaling intermediaries including Rho, Rho-associated kinase (ROCK), mDia, myosin-light chain kinase, tyrosine kinase Src, stretch activated ion channels, and calcium ion influx. These events orchestrate F-actin polymerization and focal adhesion assembly at the site of force application leading to a net increase in traction. In direct contrast stands fluidization, exemplified by depolymerization of F-actin, and ablation of cell traction. Effecting these changes are physical mechanisms that appear considerably more generic and substantially less detailed compared with known molecular signaling cascades. Are the opposing responses of reinforcement and fluidization at play all the time, or in some physiological circumstances might one response be dominant?

To resolve this paradox, Krishnan et al. [5] developed a novel embodiment of TM called cell mapping rheometry (CMR). Implementation of CMR in the isolated human airway smooth muscle (HASM) cell that is subjected to a transient biaxial stretch-unstretch perturbation of 4 s duration is shown in Fig. 4.3. The tractions showed a dramatic and prompt decrease followed by a slow recovery, which over a longer time scale recovered to the pre-stretch baseline value. For graded stretch magnitudes within the physiological range (2.5%, 5%, and 10% isotropic biaxial strains, respectively, with 0% strain corresponding to a time-control), as the magnitude of the stretch was increased the ablation of traction became progressively larger ($p < 0.005$, two-tailed unpaired Student's *t* test). Traction recovery also varied in a load-dependent manner with the largest stretch magnitude showing the fastest recovery.

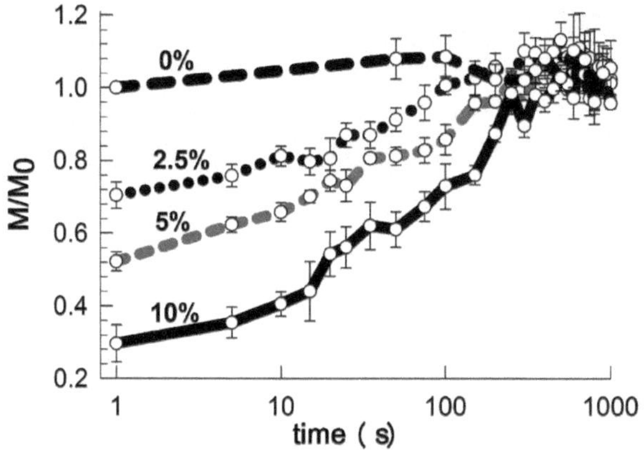


**Figure 4.3** In response to a transient biaxial stretch, the tractions represented by the contractile moment, *M*, promptly ablate in a dose-dependent manner (0 to 10%). This tendency of the cell is consistent with the physical response called fluidization. The fluidized cell subsequently recovers its traction to the pre-stretch baseline value, $M_0$. Reprinted with permission from Krishnan et al. [5].

When a similar protocol as describe above was administered through a localized cell stretch as opposed to a homogeneous biaxial stretch, the cell elicited a dramatically different response. While it promptly fluidized much to the same extent as in the case of biaxial stretch, the subsequent recovery differed dramatically. In the long time limit, tractions exceeded baseline values by as much as 92% ($p$ = 0.04), which is a response consistent with reinforcement [83–85].

These findings showed that while the reinforcement response is confined to non-homogeneous cell deformation as would be imposed by needle poking or bead pulling, it does not scale up to the case of repeated homogeneous cell stretch as would be associated with breathing or pulsatile blood flow. As regards the latter, recent findings implicate fluidization as a crucial mechanism for stretch regulation in the intact bronchial airway [90], and, during reorientation of the endothelial cell [4].

### 4.4.5 Mechanical Stress Associated with Fluid Flow in the Extracellular Matrix

As blood circulates in the body, the endothelial cells lining the inner side of blood vessels experience significant fluid shear stress. In response to comparable stress in vitro, the endothelial cell contracts, remodels, and realigns itself along the flow direction [91]. To elucidate how tractions might underlie these changes, exciting methodologies have now emerged, as is described below.

Among the first to investigate the effects of fluid shear stress on cell traction was Shiu et al. [92]. Using a parallel-plate flow chamber system, they imposed a laminar fluid shear stress of 12 dynes/cm$^2$ for 30 min on bovine aortic endothelial cells. In response, the cells increased their tractions via a Rho-p160 ROCK mediated pathway. These studies were confined to laminar flow and to the single isolated cell in culture, however. To overcome these limitations, Ting et al. [93] incorporated both laminar flow (1.7 Pa) and disturbed flow (–2.4 to 1.9 Pa) and expanded traction measurements to micropatterned human pulmonary artery endothelial cell monolayers. They reported that while laminar flow increased cellular tractions, disturbed flow decreased them. These findings have been recently challenged, however; Conway et al. [94] showed that when bovine aortic endothelial cells were exposed to a laminar shear stress that is comparable in magnitude to that imposed by Ting et al. [93], tractions decreased. This decrease was associated with a decrease in tension within the cell adhesion molecule VE-Cadherin. A potentially important difference in the experimental protocol of Ting et al. [93] and Conway et al. [94] is the flow duration. While the former measurements imposed 14 h of flow, the latter imposed 24 h of flow. As such, how flow duration, flow magnitude, and loading frequency interact to impact cellular traction remains an open question.

### 4.4.6 Mechanical Stress Associated with Neighboring Cellular Interactions

#### 4.4.6.1 Cell–cell force measured from cell–substrate tractions

In addition to exerting tractions on their substrate, cells in a monolayer also exert forces on one other. For a typical adherent sheet of cells, in accordance with Newton laws, the cell-cell forces and the cell-substrate forces together establish a state of mechanical force equilibrium. This force equilibrium allows recovery of cell-cell forces from cell-substrate tractions. A similar concept was used to obtain a 1D scalar prestress number for a single isolated cell [95–97]; a 2D intracellular stress map in a single isolated micropatterned cell [98]; a 1D scalar cell-cell force number across an isolated, micropatterned cell pair [32]; and a map of 2D tensor monolayer stress (stresses within and between the cells) map across a cell sheet [2]. For the technique used to recover monolayer stresses (Monolayer Stress Microscopy, MSM), detailed analysis of accuracy of the measurement and validation of the underlying assumptions established MSM as a robust technique applicable in variety of in vitro assays [12].

#### 4.4.6.2 Plithotaxis: The principle of collective cell guidance

The advantage of tensorial methods such as MSM is the ability to study not just the magnitude of intercellular stress, but also the magnitude and orientation of anisotropy that may be present. At any given point in a monolayer of cells, for example, the 2D state of cell-cell stress can be expressed in components acting along a cell junction (normal stress) and a component acting tangential to it (Fig. 4.4). Two hidden features of the intercellular stress landscape now become apparent. First, the intercellular stress landscape is extremely heterogeneous. Second, the intercellular stress is seen to mostly align with the cell body along the direction where the normal stress is greatest, or equivalently, where the tangential stress vanishes. During collective cell migration, it is the local orientation of the minimum value of tangential stress that defines the tendency of each individual cell in a monolayer to migrate. This feature of intercellular stress guidance is called plithotaxis [2,99], and implies that the cell–cell junction is quite capable of bearing appreciable normal stress but is either incapable of bearing tangential stress, or orients such that the tangential stress is minimized.

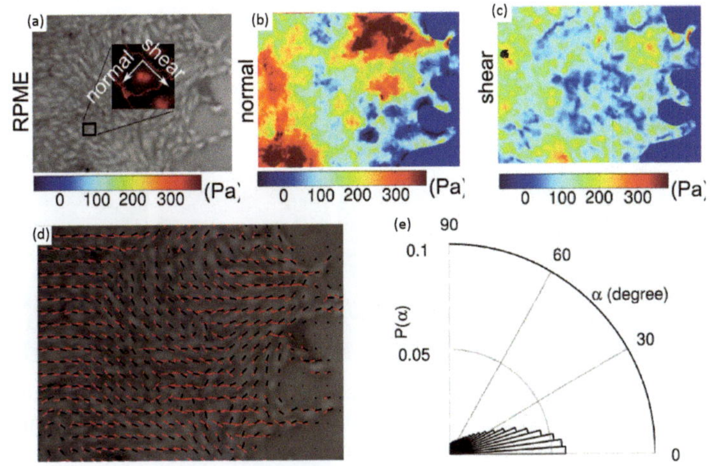


**Figure 4.4** Maps of intercellular stress. (a) Bright field image of the monolayer. Inset: at each point in the cell-cell junction, the intercellular stress has a normal and tangential/shear component. Over the entire monolayer, these components are represented in magnitude as (b) the map of local average normal stress and (c) the map of local maximum shear stress. Seen are large numbers of stress hotspots. (d) To represent stress directions, it is customary in continuum mechanics to report stresses in a coordinate system rotated such that the only stresses are normal, and the shear stress vanishes. The local orientation of the maximum value of normal stress (in red) coincides with the local orientation of the cell (in black). (e) This qualitative observation in (d) is quantified by the narrow probability distribution of the angle, $\alpha$, between the red and black orientations. Vertical size of the images of cell monolayers is 545 μm. Reprinted with permission from Tambe et al. [2].

### 4.4.6.3 Kenotaxis: Traction regulation to fill unfilled space

During wound healing, growth, pattern formation, or tumor cell invasion, cells migrate collectively until available space is filled. Might this behavior by guided by cell traction? To answer this question, Kim et al. [11] implemented TM in a migrating sheet of

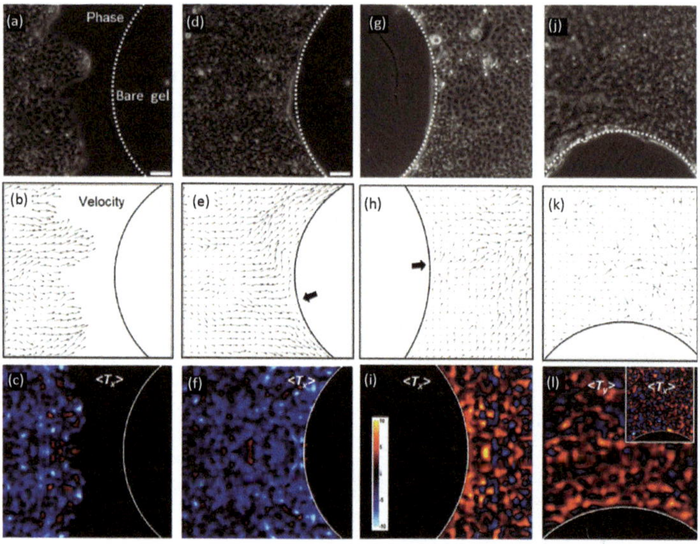


**Figure 4.5** The orientations of velocity and traction as a monolayer of Madine Darby canine kidney (MDCK) epithelial cell monolayer envelops a circular obstacle. (a, d: upstream region), (g: downstream region), (j: lateral region) are phase contrast images of the cells at various times before, during, and after the encounter with the obstacle. (b), (e), (h) and (k) are the respective velocity fields, and, (c), (f), (i), and (l) are the respective tractions. The tractions are ensemble averaged along the $x$ component, $<T_x>$, or the $y$ component, $<T_y>$. Regardless of the position of the cells near the obstacle, tractions pull towards the obstacle. Note that the tractions in this representation are those of the substrate on the cell; these are the negatives of the tractions of cell on substrate. Here, for example, positive values of $T_x$ are the tractions that the substrate exerts on the cell in the direction from east to west and positive values of $T_y$ are the tractions that the substrate exerts on the cell in the direction from north to south. Scale bar in (a) and (d), 100 μm. Reprinted with permission from Kim et al. [11].

endothelial and epithelial cells that are confronted by an obstacle. They observed that constituent cells within the sheet exert tractions to pull toward unfilled space, whether approaching the space, advancing toward it, or even receding away from it. This tendency of the cellular collective is called kenotaxis [11]. In the vicinity of the advancing free edge, kenotactic tractions were found to align with the direction of local cellular velocities. In the vicinity of the obstacle, however, TM measurements revealed a paradoxical uncoupling; tractions pulled in the direction that is nearly perpendicular to the local cellular velocities or even contrariwise (Fig. 4.5). Based upon this evidence, the authors suggest that kenotaxis might comprise a patterning motif that provides the cellular collective the necessary guidance during coordinated motion. During tissue engineering, wound healing, or organ development kenotaxis is likely to have far reaching implications, but these remain to be elucidated.

## 4.5 Concluding Remarks

A renaissance in the field of mechanobiology has enabled exciting advances in our understanding of cellular function. Much of this progress has been made possible by the various force imaging techniques of TM. TM measurements now encompass the single cell, the micropatterned cell or monolayer, the continuous monolayer cell sheet, and even 3D substrates, but in each case, the cell is studied in isolation from its native milieu in vivo. Accordingly, while current measurement techniques comprise a valuable tool, we must remain cognizant of their limitations to the dish, and the extent to which they truly reflect cellular behaviors in vivo, both in normal physiology and in the pathophysiologic context. A critical and currently unmet challenge is to develop next-generation TM technologies that bridge the measurement gap between the isolated cell or cell sheet in culture, the tissue strip, and the whole animal, the latter hopefully recapitulating human physiology. Overcoming this challenge will require a concerted effort among physiologists, engineers, biologists, and cell mechanists, and together will define the path forward toward understanding the role of mechanical forces in biology.

## References


1. Butler, J. P., et al., Traction fields, moments, and strain energy that cells exert on their surroundings. *Am. J. Physiol. Cell Physiol.*, 2002. **282**(3), C595–C605.
2. Tambe, D. T., et al., Collective cell guidance by cooperative intercellular forces. *Nat. Mater.*, 2011. **10**(6), 469–475.
3. Engler, A. J., et al., Matrix elasticity directs stem cell lineage specification. *Cell*, 2006. **126**(4), 677–689.
4. Krishnan, R., et al., Fluidization, resolidification, and reorientation of the endothelial cell in response to slow tidal stretches. *Am. J. Physiol. Cell Physiol.*, 2012. **303**(4), C368–C375.
5. Krishnan, R., et al., Reinforcement versus fluidization in cytoskeletal mechanoresponsiveness. *PLoS One*, 2009. **4**(5), e5486.
6. Balaban, N. Q., et al., Force and focal adhesion assembly: A close relationship studied using elastic micropatterned substrates. *Nat. Cell Biol.*, 2001. **3**(5), 466–472.
7. Paszek, M. J., et al., Tensional homeostasis and the malignant phenotype. *Cancer Cell*, 2005. **8**(3), 241–254.
8. An, S. S., et al., Do biophysical properties of the airway smooth muscle in culture predict airway hyperresponsiveness? *Am. J. Respir. Cell Mol. Biol.*, 2006. **35**(1), 55–64.
9. Krishnan, R., et al., Substrate stiffening promotes endothelial monolayer disruption through enhanced physical forces. *Am. J. Physiol. Cell Physiol.*, 2011. **300**(1), C146–C154.
10. Prager-Khoutorsky, M., et al., Fibroblast polarization is a matrix-rigidity-dependent process controlled by focal adhesion mechanosensing. *Nat. Cell Biol.*, 2011. **13**(12), 1457–1465.
11. Kim, J. H., et al., Propulsion and navigation within the advancing monolayer sheet. *Nat. Mater.*, 2013. **12**(9), 856–863.
12. Tambe, D. T., et al., Monolayer stress microscopy: Limitations, artifacts, and accuracy of recovered intercellular stresses. *PLoS One*, 2013. **8**(2), e55172.
13. Hardin, C., et al., Glassy dynamics, cell mechanics, and endothelial permeability. *J. Phys. Chem. B*, 2013. **117** (42), 12850–12856.
14. Wozniak, M. A., and Chen, C. S. Mechanotransduction in development: a growing role for contractility. *Nat. Rev. Mol. Cell Biol.*, 2009. **10**(1), 34–43.

15. Eyckmans, J., et al., A hitchhiker's guide to mechanobiology. *Dev. Cell*, 2011. **21**(1), 35–47.
16. Chen, C. S., Mechanotransduction—a field pulling together? *J. Cell Sci.*, 2008. **121**(Pt 20), 3285–3292.
17. Schwarz, U. S., and Gardel, M. L. United we stand: integrating the actin cytoskeleton and cell-matrix adhesions in cellular mechanotransduction. *J. Cell Sci.*, 2012. **125**(Pt 13), 3051–3060.
18. Kraning-Rush, C. M., et al., Quantifying traction stresses in adherent cells. *Methods Cell Biol.*, 2012. **110**, 139–178.
19. Brenner, M. D., Zhou, R., and Ha, T. Forcing a connection: Impacts of single-molecule force spectroscopy on in vivo tension sensing. *Biopolymers*, 2011. **95**(5), 332–344.
20. Wang, J. H., and Lin, J. S. Cell traction force and measurement methods. *Biomech. Model Mechanobiol.*, 2007. **6**(6), 361–371.
21. Li, B., and Wang, J. H. Application of sensing techniques to cellular force measurement. *Sensors (Basel)*, 2010. **10**(11), 9948–9962.
22. Sen, S., Engler, A. J., and Discher, D. E. Matrix strains induced by cells: computing how far cells can feel. *Cell. Mol. Bioeng.*, 2009. **2**, 39–48.
23. Mierke, C. T., et al., Contractile forces in tumor cell migration. *Eur. J. Cell Biol.*, 2008. **87**(8–9), 669–676.
24. Chown, M. G., and Kumar, S. Imaging and manipulating the structural machinery of living cells on the micro- and nanoscale. *Int. J. Nanomed.*, 2007. **2**(3), 333–344.
25. Wang, J. H., Cell traction forces (CTFs) and CTF microscopy applications in musculoskeletal research. *Oper. Tech. Orthop.*, 2010. **20**(2), 106–109.
26. Bereiter-Hahn, J., Mechanics of crawling cells. *Med. Eng. Phys.*, 2005. **27**(9), 743–753.
27. Califano, J. P., and Reinhart-King, C. A. Substrate stiffness and cell area predict cellular traction stresses in single cells and cells in contact. *Cell Mol. Bioeng.*, 2010. **3**(1), 68–75.
28. Han, S. J., et al., Decoupling substrate stiffness, spread area, and micropost density: A close spatial relationship between traction forces and focal adhesions. *Biophys. J.*, 2012. **103**(4), 640–648.
29. Hu, S., Chen, J., and Wang, N. Cell spreading controls balance of prestress by microtubules and extracellular matrix. *Front Biosci.*, 2004. **9**, 2177–2182.
30. Chen, C., et al., Fluidization and resolidification of the human bladder smooth muscle cell in response to transient stretch. *PLoS One*, 2010. **5**(8), e12035.

31. Maruthamuthu, V., et al., Cell-ECM traction force modulates endogenous tension at cell–cell contacts. *Proc. Natl. Acad. Sci. U. S. A.*, 2011. **108**(12), 4708–4713.

32. Liu, Z., et al., Mechanical tugging force regulates the size of cell–cell junctions. *Proc. Natl. Acad. Sci. U. S. A.*, 2010. **107**(22), 9944–9949.

33. Plotnikov, S. V., et al., Force fluctuations within focal adhesions mediate ECM-rigidity sensing to guide directed cell migration. *Cell*, 2012. **151**(7), 1513–1527.

34. Harris, A. K., Wild, P., and Stopak, D. Silicone rubber substrata: A new wrinkle in the study of cell locomotion. *Science*, 1980. **208**(4440), 177–179.

35. Harris, D. E., and Warshaw, D. M. Length vs. active force relationship in single isolated smooth muscle cells. *Am. J. Physiol.*, 1991. **260**(5 Pt 1), C1104–C1112.

36. Harris, A. K., Cell motility and the problem of anatomical homeostasis. *J. Cell Sci. Suppl*, 1987. **8**, 121–140.

37. Hinz, B., et al., Myofibroblast development is characterized by specific cell–cell adherens junctions. *Mol. Biol. Cell*, 2004. **15**(9), 4310–4320.

38. Hinz, B., et al., Mechanical tension controls granulation tissue contractile activity and myofibroblast differentiation. *Am. J. Pathol.*, 2001. **159**(3), 1009–1020.

39. Hinz, B., Gabbiani, G., and Chaponnier, C. The $NH_2$-terminal peptide of alpha-smooth muscle actin inhibits force generation by the myofibroblast in vitro and in vivo. *J. Cell Biol.*, 2002. **157**(4), 657–663.

40. Hinz, B., and Gabbiani, G. Mechanisms of force generation and transmission by myofibroblasts. *Curr. Opin. Biotechnol.*, 2003. **14**(5), 538–546.

41. Hinz, B., and Gabbiani, G. Cell-matrix and cell–cell contacts of myofibroblasts: role in connective tissue remodeling. *Thromb. Haemost.*, 2003. **90**(6), 993–1002.

42. Hinz, B., et al., Alpha-smooth muscle actin is crucial for focal adhesion maturation in myofibroblasts. *Mol. Biol. Cell*, 2003. **14**(6), 2508–2519.

43. Hinz, B., et al., Alpha-smooth muscle actin expression upregulates fibroblast contractile activity. *Mol. Biol. Cell*, 2001. **12**(9), 2730–2741.

44. Hinz, B., Formation and function of the myofibroblast during tissue repair. *J. Invest. Dermatol.*, 2007. **127**(3), 526–537.

45. Hinz, B., Masters and servants of the force: the role of matrix adhesions in myofibroblast force perception and transmission. *Eur. J. Cell Biol.*, 2006. **85**(3–4), 175–181.

46. Dembo, M., and Wang, Y. L. Stresses at the cell-to-substrate interface during locomotion of fibroblasts. *Biophys. J.*, 1999. **76**(4), 2307–2316.

47. Galbraith, C. G., and Sheetz, M. P. A micromachined device provides a new bend on fibroblast traction forces. *Proc. Natl. Acad. Sci. U. S. A.*, 1997. **94**(17), 9114–918.

48. Tan, J. L., et al., Cells lying on a bed of microneedles: an approach to isolate mechanical force. *Proc. Natl. Acad. Sci. U. S. A.*, 2003. **100**(4), 1484–1489.

49. du Roure, O., et al., Force mapping in epithelial cell migration. *Proc. Natl. Acad. Sci. U. S. A.*, 2005. **102**(7), 2390–2395.

50. Sniadecki, N. J., et al., Magnetic microposts as an approach to apply forces to living cells. *Proc. Natl. Acad. Sci. U. S. A.*, 2007. **104**(37), 14553–1458.

51. Rabodzey, A., et al., Mechanical forces induced by the transendothelial migration of human neutrophils. *Biophys. J.*, 2008. **95**(3), 1428–1438.

52. Landau, L. D., and E. M. Lifshitz, in *Mechanics. Course of Theoretical Physics*, vol. 1. 1960.

53. Stricker, J., et al., Optimization of traction force microscopy for micron-sized focal adhesions. *J. Phys. Condens. Matter.*, 2010. **22**(19), 194104.

54. Sabass, B., et al., High resolution traction force microscopy based on experimental and computational advances. *Biophys J.*, 2008. **94**(1), 207–220.

55. Gardel, M. L., et al., Traction stress in focal adhesions correlates biphasically with actin retrograde flow speed. *J. Cell Biol.*, 2008. **183**(6), 999–1005.

56. Polio, S. R., et al., A micropatterning and image processing approach to simplify measurement of cellular traction forces. *Acta Biomater.*, 2012. **8**(1), 82–88.

57. Marinkovic, A., et al., Improved throughput traction microscopy reveals pivotal role for matrix stiffness in fibroblast contractility and TGF-beta responsiveness. *Am. J. Physiol. Lung Cell Mol. Physiol.*, 2012. **303**(3), L169–L180.

58. Liu, K., et al., Improved-throughput traction microscopy based on fluorescence micropattern for manual microscopy. *PLoS One*, 2013. **8**(8), e70122.

59. Yang, Z., et al., Determining substrate displacement and cell traction fields—a new approach. *J. Theor. Biol.*, 2006. **242**(3), 607–616.

60. Nelson, C. M., et al., Vascular endothelial-cadherin regulates cytoskeletal tension, cell spreading, and focal adhesions by stimulating RhoA. *Mol. Biol. Cell*, 2004. **15**(6), 2943–2953.

61. Mammoto, A., et al., A mechanosensitive transcriptional mechanism that controls angiogenesis. *Nature*, 2009. **457**(7233), 1103–1108.

62. Ghosh, K., et al., Tumor-derived endothelial cells exhibit aberrant Rho-mediated mechanosensing and abnormal angiogenesis in vitro. *Proc. Natl. Acad. Sci. U. S. A.*, 2008. **105**(32), 11305–11310.

63. Kumar, S., et al., Viscoelastic retraction of single living stress fibers and its impact on cell shape, cytoskeletal organization, and extracellular matrix mechanics. *Biophys. J.*, 2006. **90**(10), 3762–3773.

64. Reinhart-King, C. A., Dembo, M., and Hammer, D. A. The dynamics and mechanics of endothelial cell spreading. *Biophys. J.*, 2005. **89**(1), 676–689.

65. Reinhart-King, C. A., Dembo, M., and Hammer, D. A. Cell–cell mechanical communication through compliant substrates. *Biophys. J.*, 2008. **95**(12), 6044–6051.

66. Califano, J. P., and Reinhart-King, C. A. The effects of substrate elasticity on endothelial cell network formation and traction force generation. *Conf. Proc. IEEE Eng. Med. Biol. Soc.*, 2009. **2009**, 3343–3345.

67. Byfield, F. J., et al., Endothelial actin and cell stiffness is modulated by substrate stiffness in 2D and 3D. *J. Biomech.*, 2009. **42**(8), 1114–1119.

68. Yeung, T., et al., Effects of substrate stiffness on cell morphology, cytoskeletal structure, and adhesion. *Cell Motil Cytoskeleton*, 2005. **60**(1), 24–34.

69. Patel, N. R., et al., Cell elasticity determines macrophage function. *PLoS One*, 2012. **7**(9), e41024.

70. Huynh, J., et al., Age-related intimal stiffening enhances endothelial permeability and leukocyte transmigration. *Sci. Transl. Med.*, 2011. **3**(112), 112ra122.

71. Engler, A., et al., Substrate compliance versus ligand density in cell on gel responses. *Biophys. J.*, 2004. **86**(1 Pt 1), 617–628.

72. Del Alamo, J. C., et al., Spatio-temporal analysis of eukaryotic cell motility by improved force cytometry. *Proc. Natl. Acad. Sci. U. S. A.*, 2007. **104**(33), 13343–13348.

73. Merkel, R., et al., Cell force microscopy on elastic layers of finite thickness. *Biophys. J.*, 2007. **93**(9), 3314–3323.

74. Maloney, J. M., et al., Influence of finite thickness and stiffness on cellular adhesion-induced deformation of compliant substrata. *Phys. Rev. E Stat. Nonlin. Soft Matter Phys.*, 2008. **78**(4 Pt 1), 041923.

75. Buxboim, A., et al., How deeply cells feel: methods for thin gels. *J. Phys. Condens. Matter.*, 2010. **22**(19), 194116.

76. Lin, Y. C., et al., Mechanosensing of substrate thickness. *Phys. Rev. E Stat. Nonlin. Soft Matter Phys.*, 2010. **82**(4 Pt 1), 041918.

77. Schwarz, U. S., et al., Calculation of forces at focal adhesions from elastic substrate data: the effect of localized force and the need for regularization. *Biophys. J.*, 2002. **83**(3), 1380–1394.

78. Maskarinec, S. A., et al., Quantifying cellular traction forces in three dimensions. *Proc. Natl. Acad. Sci. U. S. A.*, 2009. **106**(52), 22108–22113.

79. Del Alamo, J. C., et al., Three-dimensional quantification of cellular traction forces and mechanosensing of thin substrata by fourier traction force microscopy. *PLoS One*, 2013. **8**(9), e69850.

80. Legant, W. R., et al., Multidimensional traction force microscopy reveals out-of-plane rotational moments about focal adhesions. *Proc. Natl. Acad. Sci. U. S. A.*, 2013. **110**(3), 881–886.

81. Notbohm, J., et al., Three-dimensional analysis of the effect of epidermal growth factor on cell–cell adhesion in epithelial cell clusters. *Biophys. J.*, 2012. **102**(6), 1323–1330.

82. Serra-Picamal, X., et al., Mechanical waves during tissue expansion. *Nat. Phys.*, 2012. **8**, 628–634.

83. Choquet, D., Felsenfeld, D. P., and Sheetz, M. P. Extracellular matrix rigidity causes strengthening of integrin-cytoskeleton linkages. *Cell*, 1997. **88**(1), 39–48.

84. Munevar, S., Wang, Y. L., and Dembo, M. Regulation of mechanical interactions between fibroblasts and the substratum by stretch-activated $Ca^{2+}$ entry. *J. Cell Sci.*, 2004. **117**(Pt 1), 85–92.

85. Matthews, B. D., et al., Cellular adaptation to mechanical stress: role of integrins, Rho, cytoskeletal tension and mechanosensitive ion channels. *J. Cell Sci.*, 2006. **119**(Pt 3), 508–518.

86. Trepat, X., et al., Universal physical responses to stretch in the living cell. *Nature*, 2007. **447**, 592–595.

87. Krishnan, R., et al., Airway smooth muscle and bronchospasm: fluctuating, fluidizing, freezing. *Respir. Physiol. Neurobiol.*, 2008. **163**(1–3), 17–24.

88. Deng, L., et al., Fast and slow dynamics of the cytoskeleton. *Nat. Mater.*, 2006, 636–640.

89. Bursac, P., et al., Cytoskeletal remodelling and slow dynamics in the living cell. *Nat. Mater.*, 2005. **4**, 557–571.

90. Lavoie, T. L., et al., Dilatation of the constricted human airway by tidal expansion of lung parenchyma. *Am. J. Respir. Crit. Care Med.*, 2012. **186**(3), 225–232.

91. Chien, S., Mechanotransduction and endothelial cell homeostasis: The wisdom of the cell. *Am. J. Physiol. Heart Circ. Physiol.*, 2007. **292**(3), H1209–H1224.

92. Shiu, Y. T., et al., Rho mediates the shear-enhancement of endothelial cell migration and traction force generation. *Biophys. J.*, 2004. **86**(4), 2558–2565.

93. Ting, L. H., et al., Flow mechanotransduction regulates traction forces, intercellular forces, and adherens junctions. *Am. J. Physiol. Heart Circ. Physiol.*, 2012. **302**(11), H2220–H2229.

94. Conway, D. E., and Schwartz, M. A. Flow-dependent cellular mechanotransduction in atherosclerosis. *J. Cell Sci.*, 2013. **126**(Pt 22), 5101–5109.

95. Stamenovic, D., et al., Effect of the cytoskeletal prestress on the mechanical impedance of cultured airway smooth muscle cells. *J. Appl. Physiol.* (1985), 2002. **92**(4), 1443–1450.

96. Stamenovic, D., Effects of cytoskeletal prestress on cell rheological behavior. *Acta Biomater.*, 2005. **1**(3), 255–262.

97. Wang, N., et al., Cell prestress. I. Stiffness and prestress are closely associated in adherent contractile cells. *Am. J. Physiol. Cell Physiol.*, 2002. **282**(3), C606–C616.

98. Park, C. Y., et al., Mapping the cytoskeletal prestress. *Am. J. Physiol. Cell Physiol.*, 2010. **298**(5), C1245–C1252.

99. Trepat, X., and Fredberg, J. J. Plithotaxis and emergent dynamics in collective cellular migration. *Trends Cell Biol.*, 2011. **21**(11), 638–646.

100. An, S. S., et al., Hypoxia alters biophysical properties of endothelial cells via p38 MAPK- and Rho kinase-dependent pathways. *Am. J. Physiol. Cell Physiol.*, 2005. **289**(3), C521–C530.

## Chapter 5

# Mechanical Regulation of Myofibroblasts: Mechanically Guided Matrix Remodeling and Prevention of Fibrosis in Regenerative Medicine

**Jenna L. Balestrini[a] and Boris Hinz[b]**

[a]*Department of Biomedical Engineering and Anesthesiology, Yale University School of Medicine, New Haven, CT 06520, USA*
[b]*Laboratory of Tissue Repair and Regeneration, Matrix Dynamics Group, Faculty of Dentistry, University of Toronto, 150 College Street, Toronto, Ontario 06511, Canada*

jenna.balestrini@yale.edu

## 5.1 Introduction

Fibroblasts within connective tissues reside in a complex and ever-changing biomechanical microenvironment. This environment provides a milieu of physical signals that guide cellular processes during development and throughout the life of an individual: during embryogenesis, cellular differentiation, tissue development, and also during wound healing and repair. Mechanical signals can be generated from a variety of external sources such as heartbeat, cytoskeletal forces from neighboring cells, or environmental cues

*Cells, Forces, and the Microenvironment*
Edited by Charles M. Cuerrier and Andrew E. Pelling

ISBN 978-981-4613-36-1 (Hardcover), 978-981-4613-37-8 (eBook)
www.panstanford.com

such as tissue elasticity. Cells respond actively to external cues by internally adjusting their tractional forces, and also by altering the biochemical and mechanical composition of their surrounding extracellular matrix (ECM) in accordance with mechanical instructions. Responses to mechanical signals are particularly important in cells that play a role in maintaining the mechanical function of the body, such as those found in load bearing connective tissues. How cells monitor and respond to alterations in mechanical signals (e.g., during aging, wound healing, or after clinical intervention) is critical in either establishing or undermining overall tissue health and function. Even seemingly minute changes in physical cues can induce cell-driven ECM remodeling toward positive outcomes such as wound closure, or toward negative outcomes such as progressive fibrosis, cell death, or even loss of organ function. Therefore, if properly understood, the relationship between mechanical conditioning and cell response represents a powerful tool that can be utilized in tissue engineering and regenerative medicine. In this chapter we will discuss how cells within connective tissues are mechanically regulated, how this mechanical regulation can play a role in fibrosis, the dynamic reciprocity between ECM and fibroblast-myofibroblast behavior, and current work dedicated toward harnessing this relationship for the purposes of regenerative medicine.

## 5.2 What Fibroblasts Feel: Stretching Is Important Not to Get Stiff

Connective tissues fibroblasts are continuously subjected to a range of dynamic forces that change the mechanical state of the tissue through either cell–cell interactions, cell–ECM interactions, or ECM–ECM interactions [1]. Different mechanical signals come from a variety of sources, including gravity, muscle movement, breathing, and heart beat [1,2]. In this chapter, we will concentrate on two mechanical signals that dominate the behavior of fibroblastic cells: stretch and ECM stiffness. Although how fibroblasts mechanosense and subsequently interpret mechanical signals is not entirely understood, it has been established that several components orchestrate mechanotransduction events in concert, including the opening of mechanosensitive ions channels, initiation of integrin-mediated mechanical pathways, binding of cell–cell linkages, and force-induced exposure to otherwise cryptic peptide sequences [3,4]. Tissue deformations, experienced by residing fibroblasts as stretch, are not only essential for the physical maintenance and management of tissues, but also guide biological responses. Responses to stretch include changes in cell shape, reorientation [5], increases in proliferation rates [6], alteration of migratory behavior [7], and changes in the expression and synthesis of a variety of contractile and regulatory proteins [6]. Stretch is particularly important for fibroblasts within connective tissues of mechanically active organs such as heart, lung, tendon, and skin. It is the task of fibroblasts to creating a stable and well-structured ECM optimized to resist external loads [8]. Therefore when mechanical loading becomes heightened or compromised from trauma, abnormal loading, or pathological onset, the way in which resident cells respond in terms of ECM reorganization and production becomes a key component in dictating healthy or diseased outcomes [9]. A paradigm for load-induced physiological and pathological ECM remodeling is the development of cardiac fibrosis in conditions of chronic hypertension [10].

In addition to stretch-dependent ECM remodeling, the effects of ECM stiffness on fibroblast biology become an area of increasing interest to researchers in the field of regenerative medicine. There is growing evidence to suggest that mechanical cues inherent to the ECM are as important as chemical composition in regulating cell behavior, and can be adjusted accordingly to manipulate specific aspects of cell behavior such as protein production or cell contractility. Stiffness-dependent cell behaviors include stem cell differentiation [11], cell migration (durotaxis) [12], morphological attributes such as focal adhesions [13], cell signaling pathways [14], and also cell proliferation, apoptosis, and myofibroblast differentiation [15]. During wound healing, disease states, or even as a consequence of aging the stiffness of the surrounding ECM can become dramatically different from native tissue. The replacement of normal tissue with stiff, collagen-dense ECM within the interstitium is considered the most problematic sequelae in connective tissue fibrosis [16]. Therefore, there a recent body of work has been dedicated toward investigating how ECM stiffness could direct fundamental long-term fibrotic behavior of cells. Characterizing the how ECM stiffness promotes either normal healing or fibrosis is crucial for developing effective therapeutic regimens to halt disease progression, and also to create functional tissue replacements and implants [17].

## 5.3 The Myofibroblast—Dedicated to Produce and Remodel ECM

Whereas fibroblastic cells remain relatively quiescent in intact tissues under conditions of normal mechanical load, they follow a series of activation steps after connective tissue injury and overload [18]. Activation includes enhanced proliferation, secretion of ECM proteins and remodeling of newly formed ECM to restore the compromised mechanical integrity of the organ. The presumptive cellular end-stage of this sequence of events is the myofibroblast (Fig. 5.1). Myofibroblasts were discovered more than 40 years ago in wound granulation tissue as fibroblastic cells with prominent endoplasmatic reticulum and contractile microfilament bundles [19]. These ultrastructural features coined the cell "myofibroblast," promote high ECM secretion and contractility, and a plethora of studies have confirmed the functional predictions made from this phenotype. The hallmark and molecular basis for the high contractile activity associated with the myofibroblast is the neo-expression of α-smooth muscle actin (α-SMA) in stress fibers [18] (Figs. 5.1 and 5.2). Strictly, α-SMA-positive cells that do not form microfilament bundles (stress fibers) are not termed myofibroblasts as they are lacking their defining contractile elements [20–23]. Conversely, α-SMA-negative fibroblastic cells that display stress fibers are contractile and are deemed "functionally myofibroblasts," at least in vivo. Since fibroblastic cells almost inevitably form stress fibers in standard cell culture conditions, i.e., stiff culture plastic and growth factor supply, "myofibroblast" usually denominates α-SMA-positive stress fiber forming cells (Fig. 5.2). The percentage of fibroblasts spontaneously acquiring the myofibroblast phenotype in standard culture is characteristic for their species and tissue origin [22].

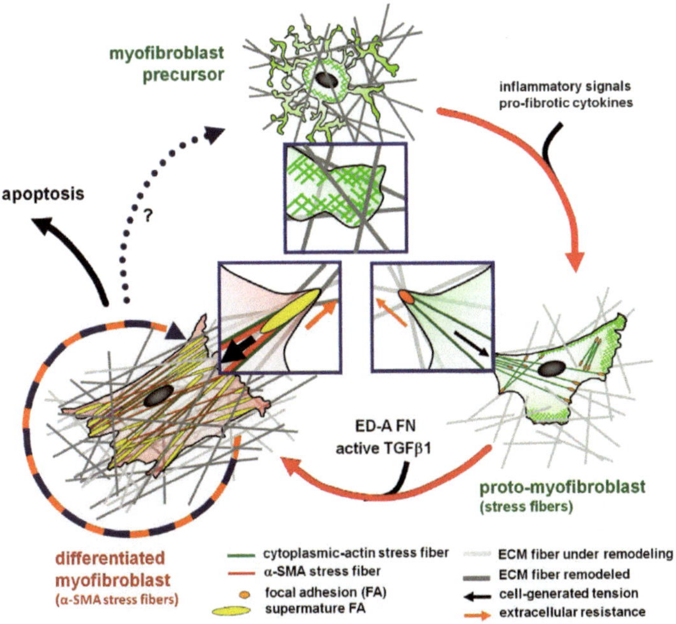


**Figure 5.1** The mechanical feedback loop in MF development. Fibroblast-to-myofibroblast differentiation is controlled by a careful balance of intracellular (contractility) and extracellular (e.g., ECM stiffness) stresses. Failure of stress-released myofibroblasts to undergo apoptosis results in their persisting activity and ultimately leads to fibrosis (adapted and reproduced with permission from [24]).

Myofibroblast activation is an almost universal feature of normal tissue repair and adaptive remodeling [24–26]. Topic of intensive studies is the contribution of myofibroblasts to the development of fibrocontractive conditions, including fibrosis of heart [27–29], liver [30], lung [31], kidney [32], muscle [33], skin [34], and the stroma reaction to tumors [35]. Because myofibroblasts drive the disease, they have been identified as possible targets in anti-fibrotic strategies and their suitability as therapeutic targets [36–40]. Whereas some of these strategies attempt to target common traits of myofibroblasts in different conditions, others aim to interfere with myofibroblast activation from their respective precursors. Our view on the origin of myofibroblasts and the extent of how different precursor cells contribute to the myofibroblast population in pathology is evolving [41,42].

Accepted myofibroblast precursors include local fibroblasts [22,35,43], fibrocytes [44–46], smooth muscle cells (SMCs) [47], pericytes [48], mesenchymal stromal cells [49,50], hepatic stellate cells [43], and epithelial and endothelial cells undergoing mesenchymal transition [51,52] (EMT and EndMT). This list is far from being complete and the concept is emerging that our body will recruit all available cell sources to generate myofibroblasts to achieve fast repair.

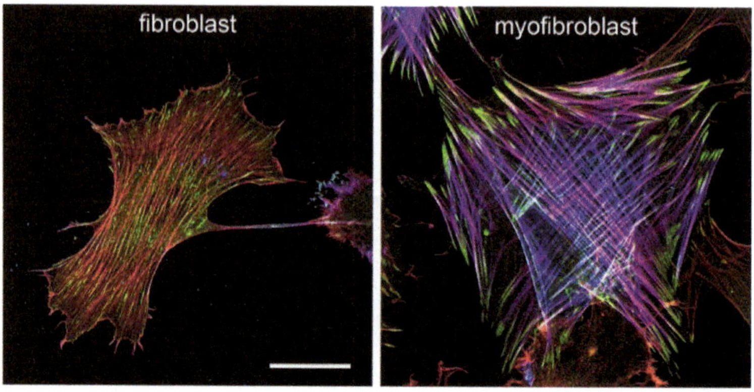


**Figure 5.2** Myofibroblast features in vitro. After 4-day culture in the absence (fibroblast) or in the presence of TGFβ1 (myofibroblast) primary fibroblasts were immunostained. Compared with fibroblasts, myofibroblasts exhibit more prominent stress fibers (F-actin-red, purple in overlay) that are positive for α-SMA (blue, purple in overlay). Myofibroblast FAs (vinculin-green) at the termini of stress fibers are larger than fibroblast cell-ECM adhesions. Scale bar: 20 μm (adapted and reproduced with permission from [156]).

It is essential to understand that the functional role of myofibroblasts and the consequences of their activities are very similar irrespective of their origin or host organ. In all fibrotic conditions, myofibroblasts replace functional tissue with a collagenous scar of a mechanical and chemical quality that dramatically differs from the parenchyma of the healthy organ. In this chapter, we will focus on the mechanical properties of the ECM but excellent recent reviews inform about the particular chemical composition of the myofibroblast ECM [53]. The sheer presence of stiff scar tissue impedes organ function, e.g., heart pumping or lung expansion and gas diffusion. In addition, disturbance of the

ECM and remodeling by myofibroblasts has profound impact on myofibroblasts (i.e., self-inflicted) themselves and neighboring cell types sharing the same microenvironment. Hence, rather than acting as a passive obstacle, the altered ECM can also contribute to the progression of the disease [54–57], partly by turning different healthy precursor cells into fibrotic myofibroblasts [24]. Myofibroblasts express α-SMA in attached stressed collagen gel cultures but not in free-floating relaxed gels [58–60]. In healing rat skin wounds, expression of α-SMA is accelerated by splinting the wound edges with plastic frames, i.e., by increasing tissue tension; tension release has the opposite effect [61]. Similar results are obtained by stretching human burn scar tissue in situ [62]. Myofibroblasts in vitro and in vivo develop their highly contractile cytoskeletal apparatus only above a certain ECM stiffness threshold. Expression of α-SMA is augmented in myofibroblasts cultured on stiff culture substrates but suppressed on soft polymer substrates [14]. In skin fibrosis, mechanical activation can affect local dermal fibroblasts [34], pericytes [63], bone marrow-derived circulating fibrocytes [64], hematopoietic cells, and mesenchymal stromal cells [65]. In addition, remodeling activities of fibroblastic cells and myofibroblasts stiffen the tissue and increase cell contractility thereby establishing a detrimental feed-forward loop that contributes to myofibroblast persistence as further discussed below (Fig. 5.1).

Myofibroblast precursor cells, including tissue-resident fibroblasts, are stress-shielded by an organized collagen ECM in normal tissues; they do not develop contractile features and cell–ECM adhesions. In chronic overload, or ECM stiffening, fibroblasts become activated to spread and migrate. Fibroblast remodeling activity leads to gradual increase in global ECM stiffness that counteracts cell traction forces. The resulting formation of small focal adhesions and stress fibers that contain only cytoplasmic actins characterize the proto-myofibroblast. Continuing ECM fiber alignment creates larger surfaces for adhesion formation; larger adhesions permit development of stronger stress fibers and generation of higher contractile forces. In the presence of active TGF-β1 and ED-A fibronectin (EDA-FN), proto-myofibroblasts begin to express α-SMA that is gradually incorporated into the pre-existing stress fibers (Fig. 5.2). The force generated by α-SMA-containing stress fiber is significantly higher compared to cytoplasmic actin stress fibers leading to adhesion supermaturation and ECM contraction, thereby establishing a mechanical feed forward loop. Myofibroblasts may exit this cycle when the original structure of the ECM is reconstituted and again takes over the mechanical load.

## 5.4 Mechanical Stretch in Wound Healing and Regenerative Medicine

From the knowledge that the physical static and dynamic states of the ECM determine myofibroblast formation, different concepts have been developed to regenerate injured tissues with minimal scarring and maximal restoration of tissue function.

### 5.4.1 Mechanical Intervention: Regulating Dermal Wound Healing through Stretch

Clinicians routinely apply mechanical stretch to healing dermal wounds to manipulate the healing rates and alter the appearance, mechanical and biochemical properties of scar tissue [66]. Some examples of mechanical conditioning include serial casting, negative pressure wound therapy, massage, dynamic splints, deep tissue massage, serial casting, z-plasty, range of motion exercises, and ambulation [67–69]. Several animal studies have also demonstrated that mechanically loading healing wounds produces a thinner, stronger, more compliant scar with a reduced incidence of contracture (pathological shortening of tissue that leads to reduced range of motion) [70]. Further, stretching techniques that utilize negative pressure wound therapy in obtaining a stable bond between the skin and percutaneous devices and accelerate the healing process by reducing edema; therefore, these systems are extensively utilized in combat medical treatments and chronic wound treatments [71,72].

### 5.4.2 Mechanical Stretch in Regenerative Medicine: Toward Building Functional Tissue Equivalents

The primary objective of any tissue engineering approach is to create patient-specific biological equivalents that ultimately restore functionality to damaged or diseased tissue. In order to restore tissue functionality, these systems must have the appropriate architecture, biocompatibility, biochemical composition, mechanical properties, and resilience for long-term usage. Functional soft connective tissue replacements would maintain their tissue-specific functions (e.g., gas exchange for lung tissue), specific repertoire of mechanical properties (e.g., blood vessels with sufficient strength to withstand forces associated with blood flow), and ability to modulate cellular responses to forces (e.g., maintain appropriate biochemical composition to promote cell anchorage and resting cytoskeletal tension) [3].

Collagen gels, fibrin gels, polymer blends, and even cell-derived matrices are often employed as tissue models for use in mechanobiological study [73–75]. In addition to use in basic research, these systems also have a strong potential for use as tissue replacements in wound healing therapy for replacing diseased or damaged tissues in regenerative medicine [76–78]. In fact, connective tissue analogs currently form the dermal component of clinically available tissue engineered skin replacements [79–81]. Despite these advances, current systems often lack sufficient mechanical integrity and durability for clinical usage in load-bearing applications [82–84]. To address this limitation, mechanical conditioning has become a primary strategy to produce tissue equivalents with superior tissue strength and has been met with some success. Similar to previous in vivo observations, researchers have found that cyclically stretching tissue models resulted in tissue with significant increases in mechanical properties as compared to statically cultured controls [8,85]. In addition, many cellular responses appear to have been specifically adapted in response to stretch orientation, magnitude, and duration.

### 5.4.3 Building Custom-Tailored Tissue Equivalents with Diverse Biomechanical Properties

In vitro stretching devices are typically grouped into three classes: uniaxial [86,87], equibiaxial [82,88], or strip biaxial [89]. These systems are composed of motor driven devices that either provide stretch in one direction (uniaxial), stretched in one axis while the transverse axis is fixed (strip biaxial stretch, or "pure uniaxial stretch"), or uniformly deformed in all directions (equibiaxial). In previous studies that utilized uniaxial stretch, a marked increase in fiber alignment in the predominant direction of stretch is observed [86], paralleling a striking increase in the stiffness [90] and tensile strength of the tissue [91–93]. Linear or circumferential alignment is a highly desirable characteristic in planar tissue substitutes such as tendons and blood vessels that are preferentially loaded in a primary direction. When building systems to replace planar tissues such as skin, pericardium, and heart valves, a highly aligned ECM would reduce the capacity of tissue to withstand multiaxially loading and could result in an even greater reduction of motion and tissue failure [94]. In planar tissues, biaxial stretching more closely models mechanical environment of planar tissues and appears to result in a more uniform angular distribution of collagen fibers in vivo [70,94,95]. Therefore, specific stretch regimens designed to capitalize on cell-induced ECM reorganization are essential for creating tissues with specific loading function.

### 5.4.4 Determining Optimal Loading Conditions for the Creation of Tissue Equivalents

Clearly, there is a plethora of information that indicates that stretch is a powerful regulator of ECM remodeling [87,96,97]. In addition to stretch orientation there is evidence to suggest that stretch-induced cell behavior is entirely dependent on the magnitude [5,98,99], duration [5,100,101], and frequency [98,102] of the mechanical conditioning. Until recently however, the bulk of research has generally focused stretch-induced ECM remodeling and therefore has utilized a single level of stretch to investigate mechanically induced cell activity. However, in order to utilize stretch as a mean to mechanically custom-tailor tissue analogs with specific requirements (e.g., strength, stiffness, and contractility), quantitative relationships between mechanical stimulation and ECM remodeling must be pursued.

The research dedicated toward creating these regimes is currently limited in terms of cell and tissue types, and only recently has been directed toward translational applications [103]. What has been shown, however, is that in terms of stretch effects, combinations of stretch magnitude and duration (i.e., continuous or intermittent stretch signaling) can be utilized to fine tune aspects of tissue strength, stiffness, contractility, and ECM production. For example, applying continuous or intermittently equibiaxial stretch at multiple magnitudes to fibroblast-populated fibrin gels resulted in strength of tissues that increased exponentially with increasing stretch magnitude [74]. Surprisingly, this increase in mechanical properties did not require large strain levels, continuous stimulation, exogenous growth factors such as TGF-β, or extended culture duration—all parameters previously thought to be required for a substantial increase in tissue strength [87,104,105]. In addition, the introduction of a "rest period" during the loading regime was shown to induce an increase in cell proliferation, ECM production, and tissue strength compared to cycling continuously (at all stretch magnitudes). Tranquillo and colleagues also found that stretch-effects on ECM strength were heightened when mechanical stretch was applied intermittently to SMC-populated fibrin gels [105].

Although it is well known that mechanical stretch regulates a variety of aspects of cell function through a variety of mechanisms (e.g., integrin production and confirmation, regulation of cell cycle and signaling pathways), it is not yet understood which underlying mechanisms are altered by increasing stretch magnitude or introducing a rest period during mechanical conditioning. It is, however, becoming clear that cells within mechanically loaded tissues undergo an "adaptation response" to continuous signaling (one magnitude, stretched continuously): Cells exposed to a constant mechanical stimuli initially demonstrate heightened activity but return to baseline levels after sustained mechanical stimulation [106,107]. Although existing dose-response curves have yet to be fine-tuned in different cell types, it is evident that (1) further research on dose-response curves in alternative cell and tissue types is warranted, and (2) stretch magnitude and duration can be used independently or in concert to manipulate different aspects of remodeling to custom tailor specific characteristics in tissue equivalents for use in tissue engineering.

### 5.4.5 Myofibroblasts in Tissue-Engineering: An Organ Specific Affair

The myofibroblast, infamous for its role in wound healing and contribution to fibrosis, is now gaining popularity in the construction of mechanically robust engineered tissue-equivalents. As discussed previously, myofibroblasts rapidly produce large amounts of ECM (mainly collagen) and large tractional forces across the ECM, and are also resistant to apoptosis under the large and often continuous mechanical loads that would be imparted on them in vivo. These characteristics make myofibroblasts ideal for recreating dynamically loaded tissues such as blood vessels [108–111], heart valves [112–114] and even visceral tissue such as the bladder and uterus [115]. Currently, (peri-)vascular myofibroblast cells represent a well-established and routinely utilized cell source for cardiovascular tissue engineering [116]. Myofibroblasts contribute to the mechanical strength and ability to withstand substantial mechanical loading and also provide a source for constant tissue maintenance—all prerequisites for tissue engineering of cardiovascular structures. When mechanical loading is imparted on myofibroblast-populated equivalents, ECM production and tissue maturation becomes even more accelerated [117]. For example, tissue-engineered heart valves have been observed to respond to mechanical conditioning by significantly increasing collagen production in a rapid fashion [118], making the combination of myofibroblast ECM production and stretch induced remodeling very appealing. Experimental approaches using ex vivo human skin or strain devices on the backs of wounded mice indicate that controlled application of stress has the therapeutic potential to improve wound healing by augmenting myofibroblast differentiation and vascularization [61,119].

Although the presence of myofibroblasts and their associated ECM production can be advantageous in tissues that require superior ECM strength, they are often detrimental in alternative tissue types that ultimately require a soft ECM with relatively little collagen type I [120]. Taking for example native lung tissue, myofibroblast presence has been directly correlated with excessive deposition of collagen and therefore could result in a non-compliant and therefore non-functional tissue equivalent [121–123]. Another example would be corneal tissue implants, where collagen deposition typically produced by myofibroblasts could result in loss of transparency of the tissue [124,125]. In these systems, stretch regimes would need to be designed such that proper ECM production would be optimized while myofibroblast presence would be tightly controlled or prevented. In alternative tissues that require less compliance and large amounts of collagen type I such as ligaments or dermal tissue, myofibroblasts could be useful in the initial fabrication of the tissue but may need to be moved or attenuated prior to tissue implantation or use.

## 5.5 Matrix Stiffness: Cues That Direct Both Cell Health and Dysfunction

Similar to mechanical conditioning via stretching, substrate rigidity is also a powerful regulator of both biological and regulatory cell behaviors involved in tissue remodeling. Recently cell behavior driven exclusively through substrate stiffness cues has gotten attention within the field of tissue engineering and regenerative medicine. In order to study stiffness-dependent cell behaviors in a simplified system, researchers have developed soft substrates to culture cells using in vitro systems that represent a variety of biological tissues and processes [126,127]. The earliest versions of these culture systems were composed of soft polymer substrates that would wrinkle as cells generated contractile force across the surface [128,129]. These systems initially gained popularity as a method to determine cell contractility, but later systems to culture cells on that more closely resembled native tissue stiffness. Subsequently, several systems have been developed using a variety of natural and synthetic polymers including collagen gels, reconstituted tissues, silicones, agarose gels, and polyacrylamides [130–133]. By adjusting the elastic moduli of these systems, ECM stiffness has been shown to guide stem cell differentiation and self-renewal [11], alter cell migration behavior (durotaxis) [12], induce morphological changes such as focal adhesions [13], differences in cell signaling pathways [14], and changes in cell proliferation, apoptosis, and myofibroblast differentiation [15]. Therefore, ECM stiffness not only has been shown to regulate fundamental biological processes essential for cell health but also governs regulatory processes such as migration and differentiation as well.

### 5.5.1 Matrix Stiffness and Fibrosis: Artifact or Active Contributor to Pathology?

The impact of ECM stiffness during wound healing is of particular consequence; in fact, a defining feature of connective tissue fibrosis is the conversion of native tissue into pathologically stiff ECM. Over the last decade, the elastic modulus of a great number of tissues and organs has been assessed at the cellular perception level using atomic force microscopy indentation [134,135] (Fig. 5.3). Very soft organs, not surprisingly, are bone marrow [136], brain (0.1–0.5 kPa) [137] and fat (1–3 kPa) [138]. Soft organs are liver (1–2 kPa) [139] and lung parenchyma (2–4 kPa) [140], whereas muscular tissues are medium stiff (10–15 kPa) [141–143]. Bone and teeth provide the stiffest structures in our body [144] (Fig. 5.3). Importantly, the elastic modulus of collagen-dense fibrotic tissue is approximately 10–100-times stiffer than seen in most normal tissues (0.5 to 15 kPa) [135,145] (Fig. 5.3). Fibroblasts embedded in a variety of tissue types are highly responsive to their resident ECM compliance and will develop either normal or fibrotic biological profiles based on their surface stiffness. Specifically, fibroblasts cultured on stiff (pathological) surfaces (>16 kPa) undergo myofibroblast activation and demonstrate an increase in both ECM production and contractility, whereas exposure to soft (healthy) substrates (5–10 kPa) results in the reduction of myofibroblast presence and associated pathological activity [14,15,130].

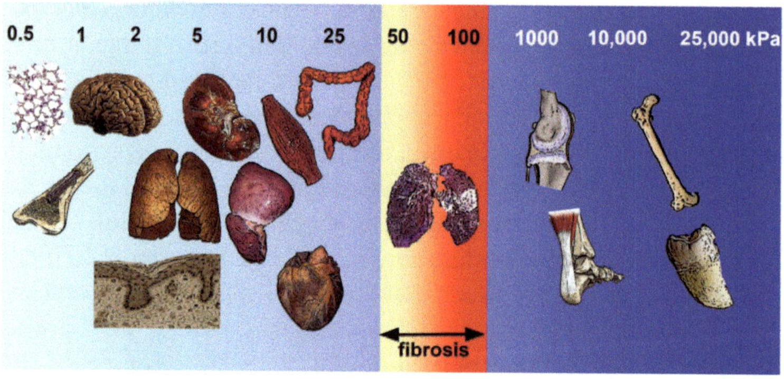


**Figure 5.3** The stiffness of our body. The stiffness range of different organs and tissues, determined by atomic force microscopy, is displayed as Young's elastic modulus in kPa. Note that fibrotic tissue (here stylized for a fibrotic lung) is always stiffer than normal tissue. References supporting this scheme are given in the text (adapted and reproduced with permission from [156]).

In the case of lung tissue, fibroblasts exposed to pathological levels of stiffness as seen during the end stages of fibrosis exhibit fibroproliferative behavior that is similar to that seen in patients with idiopathic pulmonary fibrosis [130,140]. It has therefore been evident for some time that stiffness propagates existing fibrotic cell behaviors among many cell types, but a more recent question has arisen regarding the role it plays in the onset of the disease [146]. In a previous experiment investigating carbon tetrachloride-induced fibrosis in a rat liver, tissue stiffening was detected prior to any significant collagen accumulation, indicating that ECM stiffness could be actively promoting disease progression as opposed to simply being artifact of disease onset [139]. These findings challenged preconceived notions of the role of ECM stiffness in disease progression, and were the first to suggest that mechanosensed stiffness information generated from local cellular environment could independently direct fibrotic behaviors of fibroblasts. The inevitable question remains: If ECM stiffness is independently driving fibroblast fibrotic behavior (i.e., myofibroblast differentiation), is this response reversible? Specifically do cells have a "mechanical memory"?

To investigate the potential for cell mechanical memory (and therefore stiffness-dependent myofibroblast reversibility), lung fibroblasts were cultured on silicone substrates (i.e., not grown on tissue culture plastic) and were exposed to substrates modeling pathologically stiff tissue (100 kPa) for several passages and subsequently placed on soft substrates (5 kPa) modeling healthy tissue for several passages (Fig. 5.4). Mechanically primed fibroblasts (those cultured on stiff substrates) were shown to have a subpopulation that maintained a sustained level of myofibroblast presence despite placing them on soft for several passages [130]. Specifically, a subset of the fibroblasts was shown to undergo an irreversible differentiation into the myofibroblast phenotype independent of a sustained fibrotic stiffness stimulus, suggesting the evidence of cell mechanical memory.

These findings were not without controversy however. In a separate study investigating stiffness-dependent myofibroblast persistence in valvular interstitial cells (VICs), the removal of the pathological stiffness stimuli after a few days of culture resulted a reduction in overall myofibroblast percentage and fibrotic activity [147]. In the same study however, they found that long-term exposure (i.e., several passages) to pathological stiffness levels did ultimately result in an irreversibility of the myofibroblast phenotype. It remains to be seen if this "mechanical memory" effect is tissue type dependent (i.e., predominantly in softer tissues where priming would be more severe), dependent on the length of stiffness exposure time, or dependent on the polymer culture system.

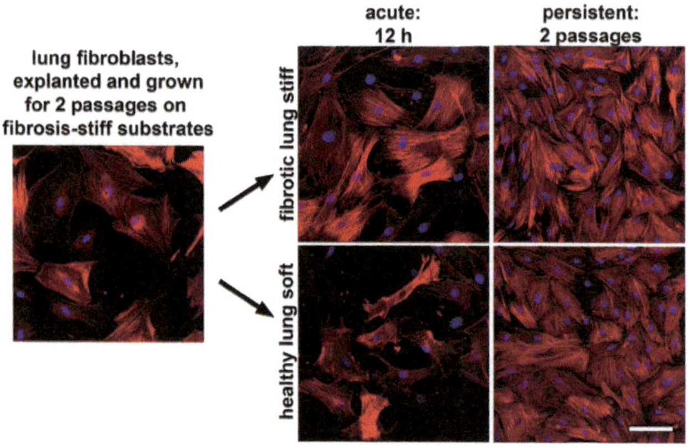


**Figure 5.4** Mechanical priming on fibrosis-stiff substrates results in persistent activation of lung myofibroblasts. Lung fibroblasts were explanted and sub-cultured for two passages on either 5 kPa healthy lung-soft or 100 kPa fibrotic lung-stiff collagen coated silicone substrates. Cells were then transferred to soft or stiff surfaces and assessed after 12 h and after 2 further passages (14 days). Cells were immunostained for α-SMA (red) and nuclei (blue). Although stiff substrate-primed myofibroblasts demonstrated acute decrease of α-SMA stress fiber organization after 12 h shift to soft substrates, myofibroblast presence was re-established and persisted even after 14 days of soft culture. Scale bar: 200 μm. Adapted and reproduced with permission from ([130]).

### 5.5.2 Controlling Stem Cell Fate: Utilizing Substrate Rigidity as a Tool in Regenerative Medicine

Stem cell renewal and differentiation is governed by local biomechanical microenvironments that span a physically diverse range of ECM stiffness [148,149]. Stiffness-dependent stem cell differentiation is typically achieved through the manipulation of focal adhesions, cytoskeletal contractility, Rho GTPase signaling, and nuclear regulation [150,151]. Recently, defined ECM stiffness in combination with specific chemical factors has become a tool to guide stem cell differentiation for use in regenerative medicine.

By simply tuning the substrate stiffness in culture systems, mesenchymal stem cells (MSCs) can be driven toward neuronal (~1 kPa), glial (~10 kPa) [152], muscle and tendon (10–20 kPa) [153,154] or osteogenic lines (40–90 kPa). In addition, MSCs are known to migrate along gradients of stiffness [155], and therefore creating implants of varying stiffness could be implemented to guide migration and differentiation into functional cell types. Although these approaches can generate specific lineages in vitro, it is not clear if mechanically determined cells will remain functional in vivo or how long the effects will last post implantation. Therefore, because disease or trauma typically results in pathological areas at the in vivo implantation site it may be necessary to create systems that can not only provide a suitable environment for the resident cells but also restore the appropriate tissue compliance and architecture. To address these limitations, scaffolds with independently tunable ECM cues have been investigated as a tool for creating complex systems to regulate stem cell fate in three dimensions and promote desirable cellular processes or tissue regeneration and have been met with some success [130]. Clearly stiffness-directed stem cell differentiation holds great promise toward providing autologously derived cell sources, and future research will be required to assess feasibility for clinical usage.

## 5.6 Conclusions and Perspective

Cells embedded have acute sensitivity to mechanical cues generated by their biomechanical microenvironment, and they respond to this mechanical signaling by creating functional changes to the surrounding tissue. Therefore, harnessing the relationship between mechanical conditioning and cell response holds great promise for use in wound healing therapies, tissue engineering, and regenerative medicine. In this chapter we have discussed two separate but equally important mechanical stimuli: external loading in the form of stretch and intrinsic loading in the form of ECM stiffness. Herein we discussed how mechanical stretch can be utilized to promote and accelerate wound healing, modify myofibroblast behavior as is appropriate, and assist in the creation of functional and custom-tailored tissue equivalents. We have also discussed how tissue compliance can be a powerful regulator of wound healing, stem cell behavior, and also a tool for tissue engineering. Together stretch and ECM compliance cues have significant implications for the design and fabrication of tissue equivalents for use tissue regeneration.

## References

1. Silver, F. H., and Siperko, L. M. Mechanosensing and mechanochemical transduction: How is mechanical energy sensed and converted into chemical energy in an extracellular matrix? *Crit. Rev. Biomed. Eng.*, 2003. **31**(4), 255–331.
2. Wang, J. H., and Thampatty, B. P. An introductory review of cell mechanobiology. *Biomech. Model Mechanobiol.*, 2006. **5**(1), 1–16.
3. Chan, M. W., Hinz, B., and McCulloch, C. A. Mechanical induction of gene expression in connective tissue cells. *Methods Cell Biol.*, 2010. **98**, 178–205.
4. Gauthier, N. C., Masters, T. A., and Sheetz, M. P. Mechanical feedback between membrane tension and dynamics. *Trends Cell Biol.*, 2012. **22**(10), 527–535.
5. Loesberg, W. A., et al., The effect of combined cyclic mechanical stretching and microgrooved surface topography on the behavior of fibroblasts. *J. Biomed. Mater. Res. A*, 2005. **75**(3), 723–732.
6. Breen, E. C., Fu, Z., and Normand, H. Calcyclin gene expression is increased by mechanical strain in fibroblasts and lung. *Am. J. Respir. Cell Mol. Biol.*, 1999. **21**(6), 746–752.
7. Raeber, G. P., Lutolf, M. P., and Hubbell, J. A. Part II: Fibroblasts preferentially migrate in the direction of principal strain. *Biomech. Model Mechanobiol.*, 2008. **7**(3), 215–225.
8. Bishop, J. E., et al., Cyclic mechanical deformation stimulates human lung fibroblast proliferation and autocrine growth factor activity. *Am. J. Respir. Cell Mol. Biol.*, 1993. **9**(2), 126–133.
9. Lundon, K., Effect of mechanical loading on soft connective tissues, in *Functional Soft Tissue Examination and Treatment by Manual Methods*, (Hammer, W., ed.), 2006, Jones and Bartlett: Sudbury, MA. p. 13–120.
10. Chen, W., and Frangogiannis, N. G. Fibroblasts in post-infarction inflammation and cardiac repair. *Biochim. Biophys. Acta*, 2013. **1833**(4), 945–953.

11. Engler, A. J., et al., Matrix elasticity directs stem cell lineage specification. *Cell*, 2006. **126**(4), 677–689.

12. Lo, C. M., et al., Cell movement is guided by the rigidity of the substrate. *Biophys. J.*, 2000. **79**(1), 144–152.

13. Pelham, R. J., Jr., and Wang, Y. Cell locomotion and focal adhesions are regulated by substrate flexibility. *Proc. Natl. Acad. Sci. U. S. A.*, 1997. **94**(25), 13661–13665.

14. Goffin, J. M., et al., Focal adhesion size controls tension-dependent recruitment of alpha-smooth muscle actin to stress fibers. *J. Cell Biol.*, 2006. **172**(2), 259–268.

15. Wipff, P. J., et al., Myofibroblast contraction activates latent TGF-beta1 from the extracellular matrix. *J. Cell Biol.*, 2007. **179**(6), 1311–1323.

16. Hinz, B., Tissue stiffness, latent TGF-beta1 activation, and mechanical signal transduction: Implications for the pathogenesis and treatment of fibrosis. *Curr. Rheumatol. Rep.*, 2009. **11**(2), 120–126.

17. Hinz, B., and Gabbiani, G. Cell-matrix and cell–cell contacts of myofibroblasts: Role in connective tissue remodeling. *Thromb. Haemost.*, 2003. **90**(6), 993–1002.

18. Tomasek, J. J., et al., Myofibroblasts and mechano-regulation of connective tissue remodelling. *Nat. Rev. Mol. Cell Biol.*, 2002. **3**(5), 349–363.

19. Gabbiani, G., Ryan, G. B., and Majno, G. Presence of modified fibroblasts in granulation tissue and their possible role in wound contraction. *Experientia*, 1971. **27**(5), 549–550.

20. Follonier Castella, L., et al., Regulation of myofibroblast activities: Calcium pulls some strings behind the scene. *Exp. Cell Res.*, 2010. **316**(15), 2390–2401.

21. Godbout, C., et al., The mechanical environment modulates intracellular calcium oscillation activities of myofibroblasts. *PLoS One*, 2013. **8**(5), e64560.

22. Hinz, B., The myofibroblast: Paradigm for a mechanically active cell. *J. Biomech.*, 2010. **43**(1), 146–155.

23. Sandbo, N., and Dulin, N. Actin cytoskeleton in myofibroblast differentiation: Ultrastructure defining form and driving function. *Transl. Res.*, 2011. **158**(4), 181–196.

24. Hinz, B., Formation and function of the myofibroblast during tissue repair. *J. Invest. Dermatol.*, 2007. **127**(3), 526–537.

25. Kis, K., Liu, X., and Hagood, J. S. Myofibroblast differentiation and survival in fibrotic disease. *Expert. Rev. Mol. Med.*, 2011. **13**, e27.
26. Friedman, S. L., et al., Therapy for fibrotic diseases: Nearing the starting line. *Sci. Transl. Med.*, 2013. **5**(167), 167sr1.
27. van den Borne, S. W., et al., Myocardial remodeling after infarction: The role of myofibroblasts. *Nat. Rev. Cardiol.*, 2010. **7**(1), 30–37.
28. Rohr, S., Myofibroblasts in diseased hearts: New players in cardiac arrhythmias? *Heart Rhythm.*, 2009. **6**(6), 848–856.
29. Krenning, G., Zeisberg, E. M., and Kalluri, R. The origin of fibroblasts and mechanism of cardiac fibrosis. *J. Cell Physiol.*, 2010. **225**(3), 631–637.
30. Hernandez-Gea, V., and Friedman, S. L. Pathogenesis of liver fibrosis. *Ann. Rev. Pathol.*, **6**, 425–456.
31. Wynn, T. A., Common and unique mechanisms regulate fibrosis in various fibroproliferative diseases. *J. Clin. Invest.*, 2007. **117**(3), 524–529.
32. Meran, S., and Steadman, R. Fibroblasts and myofibroblasts in renal fibrosis. *Int. J. Exp. Pathol.*, 2011. **92**(3), 158–167.
33. Serrano, A. L., et al., Cellular and molecular mechanisms regulating fibrosis in skeletal muscle repair and disease. *Curr. Top. Dev. Biol.*, 2011. **96**, 167–201.
34. Varga, J., and Abraham, D. Systemic sclerosis: A prototypic multisystem fibrotic disorder. *J. Clin. Invest.*, 2007. **117**(3), 557–567.
35. De Wever, O., et al., Stromal myofibroblasts are drivers of invasive cancer growth. *Int. J. Cancer*, 2008. **123**(10), 2229–2238.
36. Hinz, B., and Gabbiani, G. Fibrosis: Recent advances in myofibroblast biology and new therapeutic perspectives. *F1000 Biol. Rep.*, 2010. **2**, 78.
37. Douglass, A., et al., Targeting liver myofibroblasts: A novel approach in anti-fibrogenic therapy. *Hepatol. Int.*, 2008. **2**(4), 405–415.
38. Sivakumar, P., et al., Into the matrix: Targeting fibroblasts in pulmonary fibrosis. *Curr. Opin. Pulm Med.*, 2012. **18**(5), 462–469.
39. Leask, A., Towards an anti-fibrotic therapy for scleroderma: Targeting myofibroblast differentiation and recruitment. *Fibrogenesis Tissue Repair.*, 2010. **3**, 8.
40. Scotton, C. J., and Chambers, R. C. Molecular targets in pulmonary fibrosis: The myofibroblast in focus. *Chest*, 2007. **132**(4), 1311–1321.
41. Hinz, B., et al., The myofibroblast: One function, multiple origins. *Am. J. Pathol.*, 2007. **170**(6), 1807–1816.

42. Hinz, B., et al., Recent developments in myofibroblast biology: Paradigms for connective tissue remodeling. *Am. J. Pathol.*, 2012. **180**(4), 1340–1355.

43. Desmouliere, A., Hepatic stellate cells: The only cells involved in liver fibrogenesis? A dogma challenged. *Gastroenterology*, 2007. **132**(5), 2059–2062.

44. Herzog, E. L., and Bucala, R. Fibrocytes in health and disease. *Exp. Hematol.*, 2010. **38**(7), 548–556.

45. Keeley, E. C., Mehrad, B., and Strieter, R. M. The role of fibrocytes in fibrotic diseases of the lungs and heart. *Fibrogenesis Tissue Repair.*, 2011. **4**, 2.

46. Bellini, A., and Mattoli, S. The role of the fibrocyte, a bone marrow-derived mesenchymal progenitor, in reactive and reparative fibroses. *Lab Invest.*, 2007. **87**(9), 858–870.

47. Coen, M., Gabbiani, G., and Bochaton-Piallat, M. L. Myofibroblast-mediated adventitial remodeling: An underestimated player in arterial pathology. *Arterioscler. Thromb. Vasc. Biol.*, 2011. **31**(11), 2391–2396.

48. Humphreys, B. D., et al., Fate tracing reveals the pericyte and not epithelial origin of myofibroblasts in kidney fibrosis. *Am. J. Pathol.*, 2010. **176**(1), 85–97.

49. Mishra, P. J., Glod, J. W., and Banerjee, D. Mesenchymal stem cells: Flip side of the coin. *Cancer Res.*, 2009. **69**(4), 1255–1258.

50. Hinz, B., The myofibroblast in connective tissue repair and regeneration, in *Regenerative Medicine and Biomaterials for the Repair of Connective Tissues* (Ralphs, C. A. J., ed.) 2010, Woodhead Publishing Ltd.: Cambridge, UK. pp. 39–82.

51. Lee, K., and Nelson, C. M. New insights into the regulation of epithelial-mesenchymal transition and tissue fibrosis. *Int. Rev. Cell Mol. Biol.*, 2012. **294**, 171–221.

52. Piera-Velazquez, S., Li, Z., and Jimenez, S. A. Role of endothelial-mesenchymal transition (EndoMT) in the pathogenesis of fibrotic disorders. *Am. J. Pathol.*, 2011. **179**(3), 1074–1080.

53. Klingberg, F., Hinz, B., and White, E. S. The myofibroblast matrix: Implications for tissue repair and fibrosis. *J. Pathol.*, 2013. **229**(2), 298–309.

54. Cox, T. R., and Erler, J. T. Remodeling and homeostasis of the extracellular matrix: Implications for fibrotic diseases and cancer. *Dis. Model Mech.*, 2011. **4**(2), 165–178.

55. Frantz, C., Stewart, K. M., and Weaver, V. M. The extracellular matrix at a glance. *J. Cell Sci.*, 2010. **123**(Pt 24), 4195–4200.

56. Hynes, R. O., The evolution of metazoan extracellular matrix. *J. Cell Biol.*, 2012. **196**(6), 671–679.

57. Ozbek, S., et al., The evolution of extracellular matrix. *Mol. Biol. Cell*, 2010. **21**(24), 4300–4305.

58. Grinnell, F., Fibroblasts, myofibroblasts, and wound contraction. *J. Cell Biol.*, 1994. **124**(4), 401–404.

59. Arora, P. D., Narani, N., and McCulloch, C. A. The compliance of collagen gels regulates transforming growth factor-beta induction of alpha-smooth muscle actin in fibroblasts. *Am. J. Pathol.*, 1999. **154**(3), 871–882.

60. Hinz, B., Masters and servants of the force: The role of matrix adhesions in myofibroblast force perception and transmission. *Eur. J. Cell Biol.*, 2006. **85**(3–4), 175–181.

61. Hinz, B., et al., Mechanical tension controls granulation tissue contractile activity and myofibroblast differentiation. *Am. J. Pathol.*, 2001. **159**(3), 1009–1020.

62. Junker, J. P., et al., Mechanical tension stimulates the transdifferentiation of fibroblasts into myofibroblasts in human burn scars. *Burns*, 2000. **34**(7), 942–946

63. Rajkumar, V. S., et al., Platelet-derived growth factor-beta receptor activation is essential for fibroblast and pericyte recruitment during cutaneous wound healing. *Am. J. Pathol.*, 2006. **169**(6), 2254–2265.

64. Bello, Y. M., Falabella, A. F., and Eaglstein, W. H. Tissue-engineered skin. Current status in wound healing. *Am. J. Clin. Dermatol.*, 2001. **2**(5), 305–313.

65. Mattoli, S., Bellini, A., and Schmidt, M. The role of a human hematopoietic mesenchymal progenitor in wound healing and fibrotic diseases and implications for therapy. *Curr. Stem. Cell Res. Ther.*, 2009. **4**(4), 266–280.

66. Hardy, M. A., The biology of scar formation. *Phys. Ther.*, 1989. **69**(12), 1014–1024.

67. Costa, A. M., et al., Mechanical forces induce scar remodeling. Study in non-pressure-treated versus pressure-treated hypertrophic scars. *Am. J. Pathol.*, 1999. **155**(5), 1671–1679.

68. Rudolph, R., Berg, J. V., and Ehrlick, H. P. Wound healing: Biochemical and clinical aspects, in *Wound Healing: Biochemical and Clinical Aspects* (I. Cohen, R. Diegelmann, and W. Lindblad, eds.). 1992, Philadelphia, PA: W. B. Saunders Co.

69. Siegel, H. J., et al., Vacuum-assisted closure for radiation-associated wound complications. *J. Surg. Oncol.*, 2007. **96**(7), 575–582.

70. Langrana, N. A., et al., Effect of mechanical load in wound healing. *Ann. Plast. Surg.*, 1983. **10**(3), 200–208.

71. Machen, S., Management of traumatic war wounds using vacuum-assisted closure dressings in an austere environment. *US Army Med. Dep. J.*, 2007, January–March, 17–23.

72. Ozer, M. T., et al., Use of vacuum-assisted closure (VAC) in high-energy complicated perineal injuries: Analysis of nine cases. *Int. Wound J.*, 2011. **8**(6), 599–607.

73. Ahlfors, J. E., and Billiar, K. L. Biomechanical and biochemical characteristics of a human fibroblast-produced and remodeled matrix. *Biomaterials*, 2007. **28**(13), 2183–2191.

74. Balestrini, J. L., et al., Applying controlled non-uniform deformation for in vitro studies of cell mechanobiology. *Biomech. Model Mechanobiol.*, 2010. **9**(3), 329–344.

75. Hinz, B., Tissue stiffness, latent TGF-beta1 activation, and mechanical signal transduction: Implications for the pathogenesis and treatment of fibrosis. *Curr. Rheumatol. Rep.*, 2009. **11**(2), 120–126.

76. Nerem, R. M., Cell-based therapies: From basic biology to replacement, repair, and regeneration. *Biomaterials*, 2007. **28**(34), 5074–5077.

77. Swartz, D. D., Russell, J. A., and Andreadis, S. T. Engineering of fibrin-based functional and implantable small-diameter blood vessels. *Am. J. Physiol. Heart Circ. Physiol.*, 2005. **288**(3), H1451–H1460.

78. Yao, L., et al., Fibrin-based tissue-engineered blood vessels: Differential effects of biomaterial and culture parameters on mechanical strength and vascular reactivity. *Tissue Eng.*, 2005. **11**(7–8), 991–1003.

79. Langer, A., and Rogowski, W. Systematic review of economic evaluations of human cell-derived wound care products for the treatment of venous leg and diabetic foot ulcers. *BMC Health Serv. Res.*, 2009. **9**, 115.

80. Gurtner, G. C., et al., Wound repair and regeneration. *Nature*, 2008. **453**(7193), 314–321.

81. Clark, R. A., Ghosh, K., and Tonnesen, M. G. Tissue engineering for cutaneous wounds. *J. Invest. Dermatol.*, 2007. **127**(5), 1018–1029.

82. Balestrini, J. L., and Billiar, K. L. Equibiaxial cyclic stretch stimulates fibroblasts to rapidly remodel fibrin. *J. Biomech.*, 2006. **39**(16), 2983–2990.

83. Balestrini, J. L., and Billiar, K. L. Magnitude and duration of stretch modulate fibroblast remodeling. *J. Biomech. Eng.*, 2009. **131**(5), 051005.

84. Grassl, E. D., Oegema, T. R., and Tranquillo, R. T. Fibrin as an alternative biopolymer to type-I collagen for the fabrication of a media equivalent. *J. Biomed. Mater. Res.*, 2002. **60**(4), 607–612.

85. Wille, J. J., Elson, E. L., and Okamoto, R. J. Cellular and matrix mechanics of bioartificial tissues during continuous cyclic stretch. *Ann. Biomed. Eng.*, 2006. **34**(11), 1678–1690.

86. Altman, G. H., et al., Cell differentiation by mechanical stress. *FASEB J.*, 2002. **16**(2), 270–272. Epub 2001 Dec 28.

87. Boerboom, R. A., et al., Effect of strain magnitude on the tissue properties of engineered cardiovascular constructs. *Ann. Biomed. Eng.*, 2008. **36**(2), 244–253.

88. Mudera, V. C., et al., Molecular responses of human dermal fibroblasts to dual cues: Contact guidance and mechanical load. *Cell Motil. Cytoskeleton.*, 2000. **45**(1), 1–9.

89. Throm Quinlan, A. M., et al., Combining dynamic stretch and tunable stiffness to probe cell mechanobiology in vitro. *PLoS One*, 2012. **6**(8), e23272.

90. Calve, S., et al. Mechanical characterization of growth in fibrin-based tendon constructs. in *Summer Bioengineering Conference*. 2005. Vail, Colorado: ASME.

91. Grenier, G., et al., Tissue reorganization in response to mechanical load increases functionality. *Tissue Eng.*, 2005. **11**(1–2), 90–100.

92. Cummings, C. L., et al., Properties of engineered vascular constructs made from collagen, fibrin, and collagen-fibrin mixtures. *Biomaterials*, 2004. **25**(17), 3699–3706.

93. Seliktar, D., et al., Dynamic mechanical conditioning of collagen-gel blood vessel constructs induces remodeling in vitro. *Ann. Biomed. Eng.*, 2000. **28**(4), 351–362.

94. Timmenga, E. J., Schoorl, R., and Klopper, P. J. Biomechanical and histomorphological changes in expanded rabbit skin. *Br. J. Plast. Surg.*, 1990. **43**(1), 101–106.

95. Timmenga, E. J., and Das, P. K. Histomorphological observations on dermal repair in expanded rabbit skin: A preliminary report. *Br. J. Plast. Surg.*, 1992. **45**(7), 503–507.

96. Boublik, J., et al., Mechanical properties and remodeling of hybrid cardiac constructs made from heart cells, fibrin, and biodegradable, elastomeric knitted fabric. *Tissue Eng.*, 2005. **11**(7–8), 1122–1132.

97. Isenberg, B. C., and Tranquillo, R. T. Long-term cyclic distention enhances the mechanical properties of collagen-based media-equivalents. *Ann. Biomed. Eng.*, 2003. **31**(8), 937–949.

98. Boccafoschi, F., et al., Dynamic fibroblast cultures: Response to mechanical stretching. *Cell Adh. Migr.*, 2007. **1**(3), 124–128.

99. Yang, G., Crawford, R. C., and Wang, J. H. Proliferation and collagen production of human patellar tendon fibroblasts in response to cyclic uniaxial stretching in serum-free conditions. *J. Biomech.*, 2004. **37**(10), 1543–1550.

100. Kratz, C., Tollback, A., and Kratz, G. Effects of continuous stretching on cell proliferation and collagen synthesis in human burn scars. *Scand. J. Plast. Reconstr. Surg. Hand Surg.*, 2001. **35**(1), 57–63.

101. Wang, J. G., et al., Uniaxial cyclic stretch induces focal adhesion kinase (FAK) tyrosine phosphorylation followed by mitogen-activated protein kinase (MAPK) activation. *Biochem. Biophys. Res. Commun.*, 2001. **288**(2), 356–361.

102. Nishimura, K., et al., Effect of different frequencies of tensile strain on human dermal fibroblast proliferation and survival. Wound Repair Regen, 2007. **15**(5), 646–656.

103. Stickler, P., et al., Cyclically stretching developing tissue in vivo enhances mechanical strength and organization of vascular grafts. *Acta Biomater.*, 2010. **6**(7), 2448–2456.

104. Clark, R. A., et al., Collagen matrices attenuate the collagen-synthetic response of cultured fibroblasts to TGF-beta. *J. Cell Sci.*, 1995. **108**(Pt 3), 1251–1261.

105. Syedain, Z. H., Weinberg, J. S., and R. T. Tranquillo, Cyclic distension of fibrin-based tissue constructs: Evidence of adaptation during growth of engineered connective tissue. *Proc. Natl. Acad. Sci. U. S. A.*, 2008. **105**(18), 6537–6542.

106. Li, Y. S., Haga, J. H., and Chien, S. Molecular basis of the effects of shear stress on vascular endothelial cells. *J. Biomech.*, 2005. **38**(10), 1949–1971.

107. Schriefer, J. L., et al., Cellular accommodation and the response of bone to mechanical loading. *J. Biomech.*, 2005. **38**(9), 1838–1845.

108. Jhun, C. S., et al., Planar biaxial mechanical behavior of bioartificial tissues possessing prescribed fiber alignment. *J. Biomech. Eng.*, 2009. **131**(8), 081006.

109. Stekelenburg, M., et al., Dynamic straining combined with fibrin gel cell seeding improves strength of tissue-engineered small-diameter vascular grafts. *Tissue Eng. Part A*, 2009. **15**(5), 1081–1089.

110. Syedain, Z. H., et al., Controlled compaction with ruthenium-catalyzed photochemical cross-linking of fibrin-based engineered connective tissue. *Biomaterials*, 2009. **30**(35), 6695–6701.

111. Weidenhamer, N. K., and Tranquillo, R. T. Influence of cyclic mechanical stretch and tissue constraints on cellular and collagen alignment in fibroblast-derived cell sheets. *Tissue Eng. Part C Methods*, 2013. **19**(5), 386–395.

112. Weber, B., et al., Tissue engineering on matrix: Future of autologous tissue replacement. *Semin. Immunopathol.*, 2011. **33**(3), 307–315.

113. Thayer, P., et al., The effects of combined cyclic stretch and pressure on the aortic valve interstitial cell phenotype. *Ann. Biomed. Eng.*, 2011. **39**(6), 1654–1667.

114. van Vlimmeren, M. A., et al., Passive and active contributions to generated force and retraction in heart valve tissue engineering. *Biomech. Model Mechanobiol.*, 2012. **11**(7), 1015–1027.

115. Campbell, G. R., et al., The peritoneal cavity as a bioreactor for tissue engineering visceral organs: Bladder, uterus and vas deferens. *J. Tissue Eng. Regen. Med.*, 2008. **2**(1), 50–60.

116. Kadner, A., et al., Human umbilical cord cells: A new cell source for cardiovascular tissue engineering. *Ann. Thorac. Surg.*, 2002. **74**(4), S1422–S1428.

117. Engelmayr, G. C., Jr., et al., A novel flex-stretch-flow bioreactor for the study of engineered heart valve tissue mechanobiology. *Ann. Biomed. Eng.*, 2008. **36**(5), 700–712.

118. Hoerstrup, S. P., et al., Tissue engineering of a bioprosthetic heart valve: Stimulation of extracellular matrix assessed by hydroxyproline assay. *Asaio J.*, 1999. **45**(5), 397–402.

119. Gurtner, G. C., et al., Improving cutaneous scar formation by controlling the mechanical environment: Large animal and phase I studies. *Ann. Surg.*, 2011. **254**(2), 217–225.

120. Fireman, E., et al., Morphological and biochemical properties of alveolar fibroblasts in interstitial lung diseases. 2001. **179**(2), 105–117.

121. Zhang, H. Y., et al., Lung fibroblast alpha-smooth muscle actin expression and contractile phenotype in bleomycin-induced pulmonary fibrosis. *Am. J. Pathol.*, 1996. **148**(2), 527–537.

122. Powell, D. W., et al., Myofibroblasts. I. Paracrine cells important in health and disease. *Am. J. Physiol.*, 1999. **277**(1 Pt 1), C1–C9.

123. King, T. E., Jr., et al., Idiopathic pulmonary fibrosis: Relationship between histopathologic features and mortality. *Am. J. Respir. Crit. Care Med.*, 2001. **164**(6), 1025–1032.

124. Carrington, L. M., and Boulton, M. Hepatocyte growth factor and keratinocyte growth factor regulation of epithelial and stromal corneal wound healing. *J. Cataract Refract. Surg.*, 2005. **31**(2), 412–423.

125. Phu, D., et al., Effect of substrate composition and alignment on corneal cell phenotype. *Tissue Eng. Part A*, 2011. **17**(5–6), 799–807.

126. Discher, D., et al., Biomechanics: Cell research and applications for the next decade. *Ann. Biomed. Eng.*, 2009. **37**(5), 847–859.

127. Georges, P. C., and Janmey, P. A. Cell type-specific response to growth on soft materials. *J. Appl. Physiol.*, 2005. **98**(4), 1547–1553.

128. Harris, A., Tissue culture cells on deformable substrata: Biomechanical implications. *J. Biomech. Eng.*, 1984. **106**, 19–24.

129. Harris, A. K., Stopak, D., and Wild, P. Fibroblast traction as a mechanism for collagen morphogenesis. *Nature*, 1981. **290**(5803), 249–251.

130. Balestrini, J. L., et al., The mechanical memory of lung myofibroblasts. *Integr. Biol.* (*Camb*), 2012. **4**(4), 410–421.

131. Engler, A., et al., Substrate compliance versus ligand density in cell on gel responses. *Biophys. J.*, 2004. **86**(1 Pt 1), 617–628.

132. Throm, A. M., et al. Effects of substrate stiffness on valvular interstitial cell activation. in *BMES*. 2005. Baltimore, MD.

133. Wang, H. B., Dembo, M., and Wang, Y. L. Substrate flexibility regulates growth and apoptosis of normal but not transformed cells. *Am. J. Physiol. Cell Physiol.*, 2000. **279**(5), C1345–C1350.

134. Buxboim, A., Ivanovska, I. L., and Discher, D. E. Matrix elasticity, cytoskeletal forces and physics of the nucleus: How deeply do cells "feel" outside and in? *J. Cell Sci.*, 2010. **123**(Pt 3), 297–308.

135. Janmey, P. A., et al., The hard life of soft cells. *Cell Motil Cytoskeleton*, 2009. **66**(8), 597–605.

136. Winer, J. P., et al., Bone marrow-derived human mesenchymal stem cells become quiescent on soft substrates but remain responsive

to chemical or mechanical stimuli. *Tissue Eng. Part A*, 2009. **15**(1), 147–54.

137. Flanagan, L. A., et al., Neurite branching on deformable substrates. *Neuroreport*, 2002. **13**(18), 2411–2415.

138. Butcher, D. T., Alliston, T., and Weaver, V. M. A tense situation: Forcing tumour progression. *Nat. Rev. Cancer*, 2009. **9**(2), 108–122.

139. Georges, P. C., et al., Increased stiffness of the rat liver precedes matrix deposition: Implications for fibrosis. *Am. J. Physiol. Gastrointest Liver Physiol.*, 2007. **293**(6), G1147–G1154.

140. Liu, F., et al., Feedback amplification of fibrosis through matrix stiffening and COX-2 suppression. *J. Cell Biol.*, 2010. **190**(4), 693–706.

141. Berry, M. F., et al., Mesenchymal stem cell injection after myocardial infarction improves myocardial compliance. *Am. J. Physiol. Heart Circ. Physiol.*, 2006. **290**(6), H2196–H2203.

142. Engler, A. J., Myotubes differentiate optimally on substrates with tissue-like stiffness: Pathological implications for soft or stiff microenvironments. *J. Cell Biol.*, 2004. **166**, 877–887.

143. Engler, A. J., et al., Microtissue elasticity: Measurements by atomic force microscopy and its influence on cell differentiation. *Methods Cell Biol.*, 2007. **83**, 521–545.

144. Ho, S. P., et al., The tooth attachment mechanism defined by structure, chemical composition and mechanical properties of collagen fibers in the periodontium. *Biomaterials*, 2007. **28**(35), 5238–5245.

145. Buxboim, A., and Discher, D. E. Stem cells feel the difference. *Nat. Methods*, 2010. **7**(9), 695–697.

146. Tschumperlin, D. J., Jones, J. C., and Senior, R. M. The fibrotic matrix in control: Does the extracellular matrix drive progression of idiopathic pulmonary fibrosis? *Am. J. Respir. Crit. Care Med.*, 2012. **186**(9), 814–816.

147. Wang, H., et al., Redirecting valvular myofibroblasts into dormant fibroblasts through light-mediated reduction in substrate modulus. *PLoS One*, 2012. **7**(7), e39969.

148. Fuchs, E., Tumbar, T., and Guasch, G. Socializing with the neighbors: Stem cells and their niche. *Cell*, 2004. **116**(6), 769–778.

149. Watt, F. M., and Hogan, B. L. Out of Eden: Stem cells and their niches. *Science*, 2000. **287**(5457), 1427–1430.

150. Yim, E. K., and Sheetz, M. P. Force-dependent cell signaling in stem cell differentiation. *Stem Cell Res. Ther.*, 2012. **3**(5), 41.

151. Trappmann, B., et al., Extracellular-matrix tethering regulates stem-cell fate. *Nat. Mater.*, 2012. **11**(7), 642–964.

152. Her, G. J., et al., Control of three-dimensional substrate stiffness to manipulate mesenchymal stem cell fate toward neuronal or glial lineages. *Acta Biomater.*, 2013. **9**(2), 5170–5180.

153. Zouani, O. F., et al., Effect of BMP-2 from matrices of different stiffnesses for the modulation of stem cell fate. *Biomaterials*, 2013. **34**(9), 2157–2166.

154. Sharma, R. I., and J. G. Snedeker, Paracrine interactions between mesenchymal stem cells affect substrate driven differentiation toward tendon and bone phenotypes. *PLoS One*, 2012. **7**(2), e31504.

155. Vincent, L. G., et al., Mesenchymal stem cell durotaxis depends on substrate stiffness gradient strength. *Biotechnol. J.*, 2013. **8**(4), 472–484.

156. Hinz, B. Matrix mechanics and regulation of the fibroblast phenotype. *Periodontology*, 2013 (2000). **63**, 14–28.

# Chapter 6

# Interstitial Fluid Flow Mechanosensing: Mechanisms and Consequences

**Arpit D. Shah and Adrian C. Shieh**

*School of Biomedical Engineering, Science and Health Systems, Drexel University, 3141 Chestnut Street, Philadelphia, PA 19104, USA*

Arpit.D.Shah@Drexel.edu

## 6.1 Introduction

Interstitial fluid flow is flow of the fluid found in interstitial space of tissues, convectively transporting nutrients, cellular waste products, and even large proteins. This constantly moving fluid, which makes up nearly 20% of an average human's body weight, has garnered much interest from researchers over recent years; in particular, the focus has been placed on understanding the physiological role this fluid flow plays on cellular behavior and the microenvironment [4].

Interstitial flow has been shown to have an important physiological role in a variety of cell processes, including invasion [50,76], differentiation [17,74], morphogenesis [51,59,74], and protein secretion [13,50]. Many tissues and cell types are sensitive to interstitial flow, including blood [51] and lymphatic endothelial

*Cells, Forces, and the Microenvironment*
Edited by Charles M. Cuerrier and Andrew E. Pelling

ISBN 978-981-4613-36-1 (Hardcover), 978-981-4613-37-8 (eBook)
www.panstanford.com

cells [7], fibroblasts [52,54], osteocytes [2,100], smooth muscle cells [14,70,95], and cancer cells [64,76,78]. This chapter focuses on the molecular mechanisms used by cells to sense and respond to interstitial fluid flow, and how these cellular responses to interstitial flow affect normal physiology and disease processes.

## 6.2 Basics of Interstitial Fluid Flow

Interstitial fluid flows through intercellular spaces of the extracellular matrix between blood and lymphatic capillaries, where it is eventually collected and drained by the lymphatic system as lymph [45,54]. Convective forces facilitate transport of molecules while simultaneously maintaining tissue fluid balance [54]. This fluid flow through the interstitium also generates biophysical forces that may directly affect cells [29]. Theoretical or computational models of interstitial flow are complicated by the nature of interstitial flow; at each location in the interstitium where fluid flow occurs, the structure and properties of the matrix can vary considerably, along with the shape and orientation of affected cells. However, despite these complexities, several governing principles can be considered to better understand interstitial fluid flow in a quantitative fashion.

### 6.2.1 Interstitial Flow Velocity

One of the fundamental ways to characterize fluid flow is by the velocity. In the case of interstitial flow, there have been a number of experimental measurements and computational models to estimate flow velocity in vivo. Using fluorescence recovery after photobleaching (FRAP) yielded a velocity range of 0.1–2 μm/s in a rabbit ear chamber with bovine serum albumin conjugated with fluorescein isothiocyanate tracer [13]. An in vivo study examined the effect of overexpression of $VEGF_{165}$ and its role on peritumoral fluid convection and lymphatic drainage. Magnetic resonance imaging (MRI) of the $VEGF_{165}$-overexpressing tumors revealed interstitial fluid flow velocities of approximately 0.2–0.3 μm/s. Overall there was increased interstitial convection, vascular permeability, and lymphatic drainage in tumors [15]. Dynamic contrast-enhanced MRI (DCE-MRI) has also been used to measure fluid flow velocity at the edge of cervical carcinoma xenografts in mice and cervical tumors in humans by wash out of gadolinium-based contrast agents. Flow at the tumor edge is of particular interest, since there is dramatic gradient in fluid pressure between the high-pressure tumor and the lower-pressure surrounding tissue. This pressure gradient resulted in an approximated flow velocity of 5 μm/s at an interstitial fluid pressure of 20 mmHg in mouse xenografts, while human cervical cancer patients with pelvic lymph node metastases showed flow velocities as high as 50 μm/s [33]. Other models have shown that the periphery of a tumor can have interstitial fluid velocities as high as 10 μm/s as well [36].

These findings still have much uncertainty with regard to determining quantitatively the flow velocity, while there are many findings that correlate it to various parameters such as convective transport of macromolecules and shear stress that is generated on the cell surface. Overall there are many challenges for current computational models and in vivo measurements of interstitial flow velocities. The internal forces and interstitial spaces of various tissues are highly heterogeneous, thus making it very difficult to determine the characteristics of interstitial flow without identifying and understanding all local parameters. Nevertheless, quantitatively determining flow velocity (and other flow characteristics) would provide a greater understanding of the effects interstitial flow may exert on cells and tissues.

## 6.2.2 Starling's Equation

To characterize interstitial fluid flow, we need to first understand the phenomenon of capillary filtration. It is widely known that oncotic (colloid osmotic) and hydrostatic pressure gradients play a major role in the movement of fluid. In 1846, Ernest Henry Starling determined that forces such as tissue fluid pressure ($P_t$), capillary fluid pressure ($P_c$), tissue colloid osmotic pressure ($\pi_t$), and plasma colloid osmotic pressure ($\pi_p$), known as Starling forces, were responsible for driving capillary filtration [79,89]. In humans typical colloid osmotic pressure is 28 mmHg for the capillaries and 8 mmHg for the interstitium, while hydrostatic pressure is 20 mmHg for the capillaries and –1 to –3 mmHg for the interstitium [30]. Oncotic pressure is an important indicator for identifying potential diseases; elevated oncotic pressure is associated with pulmonary edema [62]. Additionally in cancer, studies in rats have shown that both oncotic and interstitial fluid pressure are elevated in solid tumors [81].

Using these hydrostatic and oncotic pressures, Starling's equation predicts fluid filtration across capillary membranes:

$$J_v = K_{f,c}[(P_c - P_t) - \sigma(\Pi_c - \Pi_t)],$$

where $J_v$ is the transcapillary fluid flow, $K_{f,c}$ is the capillary filtration coefficient, and $\sigma$ is the capillary protein reflection coefficient. The reflection coefficient represents the degree to which the capillary is impermeable to the proteins generating the oncotic pressure gradient [42]. If the capillary was permeable to all plasma proteins, then the reflection coefficient would be 0, whereas a completely impermeable capillary would have a reflection coefficient of 1. A majority of the fluid filtered through the capillary wall is reabsorbed by downstream post-capillary venules; however a small fraction of this fluid remains within the tissue interstitium [57,98]. This fluid is ultimately converted into lymph once it is drained from the interstitium by the initial lymphatic capillaries. However, while Starling's equation describes capillary fluid filtration, it does not model interstitial flow within the tissue. For this, we turn to civil engineering and models of flow through porous media.

### 6.2.3 Darcy's Law and Brinkman's Equation

Interstitial fluid flow depends on many factors, such as pressure gradients within the interstitium and the structure and properties of the tissue. Darcy's Law was developed to characterize fluid flowing through a porous medium. In 1856 Henry Darcy examined the flow of groundwater through sand-beds that assisted in filtration [98]:

$$\overline{v} = -\frac{k}{\mu}\nabla P$$

The equation is comprised of $\overline{v}$, the bulk-averaged velocity; $k$, the permeability (measured in $m^2$); $\mu$, the fluid viscosity; and $P$, the fluid pressure [82]. Collagen gels have been shown to have permeability values ranging from $10^{-7}$ to $10^{-8}$ $cm^2$, with gel permeability increasing with increasing culture time [27]. The permeability of gels made from Matrigel was found to be approximately $10^{-8}$ cm$^2$ [53]. Tissue permeability ranges between $10^{-8}$ and $10^{-12}$ cm$^2$ [45].

The equation that Darcy developed for one-dimensional flow is helpful in determining average flow velocity; however it is not sufficient in understanding complex permeability and flow in tissues. Brinkman's equation was developed by Henry Brinkman in 1949, when he modified Darcy's equation to take into consideration changes in viscosity, viscous shear effects, and the presence of a boundary layer [29]:

$$\nabla P = -\frac{\mu}{k}\overline{v} + \mu\nabla^2\overline{v}$$

The Darcy-Brinkman equation helps characterize flow of fluid through porous media where fluid viscosity is factored in with boundary layer effects. These fundamental equations allow for analysis of fluid flow through the interstitium due to pressure gradients that are developed. However, these equations normally assume constant matrix permeabilities, and do not consider the heterogeneity in tissue structure and properties [16]. This can result in large local variations in interstitial flow. Regardless of these limitations these two equations have helped create a fundamental understanding of how interstitial fluid flows through tissues.

### 6.2.4 Shear Stress

Interstitial fluid flow generates direct stress on cells within a microenvironment along with indirect forces transmitted via the matrix [29]. These direct and indirect forces result in biochemical and biophysical cues that can lead to profound changes to the cell and its normal function [2,11,13,24,37,39,44,46,57,63,65,92]. Forces such as fluid shear stress are exerted on surfaces due to a tangential stress caused by fluid viscosity and flow [64]. However the intensity of the shear stress is dramatically less on the cell due to extracellular matrix fibers that shield the cell from direct fluid shear [57]. Fluid shear stress that occurs on cells may be transmitted through structures such as the glycocalyx—an extracellular coat of glycoproteins, glycolipids, and proteoglycans that has been shown to transduce fluid flow-induced shear stress in endothelial cells [3]. Given estimated interstitial flow velocities, it is generally assumed that the resulting shear stresses are comparatively low. It has been hypothesized that interstitial flow can result in shear stresses of 0.1 dyne $cm^{-2}$ or even lower; however, this does not mean that they can be neglected [69]. Computational fluid dynamic simulations of flow through in vitro collagen gels have shown that maximum shear stresses are three times higher and average shear stress two times higher than values predicted by Brinkman's equation [57]. Average shear stress can be calculated with consideration of high porosity matrices, cell profile, and fibers on the cell membrane [57]:

$$\tau_{\text{avg}} = \frac{3\pi}{8} \cdot \frac{(U_0\mu)}{a} \cdot \left(1 + \frac{1}{\sigma}\right)$$

The average shear stress is denoted by $\tau_{\text{avg}}$, free stream velocity is noted as $U_0$, viscosity is $\mu$, the cell radius is represented by $a$, and $\sigma$ is the dimensionless permeability [57]. Shear stress plays a major role in triggering various biochemical signals [6,12,18,46,66,84,94].

### 6.2.5 Convective Mass Transport

Convection has been examined in many models, from transport of fluid through sediment to convective transport in tumors [5,90]. As noted earlier, gradients created by interstitial pressure play a role in transcapillary exchange of fluid and macromolecules. The dimensionless Peclet number can be used to determine if convection can be neglected for mass transport in a particular setting or system. The Peclet number ($P_e$) is a dimensionless ratio that determines the relative importance of convective versus diffusive mass transport of a species and is defined as follows:

$$P_e = \frac{Lu_i}{D_t}$$

$L$ is the characteristic length, $u_i$ is the bulk convective velocity of the solute, and $D_t$ is the diffusion coefficient [82]. When $P_e << 1$, diffusion is the dominant form of transport, while for $P_e >> 1$, convection is the driving force for transport. The utilization of the Peclet number in the study showed that when there was a high Peclet ratio with increased velocities or decreased diffusion coefficients of secreted proteins, a biased transcellular gradient formed [21]. However, even at low Peclet numbers (less than 1), interstitial flow may still generate biologically relevant, asymmetric pericellular morphogen gradients [21].

## 6.3 Mechanisms of Interstitial Fluid Flow Mechanosensing

Many studies have shown that biomechanical forces indisputably play a significant role within the cellular microenvironment; interstitial fluid flow is one of these forces that can drive biochemical and biophysical changes within the cellular microenvironment [10,15,21,32,34,54,70,75,80,93,99]. The mechanisms by which cells sense and respond to interstitial flow remain elusive; currently, there are three major hypothesized mechanisms for interstitial flow mechanosensing: shear stress sensing via the glycocalyx, autologous gradient formation, and integrin activation due to matrix tension [20,21,35,39,57,59,87].

### 6.3.1 Glycocalyx Shear Stress Sensing

Cells subjected to interstitial flow may be stimulated by the shear stresses acting on their surface [57]. It is typically assumed that the shear stresses resulting from interstitial flow would be relatively low, especially compared to better understand systems such as blood vessels; however, heterogeneous matrix architecture and cell orientation could result in localized elevated stresses. One of the ways that cells may sense shear stress is through a structure known as the glycocalyx. The glycocalyx is a membrane-bound coating of proteoglycans, glycoproteins, hyaluronic acid, and glycolipids [97]. In endothelial cells, shear stress can be sensed via the glycocalyx resulting from blood flow. The stress is transmitted via core proteins of the glycocalyx, which in turn can initiate signals for production of nitric oxide, leading to many downstream affects [87]. These core proteins are part of a "bush-like" structure that experiences the drag force at the core protein tips, which is amplified on the cortical cytoskeleton due to its length (Fig. 6.1). Prior studies in vascular smooth muscle cells suggest that heparan sulfate proteoglycans in the glycocalyx (i.e., syndecans and glypicans) are critical to cellular responses to interstitial flow, mediated by integrins, focal adhesion kinase (FAK), and extracellular signal-related kinase (ERK) [67,71]. These findings are supported by modeling of interstitial flow and shear stress on the glycocalyx, where the solid shear stress transmitted by the glycocalyx can be as much as one to two orders of magnitude greater than the predicated fluid shear stress applied to the cell [88]. The model also predicts that glycocalyx thickness and permeability would strongly affect the solid shear stress, raising the possibility that shear stress sensing could be tuned or disrupted by modifying the structure and composition of the glycocalyx. These findings suggest that while the fluid shear stresses applied by interstitial flow may be relatively low, the glycocalyx could amplify and transmit those stresses to the cell [88].

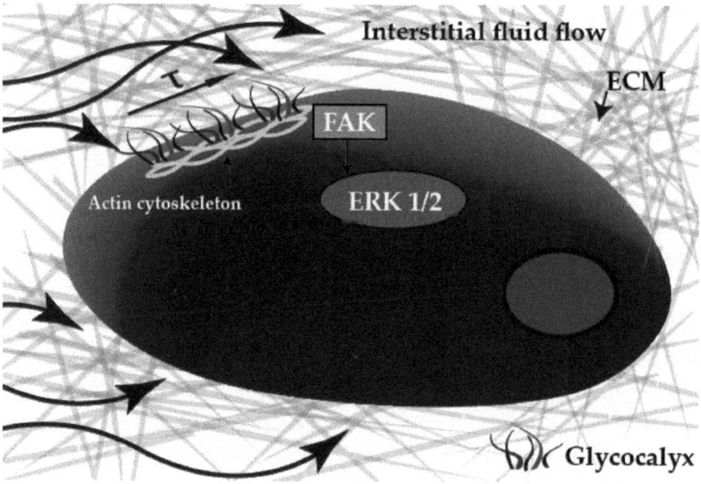


**Figure 6.1** Glycocalyx-mediated shear stress sensing by a cell exposed to interstitial fluid flow.

## 6.3.2 Autologous Gradient Formation

Another potential mechanism for interstitial flow sensing revolves around the convective mass transport associated with flow. During fluid flow, molecules are transported through the tissue interstitium where they are needed; this can cause the formation of extracellular gradients, to which cells are exquisitely sensitive [21,102]. Fluid flow can be sensed by solute transport in intervertebral discs; this convective transport of molecules provides nutritional needs to the cells allowing them to function normally [20]. Similarly, joint loading alters soluble macromolecule distribution in articular cartilage via fluid flow. This redistribution of macromolecules alters the regulation of matrix protein formation and degradation of the extracellular matrix of cartilage [25].

Models of interstitial flow and autocrine morphogen secretion of a single cell demonstrate that convective forces alone can establish transcellular autologous gradients and bias protein concentrations that can change cellular behavior (Fig. 6.2). These gradients can be amplified if the secreted morphogen is matrix binding, these morphogens can then be liberated or activated by cell-secreted proteases. Under interstitial flow conditions these proteins released from the extracellular matrix result in amplified gradients downstream compared to the soluble form of the same protein [21]. Amplified gradients of these morphogens possess capabilities of signaling via receptor tyrosine kinases and G-protein coupled receptors [40]. The model demonstrated that biologically relevant gradients can result from interstitial fluid flow velocities ranging from 0.12–6.0 μm/s, even if the Peclet number suggests that diffusion is the dominant mechanism of mass transport. If the secreted factor is a chemoattractant, these autologous gradients can drive directed cell migration via a mechanism called autologous chemotaxis [50,76]. While sensing of autologous gradients generated by interstitial flow may not fit the traditional definition of mechanosensing (which generally

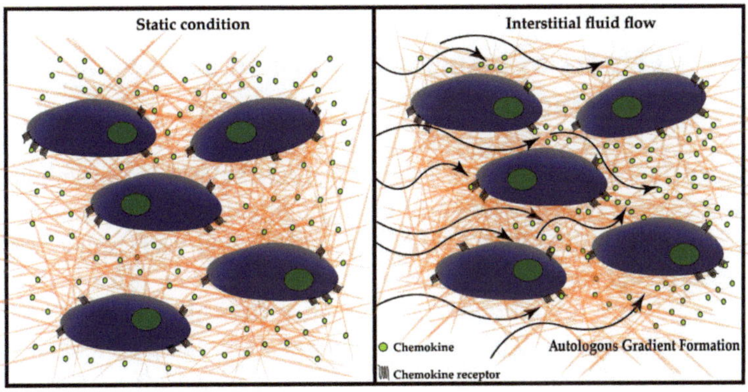


**Figure 6.2** Transcellular autologous gradient generated by convective forces (right). Under static conditions, no autologous gradient is generated (left).

involves transmission of stresses or strains), it nonetheless represents a mechanism by which cells can transduce the physical phenomenon of interstitial flow into an intracellular biochemical signal.

### 6.3.3 Matrix Tension and Integrin Activation

Interstitial fluid flow is known to create pressure forces on cells within a three-dimensional environment, leading to transcellular stress gradients and asymmetric forces between the cell and extracellular matrix [58,59]. Interstitial fluid flow can also create drag forces on matrix fibers found on the extracellular matrix [57]. The forces can generate relatively high shear stresses on these fibers that can, in turn, be transmitted to engaged integrin receptors. This may represent a distinct mechanism from the activation of integrins by shear stress transmitted via the glycocalyx, as described above. Integrins have been shown to play important roles in mechanosensing in various systems, such as endothelial cells [77] and tumor cells [56]. Cells can sense this fluid drag-induced matrix tension and respond via increased integrin and FAK activation [59]. Interestingly, the glycocalyx may potentially play a role in this mechanism as well, as models have suggested that integrin clustering can be regulated by the glycocalyx [55]. However, compared to glycocalyx shear stress sensing and autologous gradient formation, there is considerably less experimental evidence for this hypothetical mechanism.

## 6.4 Effects of Interstitial Flow in Biological Systems

Interstitial fluid flow plays a significant role in various physiological and disease systems. This section highlights examples of the effects of interstitial fluid flow in three systems: blood vessels, lymphatic vessels, and cancer. Blood and lymphatic vessels are constantly exposed to mechanical forces and interstitial fluid flow plays an important role in cellular behavior, signaling, lymphangiogenesis, and angiogenesis [7,51,64,69,70,95]. In cancer, interstitial fluid flow may cause reorganization of extracellular matrices and promote tumor cell invasion mediated by autologous chemotaxis [72,85].

We would like to note that one system where interstitial fluid flow mechanotransduction has been extensively studied is in bone. Unlike many other tissues described here, where interstitial flow is driven primarily by the pressure differences between blood capillaries, the interstitium, and lymphatic vessels, interstitial flow in bone occurs in response to external mechanical loading. These external loads result in transient pressure gradients that drive dynamic and often oscillatory interstitial flow through the extensive lacunae-canaliculi network, simulating osteocyte and osteoblast cell processes. There are many nuances to interstitial flow mechanosensing that are qualitatively and quantitatively different from those in other tissues, and extensive research has been conducted in this area. We point the reader to specific treatments of this topic [9,23,32,39,41,47,49,86,93,96].

### 6.4.1 Blood Vessels

Pressure differences across the vascular wall can result in transmural interstitial flow, potentially affecting the behavior of vascular smooth muscle cells and adventitial fibroblasts. Interstitial flow has shown to play an intimate yet indirect role on smooth muscle cell biochemical signaling and responsiveness [70,95]. Transmural pressure differentials drive interstitial flow velocities of nearly 0.1–1.0 μm/s [14]. The fluid shear stress generated by interstitial fluid flow can be 100 times greater near fenestral pores in the vicinity of smooth muscle cells. This increased shear stress can trigger changes such as increased production of prostaglandin $I_2/E_2$ by smooth muscle cells in collagen gels [95]. Physiologically interstitial fluid flow was also shown to initiate signaling crosstalk between the blood vessel lumen and smooth muscle cells. These results demonstrated that interstitial fluid flow and shear stress may affect smooth muscle cell proliferation and local vascular tone [95]. Further work suggested that interstitial flow can directly govern contraction and relaxation of smooth muscle. It was determined that smooth muscle cells which were exposed to 11 dyn/cm$^2$ shear stress for 15 min exhibited increased contractility via a $Ca^{2+}$-independent, Rho/Rho-kinase-dependent mechanism. This study was one of the first of its kind to reveal that interstitial flow shear stresses could affect myogenic responses in smooth muscle cells [14].

In addition to effects on smooth muscle cell function, interstitial flow may also affect the motility of vascular cells via the upregulation of matrix metalloproteinase 1 (MMP-1) [69]. Exposure to an approximate interstitial flow velocity of 0.5 μm/s (under a pressure of 1 cm $H_2O$) increased migration of myofibroblasts, fibroblasts, and smooth muscle cells [69]. In this experiment, cells were initially stimulated by interstitial flow, and then exposed to a chemotactic gradient (without flow) to quantify changes in cell motility. Flow-induced increases in cell migration were related to increased MMP-1 production and activity in response to interstitial flow in smooth muscle cells and fibroblasts [68,69]. Furthermore, it was concluded that upregulation of MMPs was dependent on downstream activity of ERK1/2 and transcription factor c-Jun activation [68]. Heparan sulfate proteoglycans are components of the glycocalyx that may serve as shear stress sensors [1,22,101]. These glycocalyx heparan sulfate proteoglycans have been shown to mediate the mechanotransduction of interstitial flow-induced forces through the FAK-ERK pathway in smooth muscle cells [71]. Modulation of smooth muscle cell marker genes under interstitial flow (1 cm $H_2O$, 0.05 dyn/cm$^2$) was similarly dependent on ERK1/2 activity and heparan sulfate proteoglycans, as demonstrated by enzymatic treatment with heparinase [67]. These results, in total, shed light on a possible mechanism of interstitial flow mechanotransduction mediated by the heparan sulfate proteoglycans in the glycocalyx, integrin and FAK activation, and downstream ERK 1/2 signaling. These findings shed light on the role of interstitial flow-induced fluid shear stress in regulating vascular wall cell phenotypes [67,68,71].

### 6.4.2 Lymphatic Vessels

Given the function of lymphatic vessels in interstitial fluid drainage, it should come as no surprise that interstitial flow can modulate the behavior of lymphatic endothelial cells. Lymphatic drainage and transport of interstitial fluid is important for maintaining tissue fluid balance and may also aid in innate and adaptive immunity [82,83]. Important physiological events such as inflammation or tissue injury generate rapid responsive measures that include increased vascular permeability, leading to elevated interstitial flow and downstream lymphatic drainage. Increased interstitial flow has been shown to upregulate intercellular adhesion molecule-1 (ICAM-1) and E-selectin on lymphatic endothelial cells, which in turn promoted dendritic cell migration in the presence of flow. There were also associated increases in secretion of the chemokine CCL21 and alterations in lymphatic endothelial cell junction proteins such as VE-cadherin. During tissue injury or inflammation the hyperpermeability of blood vessels and concomitant increases in interstitial fluid flow can affect lymphatic endothelial cells by increasing chemokine secretion, up-regulating of adhesion molecules, and facilitating dendritic cell transmigration [48].

Interstitial fluid flow also appears to have a significant role in the process of lymphangiogenesis. Lymphatic endothelial cells are capable of systematically organizing into working lymphatic capillaries, which is crucial for lymphangiogenesis. This study was conducted in vivo in mice; a small 2-mm-wide band of skin was removed from the tail and replaced with a collagen dermal equivalent and sealed with surgical glue. Fluorescently labeled macromolecules were used to trace fluid flow channels and lymphatic capillaries via fluorescence microlymphangiography. Using these approaches, it was observed that fluid channeling occurred prior to formation a new lymphatic capillary network [7]. In addition, cell migration, VEGF-C expression, and ultimately lymphatic capillary formation appeared driven by primarily in the direction of interstitial flow, suggesting that flow was required to dictate the organization of new lymphatic vessels. These findings were corroborated by in vitro results that showed that interstitial fluid flow had morphological consequences on blood and lymphatic endothelial cells [51]. Interstitial flow (10 μm/s flow velocity) stimulated formation of vacuoles associated with geodesic actin networks and dendritic extensions in lymphatic endothelial cells, while in blood endothelial cells interstitial flow caused enhanced multicellular tubule formation and cell alignment. Endothelial cell organization and structure formation was further enhanced by interstitial flow when cells were seeded in fibrin gels with fibrin-bound VEGF, suggesting an interstitial flow-induced gradient formation mechanism [31]. In addition to understanding how interstitial flow might affect lymphangiogenesis and angiogenesis, these findings are important in the broader sense that interstitial fluid flow could be utilized to organize endothelial cells into a functional network as part of in vitro and in vivo regenerative medicine strategies [31,51].

### 6.4.3 Cancer

The tumor microenvironment is very different from normal healthy tissue structures. Tumor growth causes physical changes to the structure of the microenvironment, resulting in matrix stresses, increased matrix stiffness, elevated fluid pressure, and alterations to fluid flow [72]. When tumors reach a critical size, they promote new blood vessel formation via secretion of VEGF [19]. The permeability of the new blood vessels and lymphatic capillaries, along with structural changes to the extracellular matrix, generate an increase in interstitial fluid pressure in tumors [8]. Tumors can possess an interstitial fluid pressure anywhere between 10 mmHg to 20 mmHg higher than normal tissue [64]. At the center of a tumor, the interstitial fluid pressure is at its highest, creating at large pressure differential across the tumor edge and driving interstitial fluid flow into the surrounding tumor stroma [5,12,15]. Thus, interstitial flow may have effects not only on the cancer cells themselves, but the various cell types (fibroblasts, endothelial cells, immune cells) in the tumor microenvironment.

Interstitial flow can potentially alter the behavior of tumor cells directly, and a seminal study demonstrated that flow and autocrine signaling worked synergistically to promote melanoma and breast cancer cell invasion to lymphatics via autologous chemotaxis [76]. The more invasive cell lines expressed higher levels of the chemokine receptor CCR7, which has been implicated in lymph node metastasis [85], and secreted elevated levels of its ligand CCL21. Under interstitial fluid flow (0.2 μm/s), this resulted in increased invasion and cell polarization in the direction of flow, which could be neutralized by function-blocking antibodies for CCR7 and CCL21. This response was further enhanced by the presence of downstream lymphatic endothelial cells, which also secrete CCL21, suggesting combined effects of autologous and paracrine gradients. Invasive glioma cells showed similar increases in invasion due to an interstitial flow and CXCR4-dependent mechanism. Invasive glioma cells under fluid flow were much more invasive due to the expression of CXCR4 and autocrine secretion of CXCL12. These in vitro findings were linked to interstitial flow patterns in the brain, which themselves have been correlated with glioma invasion in vivo [26]. These studies demonstrate that interstitial fluid flow plays an active role in tumor cell invasion, potentially through the mechanism of autologous chemotaxis.

In addition to autologous chemotaxis, there is evidence that interstitial flow may drive cancer cell invasion via other mechanisms. In prior work using a microfluidic cell culture system, changes in cell migration and morphology were observed in response to low velocity interstitial fluid flow [59]. Interestingly, migration both with and against the direction of flow was observed, and each seemed to involve different mechanisms. While migration downstream was CCR7-dependent, migration upstream was mediated by integrins and driven by FAK activation [59]. Models incorporating interstitial flow and extracellular matrix fiber architecture suggest that FAK activation could be the result of fluid drag-induced matrix tension transmitted to integrins [58]. Another mechanism for regulating tumor invasion may involve degradation and remodeling of the extracellular matrix via MMP activity, as demonstrated in smooth muscle cells and fibroblasts [68,69,73]. FAK-ERK pathway activation has been shown to upregulate MMP expression in response to flow, resulting in enhanced renal carcinoma cell migration [60]. However, there is also evidence that interstitial flow can suppress migration of CNS-1, U87, and U251 glioma cells by downregulating MMP-1 and MMP-2. These results are contrary to most findings that interstitial flow promotes cancer cell invasion, but it brings to light the potential for a spectrum of tumor cell responses to interstitial fluid flow [61].

In healthy tissues, fibroblasts are dynamic contributors to extracellular matrix remodeling [91]. In cancer, fibroblasts also drive matrix remodeling and are key contributors to tumor progression [38]. Thus, considering the responses of non-tumor cells in the microenvironment will also be highly relevant to understanding the role of interstitial flow in cancer progression. In previous work, fibroblasts were seeded in collagen gels and exposed to interstitial fluid flow in a radial flow chamber, resulting in varying average flow velocities across the chamber (average fluid velocity of 13 μm/s near the inlet and 3.6 μm/s at the outlet). Exposure to interstitial flow for 72 h caused fibroblasts to align perpendicular to the flow direction and adopt a more spindle-shaped morphology [54]. Changes in cell alignment were matched by changes in matrix alignment as well [54]. These morphological changes that resulted from the alignment of fibroblasts and the matrix could be a major component in controlling permeability within the microenvironment. In addition to cell and matrix alignment, interstitial flow also induced myofibroblast differentiation via a transforming growth factor (TGF)-β1-dependent mechanism [52]. When interstitial fluid flow rates are elevated, e.g., during inflammation and cancer, fibroblasts may differentiate or begin to reorganize the extracellular matrix. The effect of interstitial flow on fibroblast-to-myofibroblast differentiation was particularly significant because it suggests a role in cancer, which is often characterized by fibrotic stroma and the presence of myofibroblasts [38,52].

More recently, studies have examined the potential interactions of tumor cells and stromal fibroblasts in the presence of interstitial flow. Experiments with metastatic melanoma cells co-cultured with dermal fibroblasts revealed that both tumor cells and fibroblasts invaded in response to interstitial flow [73]. Fibroblast invasion was driven largely by increased activation of TGF-β1 and increased degradation of collagen by MMPs [73]. Tumor cell invasion, in turn, was facilitated by fibroblast invasion and simultaneous "matrix priming" via Rho-dependent fibroblast contractility. This matrix priming would lead to local remodeling of the architecture of the cellular microenvironment and promote tumor cell invasion.

There are many clinical benefits that could potentially come out from examining the role of interstitial fluid flow in cancer progression, given the growing evidence of its effects on tumor cell invasion. There is already clinical evidence that elevated interstitial flow velocities at the periphery of cervical tumors correlates with increased lymph node metastasis, suggesting the potential use interstitial flow as a diagnostic or prognostic indicator [33]. Novel therapeutic agents could prevent or reduce migration of cancer cells by specifically targeting pathways stimulated by interstitial flow. More broadly, better understanding the context in which a tumor grows and progresses, which necessarily must include the contributions of interstitial flow, could lead to advances in a wide array of detection and treatment methods for cancer.

## 6.5 Conclusion

Interstitial fluid flow is a near-ubiquitous biophysical force that can have profound effects in a variety of systems and on an array of cellular phenotypes. Understanding the microenvironment is crucial for identifying the cellular behaviors that potentially occur due to mechanical forces. Many of the studies described the effect interstitial fluid flow has on cellular invasion; the ability to control or even regulate that with mechanical forces could provide greater understanding of the invasion of cancer cells. This could potentially aid in developing therapeutic approaches to control and better understand metastasis formation. However, stimulating migration and invasion could also be a tremendous boon for engineering or regenerating tissues in vivo and in vitro. Furthermore, interstitial fluid flow has been shown to play a role in development of blood and lymphatic vessels; building upon this knowledge could lead to more effective strategies to generate functional vascular networks in these engineered tissues.

One of the major challenges in elucidating the biological effects of interstitial flow, the mechanisms underlying these effects, and their physiological relevance, is simply mimicking accurately the interstitial fluid flow environment that cells actually encounter in vivo. To achieve this goal, we need both (1) accurate characterization of in vivo interstitial flow velocities (and other flow characteristics) under a variety of physiological and pathophysiological conditions, and (2) improved experimental systems for controlling interstitial flow both in vitro and in vivo.

As described earlier, measurements of interstitial fluid flow velocity in vivo have yielded a velocity range of 0.1–50 μm/s for normal tissues and tumors. Interpreting these measurements is complicated by the fact that different techniques for measuring interstitial flow rely on different assumptions and have various advantages and disadvantages. In addition, these flow velocity measurements represent bulk average values; the fluid flow velocity profiles and shear stresses experienced by cells may be dramatically different, depending on the architecture of the extracellular matrix. Developing new methodologies or adapting existing techniques to study flows and forces at the cellular length scale would provide a significant benefit to current and future studies of the role of interstitial flow in the cellular microenvironment. These experimental approaches can be complemented by the formulation of new models and refinement of current models of interstitial flow.

Currently many in vitro studies utilize modified versions of the venerable Boyden chamber assay with cell-seeded collagen, Matrigel and/or hyaluronic acid gels [61,73,75,76]. Flow rates and calculated flow velocities are generally determined by measuring the passage of medium through the gel over time. These assays are straightforward to use and have yielded important results, but offer limited control over the cellular microenvironment and do not permit direct visualization of cells in real time. To overcome this, microfluidic devices [28,59] can also be used to apply interstitial flow to cells and monitor their responses. Continued refinement of these experimental systems will only improve our ability to study the effects of interstitial flow. However, the next major frontier for experimental studies lies in novel in vivo models. While interstitial flow velocities have been measured in mouse tumor models and humans with cervical cancer [15,33], suggesting a correlation between interstitial flow and cancer metastasis, there is a need for in vivo models where interstitial flow can be modulated in an effective manner to directly test hypotheses. Overhydration and direct enhancement of interstitial pressure [43,48] have been used to enhance fluid flow in vivo, but much more development is needed before we have models that offer the type of control required to determine the relevance of interstitial flow mechanotransduction to in vivo physiological processes.

Development of systems that can examine interstitial fluid flow in various physiological and disease models, combined with more accurate measurements and models of interstitial flow in vivo, will allow us to better understand the role fluid flow as a mechanical force plays in the cellular microenvironment. These capabilities would provide greater background in mechanisms that are involved in interstitial fluid flow mechanosensing and ultimately lead to better approaches to treat various diseases, more effectively regenerate tissues, and in general better understand the complexities of a variety of physiological processes.

## Acknowledgments

We would like to thank the members of the Laboratory of Tumor Mechanobiology and Microenvironment at Drexel University for their feedback and support during the writing process.

## References

1. Ainslie, K. M., Garanich, J. S., Dull, R. O., and Tarbell, J. M. (2005). Vascular smooth muscle cell glycocalyx influences shear stress-mediated contractile response, *J. Appl. Physiol. (1985)*, **98**, 242–249.
2. Ajubi, N. E., Klein-Nulend, J., Nijweide, P. J., Vrijheid-Lammers, T., Alblas, M. J., and Burger, E. H. (1996). Pulsating fluid flow increases prostaglandin production by cultured chicken osteocytes—a cytoskeleton-dependent process, *Biochem. Biophys. Res. Commun.*, **225**, 62–68.
3. Arisaka, T., Mitsumata, M., Kawasumi, M., Tohjima, T., Hirose, S., and Yoshida, Y. (1995). Effects of shear stress on glycosaminoglycan synthesis in vascular endothelial cells, *Ann. N. Y. Acad. Sci.*, **748**, 543–554.
4. Aukland, K., and Reed, R. K. (1993). Interstitial-lymphatic mechanisms in the control of extracellular fluid volume, *Physiol. Rev.*, **73**, 1–78.
5. Baxter, L. T., and Jain, R. K. (1989). Transport of fluid and macromolecules in tumors. I. Role of interstitial pressure and convection, *Microvasc. Res.*, **37**, 77–104.
6. Blatnik, J. S., Schmid-Schonbein, G. W., and Sung, L. A. (2005). The influence of fluid shear stress on the remodeling of the embryonic primary capillary plexus, *Biomech. Model Mechanobiol.*, **4**, 211–220.
7. Boardman, K. C., and Swartz, M. A. (2003). Interstitial flow as a guide for lymphangiogenesis, *Circ. Res.*, **92**, 801–808.
8. Boucher, Y., and Jain, R. K. (1992). Microvascular pressure is the principal driving force for interstitial hypertension in solid tumors: Implications for vascular collapse, *Cancer Res.*, **52**, 5110–5114.
9. Brighton, C. T., and McCluskey, W. P. (1986). Cellular response and mechanisms of action of electrically induced osteogenesis, *Bone and Mineral Research/4*. Amsterdam, The Netherlands: Elsevier Science Publishers BV, 213–254.

10. Brown, L. F., Detmar, M., Claffey, K., Nagy, J. A., Feng, D., Dvorak, A. M., and Dvorak, H. F. (1996). Vascular permeability factor/vascular endothelial growth factor: A multifunctional angiogenic cytokine, *EXS*, **79**, 233–269.

11. Buschmann, M. D., Kim, Y. J., Wong, M., Frank, E., Hunziker, E. B., and Grodzinsky, A. J. (1999). Stimulation of Aggrecan Synthesis in Cartilage Explants by Cyclic Loading Is Localized to Regions of High Interstitial Fluid Flow 1, *Arch. Biochem. Biophys.*, **366**, 1–7.

12. Butler, T. P., Grantham, F. H., and Gullino, P. M. (1975). Bulk transfer of fluid in the interstitial compartment of mammary tumors, *Cancer Res.*, **35**, 3084–3088.

13. Chary, S. R., and Jain, R. K. (1989). Direct measurement of interstitial convection and diffusion of albumin in normal and neoplastic tissues by fluorescence photobleaching, *Proc. Natl. Acad. Sci. U. S. A.*, **86**, 5385–5389.

14. Civelek, M., Ainslie, K., Garanich, J. S., and Tarbell, J. M. (2002). Smooth muscle cells contract in response to fluid flow via a $Ca^{2+}$-independent signaling mechanism, *J. Appl. Physiol.*, **93**, 1907–1917.

15. Dafni, H., Israely, T., Bhujwalla, Z. M., Benjamin, L. E., and Neeman, M. (2002). Overexpression of vascular endothelial growth factor 165 drives peritumor interstitial convection and induces lymphatic drain: Magnetic resonance imaging, confocal microscopy, and histological tracking of triple-labeled albumin, *Cancer Res.*, **62**, 6731–6739.

16. de Marsily, G., *Quantitative Hydrogeology: Groundwater Hydrology for Engineers*. 1986: Academic Press.

17. Desmouliere, A., Geinoz, A., Gabbiani, F., and Gabbiani, G. (1993). Transforming growth factor-beta 1 induces alpha-smooth muscle actin expression in granulation tissue myofibroblasts and in quiescent and growing cultured fibroblasts, *J. Cell Biol.*, **122**, 103–111.

18. Dewey, C. F., Jr., Bussolari, S. R., Gimbrone, M. A., Jr., and Davies, P. F. (1981). The dynamic response of vascular endothelial cells to fluid shear stress, *J. Biomech. Eng.*, **103**, 177–185.

19. Dvorak, H. F., Brown, L. F., Detmar, M., and Dvorak, A. M. (1995). Vascular permeability factor/vascular endothelial growth factor, microvascular hyperpermeability, and angiogenesis, *Am. J. Pathol.*, **146**, 1029–1039.

20. Ferguson, S. J., Ito, K., and Nolte, L. P. (2004). Fluid flow and convective transport of solutes within the intervertebral disc, *J. Biomech.*, **37**, 213–221.

21. Fleury, M. E., Boardman, K. C., and Swartz, M. A. (2006). Autologous morphogen gradients by subtle interstitial flow and matrix interactions, *Biophys. J.*, **91**, 113–121.

22. Florian, J. A., Kosky, J. R., Ainslie, K., Pang, Z., Dull, R. O., and Tarbell, J. M. (2003). Heparan sulfate proteoglycan is a mechanosensor on endothelial cells, *Circ. Res.*, **93**, e136–e142.

23. Fukada, E., and Yasuda, I. (1957). On the piezoelectric effect of bone, *J. Phys. Soc. Japan.*, **12**, 1158–1162.

24. Galbraith, C. G., Skalak, R., and Chien, S. (1998). Shear stress induces spatial reorganization of the endothelial cell cytoskeleton, *Cell Motil Cytoskeleton.*, **40**, 317–330.

25. Garcia, A. M., Frank, E. H., Grimshaw, P. E., and Grodzinsky, A. J. (1996). Contributions of fluid convection and electrical migration to transport in cartilage: Relevance to loading, *Arch. Biochem. Biophys.*, **333**, 317–325.

26. Geer, C. P., and Grossman, S. A. (1997). Interstitial fluid flow along white matter tracts: A potentially important mechanism for the dissemination of primary brain tumors, *J. Neurooncol.*, **32**, 193–201.

27. Gentleman, E., Nauman, E. A., Dee, K. C., and Livesay, G. A. (2004). Short collagen fibers provide control of contraction and permeability in fibroblast-seeded collagen gels, *Tissue Eng.*, **10**, 421–427.

28. Haessler, U., Kalinin, Y., Swartz, M. A., and Wu, M. (2009). An agarose-based microfluidic platform with a gradient buffer for 3D chemotaxis studies, *Biomed. Microdevices.*, **11**, 827–835.

29. Hamdan, M., Kamel, M., Siyyam, H., Perlovsky, L., Dionysiou, D., Zadeh, L., Kostic, M., Gonzalez-Concepcion, C., Jaberg, H., and Mastorakis, N. A permeability function for Brinkman's equation. in *WSEAS International Conference. Proceedings. Mathematics and Computers in Science and Engineering*. 2009. WSEAS.

30. Heldin, C.-H., Rubin, K., Pietras, K., and Östman, A. (2004). High interstitial fluid pressure—an obstacle in cancer therapy, *Nat. Rev. Cancer.*, **4**, 806–813.

31. Helm, C. L., Fleury, M. E., Zisch, A. H., Boschetti, F., and Swartz, M. A. (2005). Synergy between interstitial flow and VEGF directs capillary morphogenesis in vitro through a gradient amplification mechanism, *Proc. Natl. Acad. Sci. U. S. A.*, **102**, 15779–15784.

32. Hillsley, M. V., and Frangos, J. A. (1994). Bone tissue engineering: The role of interstitial fluid flow, *Biotechnol. Bioeng.*, **43**, 573–581.

33. Hompland, T., Ellingsen, C., Ovrebo, K. M., and Rofstad, E. K. (2012). Interstitial fluid pressure and associated lymph node metastasis revealed in tumors by dynamic contrast-enhanced MRI, *Cancer Res.*, **72**, 4899–4908.

34. Jain, R. K. (1987). Transport of molecules in the tumor interstitium: A review, *Cancer Res.*, **47**, 3039–3051.

35. Jain, R. K., and Baxter, L. T. (1988). Mechanisms of heterogeneous distribution of monoclonal antibodies and other macromolecules in tumors: Significance of elevated interstitial pressure, *Cancer Res.*, **48**, 7022–7032.

36. Jain, R. K., Tong, R. T., and Munn, L. L. (2007). Effect of vascular normalization by antiangiogenic therapy on interstitial hypertension, peritumor edema, and lymphatic metastasis: Insights from a mathematical model, *Cancer Res.*, **67**, 2729–2735.

37. Johnson, D. L., McAllister, T. N., and Frangos, J. A. (1996). Fluid flow stimulates rapid and continuous release of nitric oxide in osteoblasts, *Am. J. Physiol.*, **271**, E205–E208.

38. Kalluri, R., and Zeisberg, M. (2006). Fibroblasts in cancer, *Nat. Rev. Cancer.*, **6**, 392–401.

39. Kapur, S., Baylink, D. J., and Lau, K. H. (2003). Fluid flow shear stress stimulates human osteoblast proliferation and differentiation through multiple interacting and competing signal transduction pathways, *Bone*, **32**, 241–251.

40. Kholodenko, B. N. (2006). Cell-signalling dynamics in time and space, *Nat. Rev. Mol. Cell Biol.*, **7**, 165–176.

41. Kufahl, R. H., and Saha, S. (1990). A theoretical model for stress-generated fluid flow in the canaliculi-lacunae network in bone tissue, *J. Biomech.*, **23**, 171–180.

42. Kvietys, P. R., in *The Gastrointestinal Circulation*. 2010: San Rafael (CA).

43. Kwon, R. Y., Meays, D. R., Tang, W. J., and Frangos, J. A. (2010). Microfluidic enhancement of intramedullary pressure increases interstitial fluid flow and inhibits bone loss in hindlimb suspended mice, *J. Bone Miner. Res.*, **25**, 1798–1807.

44. Leu, A. J., Berk, D. A., Yuan, F., and Jain, R. K. (1994). Flow velocity in the superficial lymphatic network of the mouse tail, *Am. J. Physiol.*, **267**, H1507–H1513.

45. Levick, J. R. (1987). Flow through interstitium and other fibrous matrices, *Q. J. Exp. Physiol.*, **72**, 409–437.

46. McAllister, T. N., Du, T., and Frangos, J. A. (2000). Fluid shear stress stimulates prostaglandin and nitric oxide release in bone marrow-derived preosteoclast-like cells, *Biochem. Biophys. Res. Commun.*, **270**, 643–648.

47. Meinel, L., Karageorgiou, V., Fajardo, R., Snyder, B., Shinde-Patil, V., Zichner, L., Kaplan, D., Langer, R., and Vunjak-Novakovic, G. (2004). Bone tissue engineering using human mesenchymal stem cells: Effects of scaffold material and medium flow, *Ann. Biomed. Eng.*, **32**, 112–122.

48. Miteva, D. O., Rutkowski, J. M., Dixon, J. B., Kilarski, W., Shields, J. D., and Swartz, M. A. (2010). Transmural flow modulates cell and fluid transport functions of lymphatic endothelium, *Circ. Res.*, **106**, 920–931.

49. Montgomery, R. J., Sutker, B. D., Bronk, J. T., Smith, S. R., and Kelly, P. J. (1988). Interstitial fluid flow in cortical bone, *Microvasc. Res.*, **35**, 295–307.

50. Munson, J. M., Bellamkonda, R. V., and Swartz, M. A. (2013). Interstitial flow in a 3D microenvironment increases glioma invasion by a CXCR4-dependent mechanism, *Cancer Res.*, **73**, 1536–1546.

51. Ng, C. P., Helm, C. L., and Swartz, M. A. (2004). Interstitial flow differentially stimulates blood and lymphatic endothelial cell morphogenesis in vitro, *Microvasc. Res.*, **68**, 258–264.

52. Ng, C. P., Hinz, B., and Swartz, M. A. (2005). Interstitial fluid flow induces myofibroblast differentiation and collagen alignment in vitro, *J. Cell Sci.*, **118**, 4731–4739.

53. Ng, C. P., and Pun, S. H. (2008). A Perfusable 3D cell–matrix tissue culture chamber for in situ evaluation of nanoparticle vehicle penetration and transport, *Biotechnol. Bioeng.*, **99**, 1490–1501.

54. Ng, C. P., and Swartz, M. A. (2003). Fibroblast alignment under interstitial fluid flow using a novel 3-D tissue culture model, *Am. J. Physiol. Heart Circ. Physiol.*, **284**, H1771–H1777.

55. Paszek, M. J., Boettiger, D., Weaver, V. M., and Hammer, D. A. (2009). Integrin clustering is driven by mechanical resistance from the glycocalyx and the substrate, *PLoS Comput. Biol.*, **5**, e1000604.

56. Paszek, M. J., Zahir, N., Johnson, K. R., Lakins, J. N., Rozenberg, G. I., Gefen, A., Reinhart-King, C. A., Margulies, S. S., Dembo, M., Boettiger, D., Hammer, D. A., and Weaver, V. M. (2005). Tensional homeostasis and the malignant phenotype, *Cancer Cell*, **8**, 241–254.

57. Pedersen, J. A., Boschetti, F., and Swartz, M. A. (2007). Effects of extracellular fiber architecture on cell membrane shear stress in a 3D fibrous matrix, *J. Biomech.*, **40**, 1484–1492.

58. Pedersen, J. A., Lichter, S., and Swartz, M. A. (2010). Cells in 3D matrices under interstitial flow: Effects of extracellular matrix alignment on cell shear stress and drag forces, *J. Biomech.*, **43**, 900–905.

59. Polacheck, W. J., Charest, J. L., and Kamm, R. D. (2011). Interstitial flow influences direction of tumor cell migration through competing mechanisms, *Proc. Natl. Acad. Sci. U. S. A.*, **108**, 11115–11120.

60. Qazi, H., Palomino, R., Shi, Z. D., Munn, L. L., and Tarbell, J. M. (2013). Cancer cell glycocalyx mediates mechanotransduction and flow-regulated invasion, *Integr. Biol. (Camb)*, **5**, 1334–1343.

61. Qazi, H., Shi, Z. D., and Tarbell, J. M. (2011). Fluid shear stress regulates the invasive potential of glioma cells via modulation of migratory activity and matrix metalloproteinase expression, *PloS One*, **6**, e20348.

62. Rackow, E. C., Fein, I. A., and Leppo, J. (1977). Colloid osmotic pressure as a prognostic indicator of pulmonary edema and mortality in the critically ill, *Chest*, **72**, 709–713.

63. Reich, K. M., Gay, C. V., and Frangos, J. A. (1990). Fluid shear stress as a mediator of osteoblast cyclic adenosine monophosphate production, *J. Cell Physiol.*, **143**, 100–104.

64. Rutkowski, J. M., and Swartz, M. A. (2007). A driving force for change: Interstitial flow as a morphoregulator, *Trends Cell Biol.*, **17**, 44–50.

65. Semino, C. E., Kamm, R. D., and Lauffenburger, D. A. (2006). Autocrine EGF receptor activation mediates endothelial cell migration and vascular morphogenesis induced by VEGF under interstitial flow, *Exp. Cell Res.*, **312**, 289–298.

66. Sheikh, S., Rainger, G. E., Gale, Z., Rahman, M., and Nash, G. B. (2003). Exposure to fluid shear stress modulates the ability of endothelial cells to recruit neutrophils in response to tumor necrosis factor-$\alpha$: A basis for local variations in vascular sensitivity to inflammation, *Blood*, **102**, 2828–2834.

67. Shi, Z. D., Abraham, G., and Tarbell, J. M. (2010). Shear stress modulation of smooth muscle cell marker genes in 2-D and 3-D depends on mechanotransduction by heparan sulfate proteoglycans and ERK1/2, *PLoS One*, **5**, e12196.

68. Shi, Z. D., Ji, X. Y., Berardi, D. E., Qazi, H., and Tarbell, J. M. (2010). Interstitial flow induces MMP-1 expression and vascular SMC migration in collagen I gels via an ERK1/2-dependent and c-Jun-mediated mechanism, *Am. J. Physiol. Heart Circ. Physiol.*, **298**, H127–H135.

69. Shi, Z. D., Ji, X. Y., Qazi, H., and Tarbell, J. M. (2009). Interstitial flow promotes vascular fibroblast, myofibroblast, and smooth muscle cell motility in 3-D collagen I via upregulation of MMP-1, *Am. J. Physiol. Heart Circ. Physiol.*, **297**, H1225–H1234.

70. Shi, Z. D., and Tarbell, J. M. (2011). Fluid flow mechanotransduction in vascular smooth muscle cells and fibroblasts, *Ann. Biomed. Eng.*, **39**, 1608–1619.

71. Shi, Z. D., Wang, H., and Tarbell, J. M. (2011). Heparan sulfate proteoglycans mediate interstitial flow mechanotransduction regulating MMP-13 expression and cell motility via FAK-ERK in 3D collagen, *PLoS One*, **6**, e15956.

72. Shieh, A. C. (2011). Biomechanical forces shape the tumor microenvironment, *Ann. Biomed. Eng.*, **39**, 1379–1389.

73. Shieh, A. C., Rozansky, H. A., Hinz, B., and Swartz, M. A. (2011). Tumor cell invasion is promoted by interstitial flow-induced matrix priming by stromal fibroblasts, *Cancer Res.*, **71**, 790–800.

74. Shieh, A. C., and Swartz, M. A. (2011). Regulation of tumor invasion by interstitial fluid flow, *Phys. Biol.*, **8**, 015012.

75. Shields, J. D., Emmett, M. S., Dunn, D. B., Joory, K. D., Sage, L. M., Rigby, H., Mortimer, P. S., Orlando, A., Levick, J. R., and Bates, D. O. (2007). Chemokine-mediated migration of melanoma cells towards lymphatics–a mechanism contributing to metastasis, *Oncogene*, **26**, 2997–3005.

76. Shields, J. D., Fleury, M. E., Yong, C., Tomei, A. A., Randolph, G. J., and Swartz, M. A. (2007). Autologous chemotaxis as a mechanism of tumor cell homing to lymphatics via interstitial flow and autocrine CCR7 signaling, *Cancer Cell*, **11**, 526–538.

77. Shyy, J. Y., and Chien, S. (2002). Role of integrins in endothelial mechanosensing of shear stress, *Circ. Res.*, **91**, 769–775.

78. Skobe, M., Hawighorst, T., Jackson, D. G., Prevo, R., Janes, L., Velasco, P., Riccardi, L., Alitalo, K., Claffey, K., and Detmar, M. (2001). Induction of tumor lymphangiogenesis by VEGF-C promotes breast cancer metastasis, *Nat. Med.*, **7**, 192–198.

79. Starling, E. H. (1894). The influence of mechanical factors on lymph production, *J. Physiol.*, **16**, 224–267.

80. Sterpetti, A. V., Cucina, A., D'Angelo, L. S., Cardillo, B., and Cavallaro, A. (1993). Shear stress modulates the proliferation rate, protein synthesis, and mitogenic activity of arterial smooth muscle cells, *Surgery*, **113**, 691–699.

81. Stohrer, M., Boucher, Y., Stangassinger, M., and Jain, R. K. (2000). Oncotic pressure in solid tumors is elevated, *Cancer Res.*, **60**, 4251–4255.

82. Swartz, M. A., and Fleury, M. E. (2007). Interstitial flow and its effects in soft tissues, *Annu. Rev. Biomed. Eng.*, **9**, 229–256.

83. Swartz, M. A., and Lund, A. W. (2012). Lymphatic and interstitial flow in the tumour microenvironment: Linking mechanobiology with immunity, *Nat. Rev. Cancer.*, **12**, 210–219.

84. Swartz, M. A., Tschumperlin, D. J., Kamm, R. D., and Drazen, J. M. (2001). Mechanical stress is communicated between different cell types to elicit matrix remodeling, *Proc. Natl. Acad. Sci. U. S. A.*, **98**, 6180–6185.

85. Takeuchi, H., Fujimoto, A., Tanaka, M., Yamano, T., Hsueh, E., and Hoon, D. S. (2004). CCL21 chemokine regulates chemokine receptor CCR7 bearing malignant melanoma cells, *Clin. Cancer Res.*, **10**, 2351–2358.

86. Tan, S. D., de Vries, T. J., Kuijpers-Jagtman, A. M., Semeins, C. M., Everts, V., and Klein-Nulend, J. (2007). Osteocytes subjected to fluid flow inhibit osteoclast formation and bone resorption, *Bone*, **41**, 745–751.

87. Tarbell, J. M., and Pahakis, M. Y. (2006). Mechanotransduction and the glycocalyx, *J. Int. Med.*, **259**, 339–350.

88. Tarbell, J. M., and Shi, Z. D. (2013). Effect of the glycocalyx layer on transmission of interstitial flow shear stress to embedded cells, *Biomech. Model Mechanobiol.*, **12**, 111–121.

89. Taylor, A. E. (1981). Capillary fluid filtration. Starling forces and lymph flow, *Circ. Res.*, **49**, 557–575.

90. Thibodeaux, L. J., and Boyle, J. D. (1987). Bedform-generated convective transport in bottom sediment, *Nature*, **325**, 341–343.

91. Tomasek, J. J., Gabbiani, G., Hinz, B., Chaponnier, C., and Brown, R. A. (2002). Myofibroblasts and mechano-regulation of connective tissue remodelling, *Nat. Rev. Mol. Cell Biol.*, **3**, 349–363.

92. Topper, J. N., and Gimbrone, M. A., Jr. (1999). Blood flow and vascular gene expression: Fluid shear stress as a modulator of endothelial phenotype, *Mol. Med. Today.* **5**, 40–46.

93. Turner, C. H., Forwood, M. R., and Otter, M. W. (1994). Mechanotransduction in bone: Do bone cells act as sensors of fluid flow?, *FASEB J.*, **8**, 875–878.

94. Tzima, E., Irani-Tehrani, M., Kiosses, W. B., Dejana, E., Schultz, D. A., Engelhardt, B., Cao, G., DeLisser, H., and Schwartz, M. A. (2005). A mechanosensory complex that mediates the endothelial cell response to fluid shear stress, *Nature*, **437**, 426–431.

95. Wang, S., and Tarbell, J. M. (2000). Effect of fluid flow on smooth muscle cells in a 3-dimensional collagen gel model, *Arterioscler. Thromb. Vasc. Biol.*, **20**, 2220–2225.

96. Weinbaum, S., Cowin, S. C., and Zeng, Y. (1994). A model for the excitation of osteocytes by mechanical loading-induced bone fluid shear stresses, *J. Biomech.*, **27**, 339–360.

97. Weinbaum, S., Tarbell, J. M., and Damiano, E. R. (2007). The structure and function of the endothelial glycocalyx layer, *Annu. Rev. Biomed. Eng.*, **9**, 121–167.

98. Whitaker, S. (1986). Flow in porous media I: A theoretical derivation of Darcy's law, *Trans. Media.*, **1**, 3–25.

99. Wilson, E., Mai, Q., Sudhir, K., Weiss, R. H., and Ives, H. E. (1993). Mechanical strain induces growth of vascular smooth muscle cells via autocrine action of PDGF, *J. Cell Biol.*, **123**, 741–747.

100. Xu, H., Zhang, J., Wu, J., Guan, Y., Weng, Y., and Shang, P. (2012). Oscillatory fluid flow elicits changes in morphology, cytoskeleton and integrin-associated molecules in MLO-Y4 cells, but not in MC3T3-E1 cells, *Biol. Res.*, **45**, 163–169.

101. Yao, Y., Rabodzey, A., and Dewey, C. F., Jr. (2007). Glycocalyx modulates the motility and proliferative response of vascular endothelium to fluid shear stress, *Am. J. Physiol. Heart Circ. Physiol.*, **293**, H1023–H1030.

102. Zigmond, S. H. (1977). Ability of polymorphonuclear leukocytes to orient in gradients of chemotactic factors, *J. Cell Biol.*, **75**, 606–616.

# Chapter 7

# Mechanical Properties of Cytoskeletal Structures and Their Response to Externally Applied Forces

**Takahiro Watanabe-Nakayama,[a,b] Shin'ichi Machida,[a] Masakazu Saito,[a] Rehana Afrin,[a] and Atsushi Ikai[a]**

[a]*Innovation Laboratory, Tokyo Institute of Technology, S2-8, 4259 Nagatsuta-cho, Midori-ku, Yokohama 226-8503, Japan*
[b]*Bio-AFM Frontier Research Center, Kanazawa University, Kakuma-machi, Kanazawa 920-1192, Japan*

aikai@bio.titech.ac.jp

## 7.1 Introduction

Mechanical properties of intracellular and extracellular environments influence cytoskeletal structures and, thereby, cell shape, contributing to cellular processes such as migration, proliferation, differentiation, and apoptosis, which are, in turn, involved in the regulation of macroscopically observable biological phenomena such as tissue development and tumor progression [1–3].

Previous studies have examined the molecular, biochemical and biophysical mechanisms that govern cellular responses to the mechanical properties of their surroundings. Total cellular

*Cells, Forces, and the Microenvironment*
Edited by Charles M. Cuerrier and Andrew E. Pelling

ISBN 978-981-4613-36-1 (Hardcover), 978-981-4613-37-8 (eBook)
www.panstanford.com

mechanical responses are known to depend mainly on two factors, the intracellular (cytoplasmic) and the cortical (consisting of anchored actomyosin cytoskeleton) viscoelasticity [4,5]. Actin cytoskeleton, actin-associated myosin motors, focal adhesions (FA), extracellular matrix (ECM), intracellular signal transduction proteins (Rho-signaling pathway) and mechanosensitive ion channels have been shown to play crucial roles in the regulation of cellular mechanical properties [6–12]. The linkages between the actin cytoskeleton and the ECM via FA containing membrane embedded ECM-receptor (integrin) enable cells to transmit mechanical signals between the ECM and themselves [7–9] (Fig. 7.1A). FAs are composed of dozens of different species of proteins, including linker proteins between the ECM and the actin cytoskeleton and signal transduction proteins [13] (Fig. 7.1B,C). Application of mechanical stimuli to cells induces activation of mechanosensitive $Ca^{2+}$ channels and conformational changes in some of the focal adhesion proteins [10,14,15]. These changes then alter activities of Rho signal transduction proteins [6,10,11,16], inducing FA reinforcement, reconstruction of actin cytoskeleton [11,16–18], and cell stiffening [6,19] via regulation of actomyosin interaction. In addition, it is also known that biological systems at various levels have feedback mechanisms that regulate their responses to external stimuli.

Quantification of input and output signals into and out of a target mechanosensory system and observation of the behavior of proteins involved in the system are essential for understanding the systems. An atomic force microscope (AFM) equipped with a fluorescence microscope is one of the most powerful tools to make such quantitative studies of cellular mechanochemical responses. The equipment allows us to apply an external force to a living cell and record mechanical and biochemical responses of the cell.

In this chapter, we introduce the following findings there of (also refer to [22]):

(1) detection and quantification of mechanical responses of the cell against externally applied stimuli using an AFM with variously functionalized cantilevers
(2) observation of the behavior of mechanosensing proteins in response to mechanical cues to actin cytoskeleton

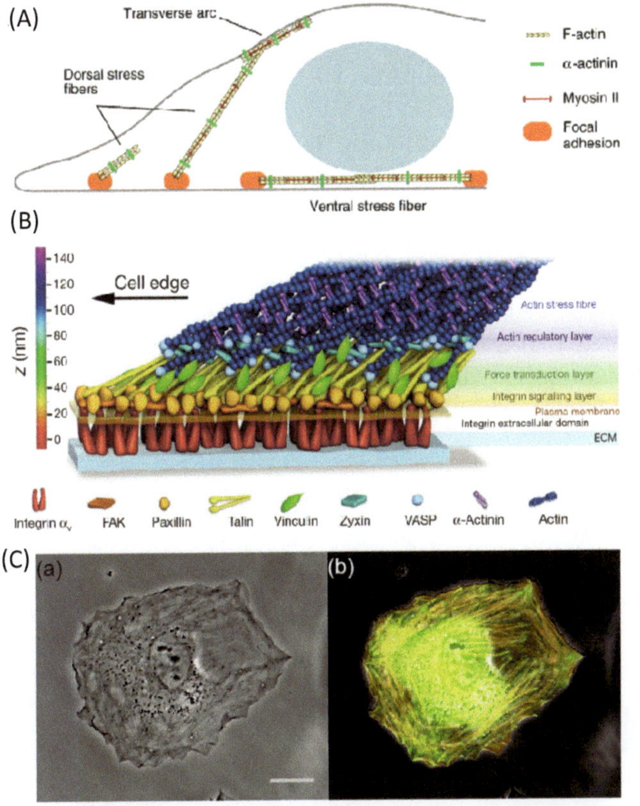


**Figure 7.1** Actin and Focal Adhesion Structures. (A) Model of stress fiber (SF) formation (cited from ref. [20] with permission). (B) Schematic model of focal adhesion molecular architecture (cited from ref. [21] with permission). (C) Microscopic images of SFs and focal adhesions inside live cell (a: phase contrast micrograph, b: merged image with fluorescence microscopic image (green: paxillin, orange: $\beta$-actin) (cited from ref. [22] with permission). Scale bar = 20 μm. Copyright (2014) The Japan Society of Applied Physics.

## 7.2 Various Methods for Measurement of Mechanical Properties in Cells Using an AFM

We have developed several methods employing an AFM to measure the mechanical properties of cytoskeletal structures in living cells.

### 7.2.1 Biochemically Modified Probes

By modifying an AFM probe with a specific ligand of a membrane receptor that is linked to the intracellular cytoskeleton, we can form a specific and continuous linkage from the probe to the cytoskeleton.

For example, Afrin et al. used AFM probes modified either with the Band 3 specific lectin, concanavalin A (Con A), or anti-F-actin antibody, and measured mechanical properties of the cytoskeleton lining the red blood cell membrane [23] (Fig. 7.2). This work revealed that spectrin-based network was compliant enough to allow a tensile extension of up to several micrometers by a very small force.

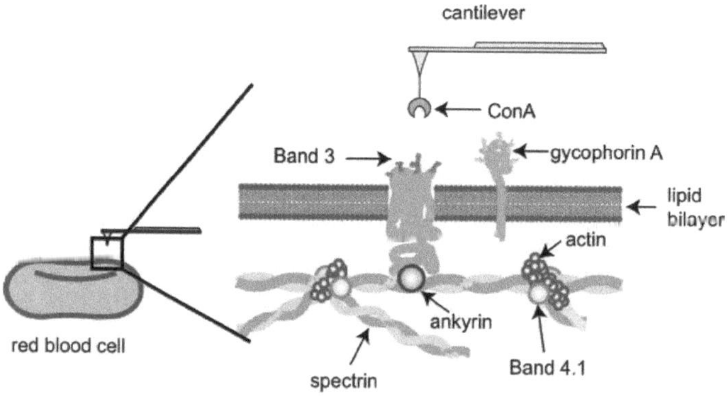


**Figure 7.2** The AFM probe coated with Con A forms a bond with Band 3 and the one coated with anti-F-actin antibody reacts with a junction node. Pulling the cytoskeleton using the two kinds of AFM probes gives indication for the possible local variations in the mechanical property of the cytoskeleton, i.e., spectrin network (cited from ref. [23] with permission).

### 7.2.2 Lateral Force Measurement

The magnitude of applicable tensile force using biochemically modified probes, however, is limited to a few tenths of nano-newtons (nN) in the case where specific non-covalent bonds are involved, and a few nano-newtons when covalent cross-linkers are used [24]. To study mechanical properties of stress fibers (SF), a tensile force as large as several tens to hundreds of nano-newtons is needed.

Hakari et al. directly applied a lateral force to individual ventral SFs using an AFM probe and recorded corresponding displacement of SFs and their tensile strength [25]. A cantilever with a pyramidal tip was inserted into a living cell by breaking through the cell membrane and placed the tip in close proximity to one of the SFs. Lateral movement of the cantilever was employed to push the fiber approximately normal to its long axis.

Figure 7.3 shows fluorescence images of SFs as they were laterally pushed. The displacement of the fibers under lateral force was described as non-linear in the sense that the displacement from the original fiber shape was not linearly dependent on the contour distance along the fiber axis. The observed nonlinear displacement of the fibers was interpreted as the result of the presence of interfiber connections mediated by the cortical actomyosin sheet.

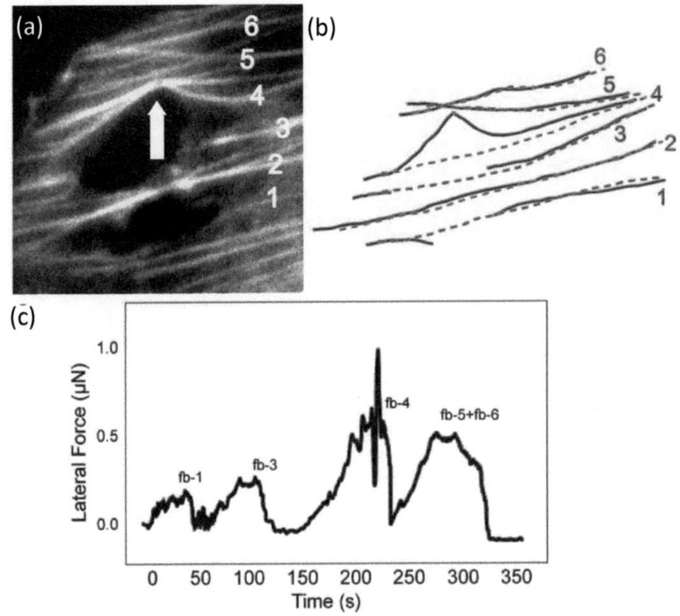


**Figure 7.3** Lateral manipulation of GFP-tagged SFs inside live cell by AFM. (a) Fluorescence micrograph of one of the major SFs (labeled #4) under a lateral force applied in the direction as indicated with an arrow. (b) Overlap of the positions of the major SFs before (dotted lines: color coded for different SF pairs) and at maximum displacement (solid lines). (c) Plot of the lateral force against the scanning time (cited from ref. [25] with permission).

### 7.2.3 Hooking Probes

Machida et al. reported the use of a fabricated AFM cantilever with a hooking function to pick up selected SFs from intact and semi-intact cells that were treated with Triton X-100 [26,27]. Figure 7.4a shows the cantilever with a hooking function in which the cutout was produced on commercial cantilevers using a focused ion beam (FIB) apparatus. Figures 7.4b,c show pulling force curves of single stress fibers obtained on intact and semi-intact cells, respectively. The stiffness of the SFs as estimated from the averaged initial slope of the force curve was clearly higher in the intact cell than that in the semi-intact cell.

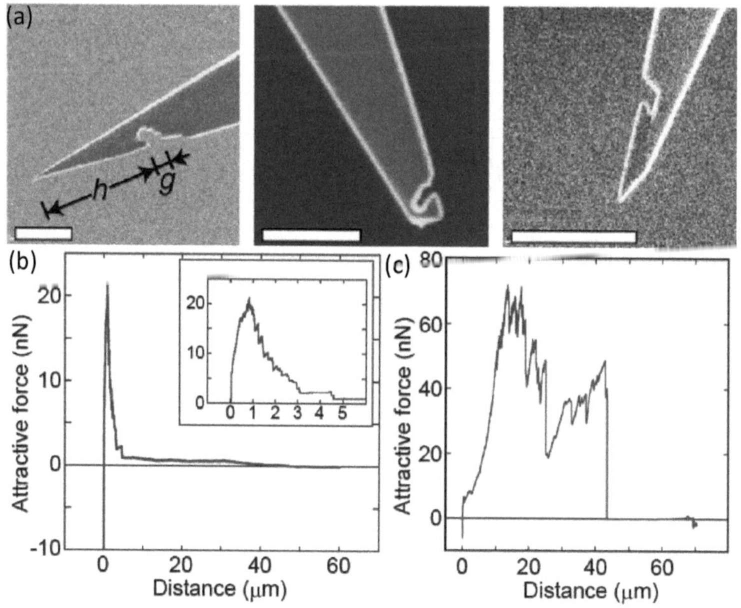


**Figure 7.4** Hooking cantilevers. (b) Scanning ion microscopy images of hook-shaped probes ($h$ = 2.2 μm, $g$ = 0.6 μm, scale bar = 1 μm). (b) Force curve obtained while hooking cantilever was pulling a single stress fiber from live intact cell. The inset shows an enlarged view of the attractive force peak. (c) Similar force curve obtained on semi-intact cell (cited from ref. [26] with permission).

Machida et al. reported that when an SF was severed in a semi-intact cell, the fiber was cut from its two ends allowing the freed fragment of the fiber to cling to the hook, whereas when it was cut in an intact cell, it was severed at the point of force application [26,27].

### 7.2.4 Bead-Glued Cantilevers

#### 7.2.4.1 Introduction

When a biochemically functionalized colloidal bead (10–100 μm in diameter) is glued on an AFM cantilever, it can form a large number of specific linkages to the intracellular cytoskeleton. After producing an extensive number of linkages to specific membrane receptors, the cantilever can be lifted under a large tensile force. This method has been used to detect a large cellular deformation and its mechanical responses [28].

#### 7.2.4.2 Overview

We modified a bead glued on a tipless AFM cantilever with the ECM proteins fibronectin. The protein has a binding capacity to the cell membrane receptor, integrin. Using this system, we studied the cellular mechanical response to a perturbation of extracellular origin [28]. We investigated the time evolution of cellular response to cyclic mechanical stresses more quantitatively than previous studies by employing an AFM technique that enabled us to simultaneously apply an external stimulus and monitor the cellular response against it. A compressive force of less than 10 nN was applied at a localized region of a cultured cell for a few minutes with the fibronectin probe. The probe was then pulled up and kept at a fixed height. During this local clamping operation of a specified duration, the probe did not detach from the cell. After the clamping period was over, the probe was returned to the pre-clamping height pressing onto the same region of the cell surface. This pressing and pulling cycle was repeated several times with simultaneous recording of the mechanical response of the cell (Fig. 7.5). We detected that the tension recovery activity of the cell was decreased as the number of pressing and pulling cycle was increased. The complex change in the tensile response of the cell during the clamping stage was analyzed in terms of stress relaxation of viscoelastic material and diffusion of cytosolic components (Fig. 7.6).

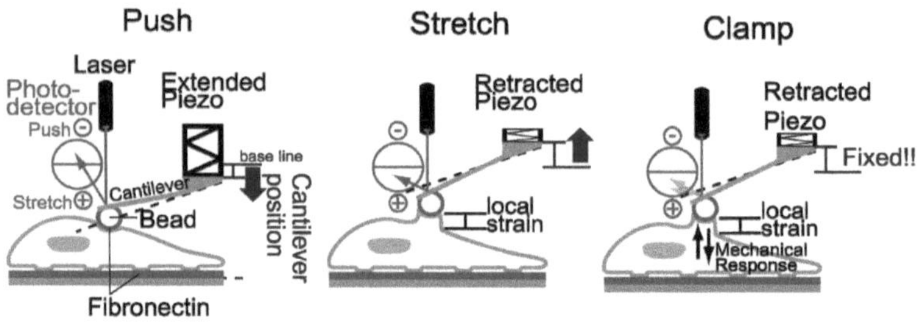


**Figure 7.5** Experimental design for the application of a local stress onto a cultured fibroblast and detection of the consequent mechanical response. (cited from ref. [28] with permission).

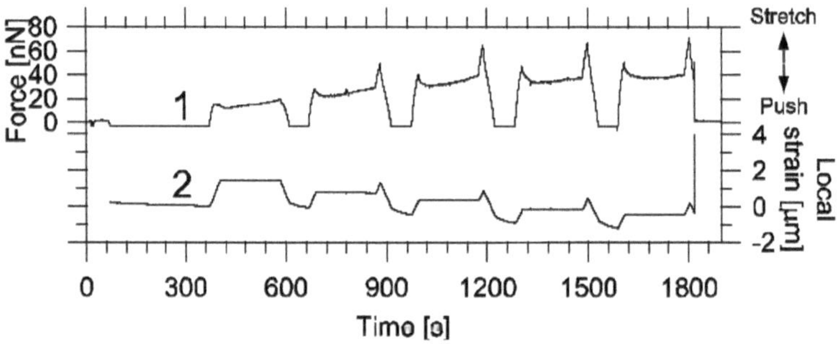


**Figure 7.6** Typical response of an intact single fibroblast to cyclic stresses. Complete time course of the mechanical response of a cell recorded as the cantilever deflection (Curve 1) and the local strain (Curve 2). See text for more detailed explanation (cited from ref. [28] with permission).

Typical results of time-dependent change of force (local cellular tension) (Curve 1) and local strain (Curve 2) during repeated cycles of compression, stretching and clamping stages are shown in Fig. 7.6. Tension first decreased rather rapidly (within ~1 min; tension relaxation phase) as a result of viscoelastic stress relaxation and then slowly increased for the rest of clamping stage (tension recovery phase).

To investigate a possible contribution of the actin–myosin interaction to the viscoelastic behavior of the cell against a local stimulus as observed above, a similar experiment was performed in the presence of cytochalasin D,(-)-blebbistatin, or (+)-blebbistatin, which are, respectively, an inhibitor of actin polymerization [29], a specific inhibitor of non-skeletal myosin II ATPase activity and an inactive enantiomer of (-)-blebbistatin

[30]. The much decreased tension recovering activity and similarly decreased magnitude of tensile force upon stretching observed for the inhibitor-treated cells was attributed to the softening of the cell caused by the treatments with the corresponding reagents [31–34].

### 7.2.4.3 Quantification of cellular mechanical response

The time dependence of the force curves was examined according to the Kelvin model of viscoelasticity [35] (Fig. 7.7a). The force–distance curve for each stretching step was fitted to the following equation, where the local tension, *F*, can be expressed in terms of the local strain, *z* [36]:

$$F(z) = k_1 \, z + \gamma v \left\{ 1\text{-}exp\left( -\frac{k_2 z}{\gamma v} \right) \right\} \tag{7.1}$$

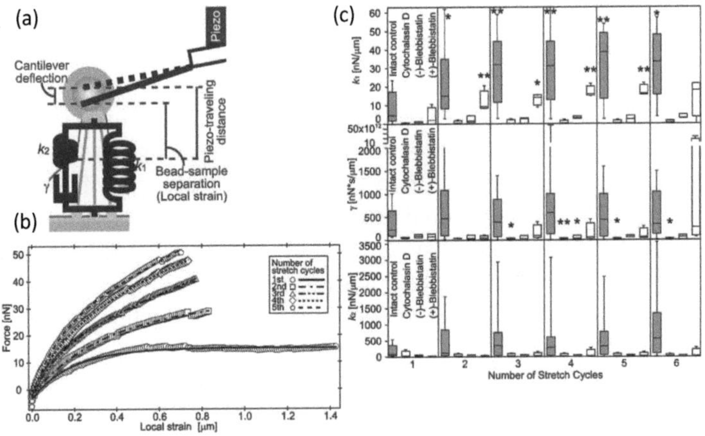


**Figure 7.7** The relation between the viscoelastic parameters of locally stretched region of the cells and the number of stretch cycles. (a) Kelvin model. (b) The *F–D* curves recorded during each stretching stage for an intact cell were fitted with Kelvin model. (c) The distribution of Kelvin model parameters are shown as the function of the number of stretch cycles. In respective experimental condition, the level of each stretch cycle was compared with the level of the first stretch (or the second stretch only in the case of (-)-blebbistatin-treated cells) by paired *t*-test analysis: **P* < 0:05; ***P* < 0:01; note that the significant enhancement in the first spring constant, k1 was observed in the intact and control, and (+)-blebbistatin-treated cells (cited from ref. [28] with permission).

where $k_1$ represents the spring constant of the spring placed in parallel to the second spring with spring constant $k_2$, $\gamma$ is the viscosity coefficient of the dashpot, and $v$ represents the retraction velocity, which is defined as $dz/dt$. According to Schmitz et al. [36], $k_1$ and $k_2$, and $\gamma$, respectively, describe the stiffness of the membrane tether pulled from the cell, the bending rigidity of the cell membrane, and the viscous contribution of the receptor-anchoring membrane. The relationship between these viscoelastic parameters obtained by fitting experimental curves to Eq. 7.1 and the number of stretching cycles is shown in Fig. 7.7b,c. In intact control cells, $k_1$ increased with the number of stretching cycles and reached a saturation level. Neither did $k_2$ nor $\gamma$ show a significant dependence on the number of stretch cycles for all cells.

#### 7.2.4.4 Model for tension recovery

We evaluated the relation between the tension recovery and the number of local stretch cycles. The initial tension relaxation can be explained by the stress relaxation of the cortical layer and anchored stress fibers. We assume that the tension recovery response is diffusion-limited, which is not an unreasonable view as the proteins associated with tension recovery processes such as myosin II should be recruited to the manipulated site from other locations [12]. We assumed that the tension recovery is promoted by the tension recovery-associated factors (TRAF) that are newly recruited during each compression and pulling cycle.

The time dependence of the local tension during clamping is explained by using Eq. 7.2 as follows:

$$F(t) = k_1 \varepsilon_0 \left\{ 1 - \left( 1 - \frac{\tau_\sigma}{\tau_\varepsilon} \right) \exp\left( -\frac{t}{\tau_\varepsilon} \right) \right\} + A\sqrt{t}, \tag{7.2}$$

where the first and the second term on the right-hand side of the equation, respectively, represent the stress relaxation of Kelvin model [35] and the diffusion-limited tension recovery. $k_1$ is already defined above. $\varepsilon_0$ represents the local cell strain as obtained from the records during clamping stages. $\tau_\sigma$ and $\tau_\varepsilon$, respectively, represents the time constant of creep and that of stress relaxation. The constant $A$ is considered a function of the following several factors: (1) the number of TRAFs that were additionally bound to actin filaments for each clamping cycle, (2) the force that an added single factor can exert, and (3) the diffusion coefficient of the cytoplasm. A, as conjectured as above, can be regarded as representing the tension recovery activity. To analyze each set of cyclic local stretching data, the time courses of the cantilever deflection was re-plotted separately for clamping stages as shown in Fig. 7.8a. Then, the time course records for local cellular tension during clamping were fitted to Eq. 7.2. The relation between the tension-recovering activity, *A*, and the number of stretching cycles is shown in Fig. 7.8b. The tension-recovering activity decreased with the increasing number of stretching cycles.

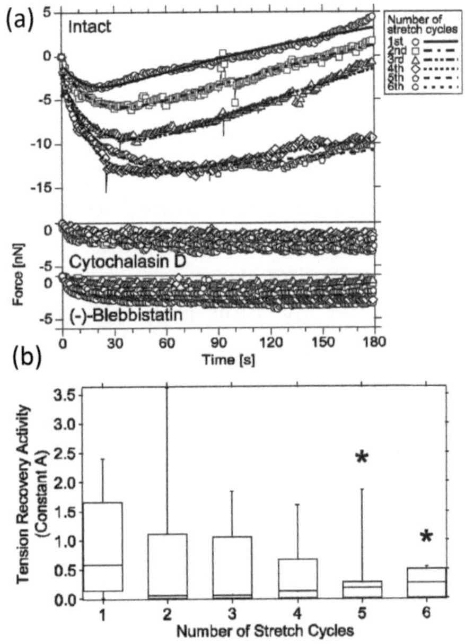


**Figure 7.8** Tension-recovering activity during clamping after stretch. (a) Each trace of cellular tension during clamping in intact control, cytochalasin D-treated, (+)-blebbistatin or (-)-blebbistatin-treated cells was collected (markers). The traces were fitted with the Eq. 7.2. (b) The relation between the tension recovering activity in the intact cells and controls, and the number of stretch cycles. The derived constant, A is the degree of the tension recovering activity and is plotted as a function of the number of stretch cycles. The level of each stretch cycle was compared with the level of the first stretch by paired *t*-test analysis: $*P < 0:05$; note that the significant attenuation was observed (cited from ref. [28] with permission).

Why does tension recovery activity decrease as the number of compression and stretch cycles increases? We speculate that this may be a form of mechanical memory in the cellular architecture. The tension recovery activity decreased as the number of compression and stretch cycle was increased because the diffusion coefficient, the number of binding sites to which the TRAFs could newly bind, and the number of diffusing TRAF most likely decreased as the stretch cycle increased.

It has been observed that the cytoplasmic fluidity is also changed under an applied mechanical force. The cytoplasm stiffened within 40 minutes under a shear stress in a Rho-signaling pathway-dependent manner [6], and the fluidity was anisotropic in response to the direction of the mechanical stress [37].

The number of newly accessible binding sites for TRAF decreased in our analysis. It is possible that the binding sites had been filled during earlier stretch cycles and the number of the TRAF that had been already bound to the sites would contribute to stiffness of the cortical layer and anchored stress fibers.

The number of the diffusible activated TRAF (i.e. myosin II) is probably regulated during the mechanical stimulation. Myosin II activity is regulated by numerous signaling pathways (i.e. Rho-signaling pathway [20]).

In our observation, the local stiffening at the stretching site was confirmed (Fig. 7.8) and the tension-recovering activity was completely inhibited by cytochalasin D and (-)-blebbstatin, suggesting that the tension recovery was promoted by actin–myosin interaction.

What is the physiological meaning of the rapid stiffening and gradual loss of the sensitivity to the cyclic mechanical stimulation? Possibly, it is a component of the cellular defense system against a severe mechanical insult. The agitation of the extracellular mechanical environment potentially could disrupt intracellular structures and cellular organization. Cells may protect themselves from harmful extracellular oscillatory mechanical forces by rapidly stiffening and gradually becoming insensitive to prolonged high-level stresses.

### 7.2.5 Micro Scoop Cantilever

In our work described above, the probe was glued on a cantilever beforehand, the throughput of measurements was low especially when it took a long time to form a strong enough bond between the probe and the cell for secure manipulation of the cell. To improve on this issue, we developed a new tool for the manipulation of micro objects using AFM and achieved high throughput measurements. A novel feature of recovering intracellular components at the bead adhesion site is also described [38].

An important aspect of the tool is in the use of a specially designed AFM-mounted cantilever with a semi-circular cut at the edge, which will be called a hole hereafter (Fig. 7.9a). The hole was cut from commercially available arrowhead-shaped cantilevers by using a FIB apparatus so that its diameter was slightly smaller than microspheres to be manipulated (Fig. 7.9b). The circular hole was intended to scoop a micro object and recover it.

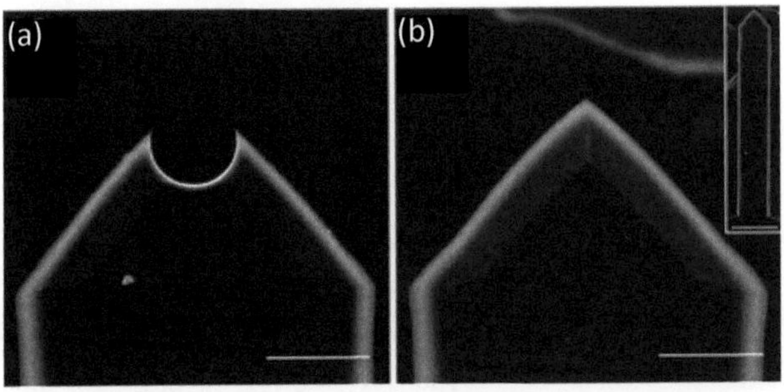


**Figure 7.9** Scanning electron micrographs of a "micro scoop" cantilever-like a pushpin remover. The edge of a fabricated cantilever (a) has a hole that is slightly smaller in diameter in comparison with a colloidal bead. The hole is cut from a commercially available arrowhead cantilever (b) by the focused ion beam technique. Inset: entire image of the arrowhead cantilever. Scale bar represents 10 μm in (a) and (b), and 50 μm in the inset (cited from ref. [38] with permission).

We analyzed the ECM adhesion force and the cell extension length when the probe was detached from cell surface [38]. Carboxylate functionalized polystyrene microspheres were covalently modified with human recombinant fibronectin type III domain 7–10 (hrFNIII7-10) and added to the culture medium of the cells and allowed to adhere on the cell surface before the manipulation experiment (Fig. 7.10A–B). We pulled up beads from the cell surface one after another using a scooping cantilever and repeated similar measurements for each bead attaining a high

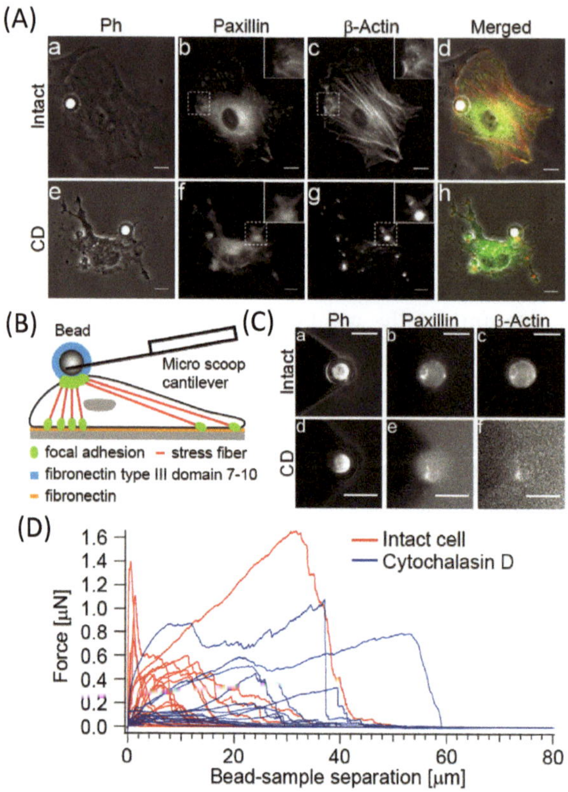


**Figure 7.10** Pickup process of beads adhered on fibroblast cells. (A) The micrographs of rat fibroblasts expressing AG1-paxillin and KO1-β-actin with (e–h) or without (a–d) cytochalasin D. The cells had adhered hrFNIII7-10-coated beads (phase contrast images (a, e), AG1-paxillin fluorescence (b, f) and KO1-$\beta$-actin fluorescence (c, g)). (d) and (h) are merged images (green and red colors in the merged images indicate AG1-paxillin fluorescence and KO1-β-actin fluorescence, respectively). The insets in (b), (c), (f) and (g) are magnified images around the bead shown as a dashed box in each image. (B) Schematic view of manipulation of a hrFNIII7-10-coated bead on a cell. (C) The micrographs of the beads after been picked up from the cells expressing AG1-paxillin and KO1-β-actin with (d–f) or without (a–c) cytochalasin D (phase contrast images (a, d), AG1-paxillin fluorescence (b, e) and KO1-β-actin fluorescence (c, f). (D) Force vs. bead-sample separation curves derived from bead-picking-up experiments for intact cells (red) and cytochalasin D-treated cells (blue) (cited from ref. [38] with permission). Scale bars = 10 μm.

overall throughput of the experiment. After a pickup manipulation, as shown in Fig. 7.10C, fluorescently labeled focal adhesion proteins and co-localized actin cytoskeleton were successfully recovered on the bead surface. We also recognized a significant difference in stiffness and viscosity between the intact cells and the cytochalasin D-treated cells (Fig. 7.10D).

## 7.3 Mechanosensing Proteins in Response to Mechanical Cues to Actin Cytoskeleton

### 7.3.1 Introduction

The transmembrane network comprising SFs and FAs is in itself a mechanosensing structure. The formation and dissociation of SFs and FAs are thus drastically changed depending on intra- and extracellular mechanical conditions. Temporary or accidental reduction in tensile stress on SFs causes SF thinning and dissociation of FA proteins [39]. In contrast, application of external strains and stresses produces FA reinforcement.

Some recent studies have revealed several candidates for mechanosensing proteins in FAs and proposed a mechanism by which these proteins convert mechanical signals into chemical ones [10,14,18,40]. The LIM protein, zyxin, is a focal adhesion protein that exhibits mechanosensing properties [17,41–43]. Zyxin translocates to severed or damaged sites on SFs [15] and recruits the phosphoprotein VASP [44], a focal adhesion protein that promotes actin polymerization [45]. Finally, $\alpha$-actinin is recruited to the repair site [44]. In this way, the zyxin-VASP system constitutes a part of the repair process of damaged SFs (Fig. 7.11). However, the mechanism underlying the assembly and disassembly of the proteins involved in this process remains to be characterized.

### 7.3.2 Transient Accumulation of Paxillin at Damaged Sites on SFs

Recently, we found that paxillin, as well as zyxin, but not $\alpha_V$-integrin nor talin, accumulates at damaged sites on SFs (Fig. 7.12) [46]. We applied an external load to fluorescently labeled

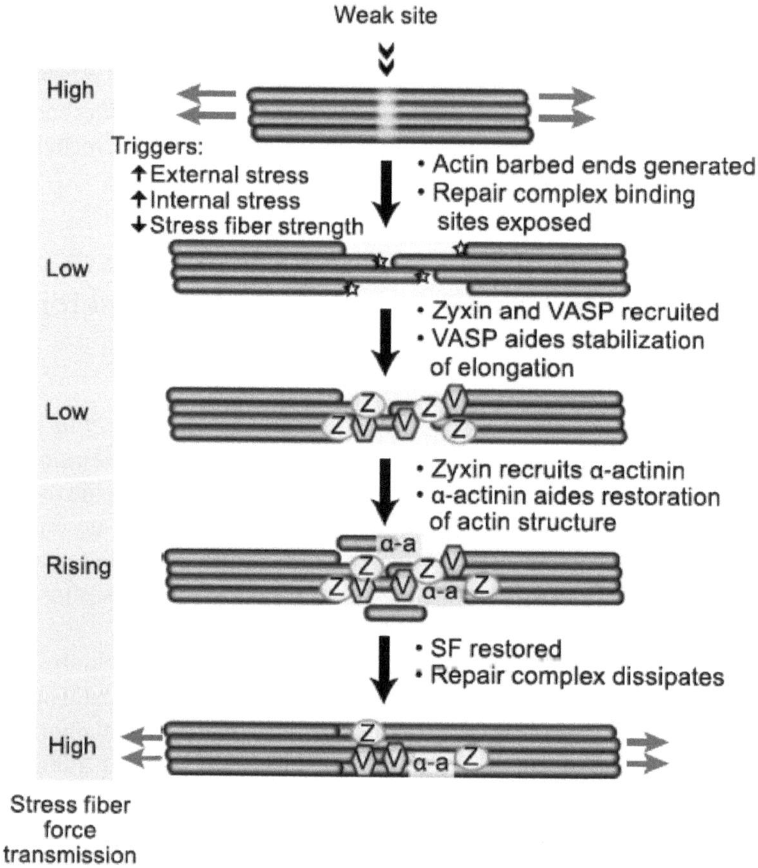


**Figure 7.11** Model for recruitment of a zyxin mediated repair complex to sites of SF strain. Stars indicate free actin barbed ends (cited from ref. [44] with permission).

SFs in rat fibroblasts using an AFM mounted on a fluorescence microscope. Kinetic analysis revealed that an accumulation and dissociation of paxillin occurred slightly earlier than zyxin. Furthermore, we found that a recombinant C-terminal fragment of paxillin comprising only the LIM domains exhibited a similar transient accumulation and dissociation pattern to that of paxillin. These results suggest that the LIM domains are essential for the mechanochemical coupling mechanism of mechanosensory LIM proteins.

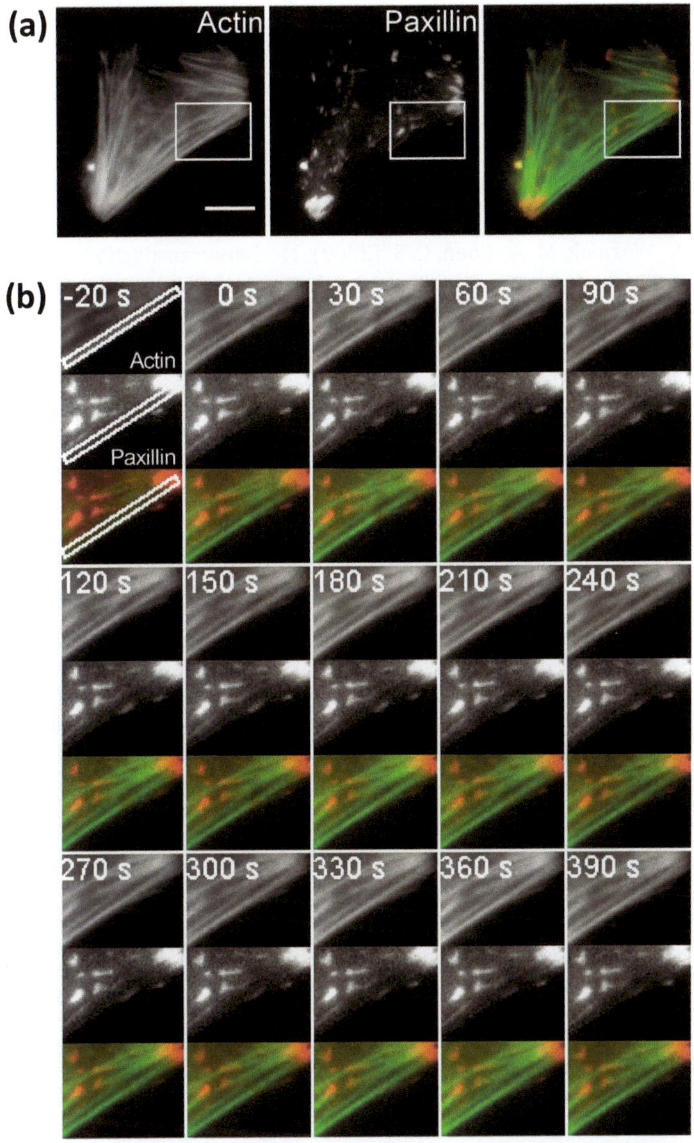


**Figure 7.12** Transient accumulation of paxillin at damaged sites on SFs. a: The micrographs of AcGFP1-actin and TagRFP-paxillin in a rat fibroblast. b: The montage image is of AcGFP1-actin and TagRFP-adhesion proteins. The montage images (b) were taken from the white boxes in the micrographs (a). Scale bar = 20 μm (cited from ref. [46] with permission).

## References

1. Butcher, D. T., Alliston, T., Weaver, V. M. (2009). A tense situation: Forcing tumour progression. *Nat. Rev. Cancer*, **9**(2), 108–22.
2. Wang, N., Tytell, J. D., Ingber, D. E. (2009). Mechanotransduction at a distance: Mechanically coupling the extracellular matrix with the nucleus. *Nat. Rev. Mol. Cell Biol.*, **10**(1), 75–82.
3. Wozniak, M. A., Chen, C. S. (2009). Mechanotransduction in development: A growing role for contractility. *Nat. Rev. Mol. Cell Biol.* **10**(1), 34–43.
4. Kole, T. P., Tseng, Y., Huang, L., Katz, J. L., Wirtz, D. (2004). Rho kinase regulates the intracellular micromechanical response of adherent cells to rho activation. *Mol. Biol. Cell*, **15**, 3475–3484.
5. Hoffman, B. D., Massiera, G., Van Citters, K. M., Crocker, J. C. (2006). The consensus mechanics of cultured mammalian cells. *Proc. Natl. Acad. Sci. U. S. A.*, **103**, 10259–10264.
6. Lee, J. S., Panorchan, P., Hale, C. M., Khatau, S. B., Kole, T. P., Tseng, Y., Wirtz, D. (2006). Ballistic intracellular nanorheology reveals ROCK-hard cytoplasmic stiffening response to fluid flow. *J. Cell Sci.*, **119** (Pt. 9), 1760–1768.
7. Choquet, D., Felsenfeld, D. P., Sheetz, M. P. (1997). Extracellular matrix rigidity causes strengthening of integrin-cytoskeleton linkages. *Cell*, **88**(1), 39–48.
8. Geiger, B., Bershadsky, A., Pankov, R., Yamada, K. M. (2001).Transmembrane crosstalk between the extracellular matrix–cytoskeleton crosstalk. *Nat. Rev. Mol. Cell Biol.*, **2**(11), 793–805.
9. Geiger, B., Bershadsky, A. (2001). Assembly and mechanosensory function of focal contacts. *Curr. Opin. Cell Biol.*, **13**(5), 584–592.
10. Giannone, G., Sheetz, M. P. (2006). Substrate rigidity and force define form through tyrosine phosphatase and kinase pathways. *Trends. Cell Biol.*, **16**(4), 213–223.
11. Kaunas, R., Nguyen, P., Usami, S., Chien, S. (2005). Cooperative effects of Rho and mechanical stretch on stress fiber organization. *Proc. Natl. Acad. Sci. U. S. A.*, **102**, 15895–15900.
12. Ren, Y., Effler, J. C., Norstrom, M., Luo, T., Firtel, R. A., Iglesias, P. A., Rock, R. S., Robinson, D. N. (2009). Mechanosensing through cooperative interactions between myosin II and the actin crosslinker cortexillin I. *Curr. Biol.*, **19**, 1421–1428.
13. Geiger, B., Spatz, J. P., Bershadsky, A. D. (2009). Environmental sensing through focal adhesions. *Nat. Rev. Mol. Cell Biol.*, **10**(1), 21–33.

14. del Rio, A., Perez Jimenez, R., Liu, R., Roca Cusachs, P., Fernandez, J. M., Sheetz, M. P. (2009). Stretching single talin rod molecules activates vinculin binding. *Science*, **323**(5914), 638–6341.

15. Colombelli, J., Besser, A., Kress, H., Reynaud, E. G., Girard, P., Caussinus, E., Haselmann, U., Small, J. V., Schwarz, U. S., Stelzer, E. H. (2009). Mechanosensing in actin stress fibers revealed by a close correlation between force and protein localization. *J. Cell Sci.*, **122**(Pt. 10), 1665–1679.

16. Riveline, D., Zamir, E., Balaban, N. Q., Schwarz, U. S., Ishizaki, T., Narumiya, S., Kam, Z., Geiger, B., Bershadsky, A. D. (2001). Focal contacts as mechanosensors: Externally applied local mechanical force induces growth of focal contacts by an mDia1-dependent and ROCK-independent mechanism. *J. Cell Biol.*, **153**(6), 1175–1186.

17. Yoshigi, M., Hoffman, L. M., Jensen, C. C., Yost, H. J., Beckerle, M. C. (2005). Mechanical force mobilizes zyxin from focal adhesions to actin filaments and regulates cytoskeletal reinforcement. *J. Cell Biol.*, **171**(2), 209–215.

18. Sawada, Y., Sheetz, M. P. (2002). Force transduction by Triton cytoskeletons. *J. Cell Biol.*, **156**(4), 609–615.

19. Mizutani, T., Kawabata, K., Koyama, Y., Takahashi, M., Haga, H. (2009). Regulation of cellular contractile force in response to mechanical stretch by diphosphorylation of myosin regulatory light chain via RhoA signaling cascade. *Cell Motil. Cytoskeleton*, **66**, 389–397.

20. Pellegrin, S., Mellor, H. (2007). Actin stress fibres. *J. Cell Sci.*, **120**, 3491–3499.

21. Kanchanawong, P., Shtengel, G., Pasapera, A. M., Ramko, E. B., Davidson, M. W., Hess, H. F., Waterman, C. M. (2010). Nanoscale architecture of integrin-based cell adhesions. *Nature*, **468**, 580–584.

22. Ikai, A., Watanabe-Nakayama, T., Machida, S., Saito, M., Afrin, R. (2011). Mechanics of Intracellular Stress Fibers: A Short Review Jpn. *J. Appl. Phys.*, **50**, 08LA04.

23. Afrin, R., Nakaji, M., Sekiguchi, H., Lee, D., Kishimoto, K., Ikai, A. (2012). Forced extension of delipidated red blood cell cytoskeleton with little indication of spectrin unfolding. *Cytoskeleton. (Hoboken.)*, **69**, 101–112.

24. Grandbois, M., Beyer, M., Rief, M., Clausen Schaumann, H., Gaub, H. E. (1999). How strong is a covalent bond? *Science*, **283**, 1727–1730.

25. Hakari, T., Sekiguchi, H., Osada, T., Kishimoto, K., Afrin, R., Ikai, A. (2011). Nonlinear displacement of ventral stress fibers under externally

applied lateral force by an atomic force microscope. *Cytoskeleton. (Hoboken.).*, **68**, 628–638.

26. Machida, S., Watanabe Nakayama, T., Harada, I., Afrin, R., Nakayama, T., Ikai, A. (2010). Direct manipulation of intracellular stress fibres using a hook-shaped AFM probe. *Nanotechnology*, **21**, 385102.
27. Machida, S., Watanabe Nakayama, T., Saito, M., Afrin, R., Ikai, A. (2012). Fabricated cantilever for AFM measurements and manipulations: Pre-stress analysis of stress fibers. Micron, **43**, 1380–1389.
28. Watanabe Nakayama, T., Machida, S., Harada, I., Sekiguchi, H., Afrin, R., Ikai, A. (2011). Direct detection of cellular adaptation to local cyclic stretching at the single cell level by atomic force microscopy. *Biophys. J.*, **100**, 564–572.
29. Cooper, J. A. (1987). Effects of cytochalasin and phalloidin on actin. *J. Cell. Biol.*, **105**(4), 1473–1478.
30. Straight, A. F., Cheung, A., Limouze, J., Chen, I., Westwood, N. J., Sellers, J. R., Mitchison, T. J. (2003). Dissecting temporal and spatial control of cytokinesis with a myosin II Inhibitor. *Science*, **299**(5613), 1743–1747.
31. Rotsch, C., Radmacher, M. (2000). Drug-induced changes of cytoskeletal structure and mechanics in fibroblasts: An atomic force microscopy study. *Biophys. J.*, **78**(1), 520–535.
32. Wakatsuki, T., Schwab, B., Thompson, N. C., Elson, E. L. (2001). Effects of cytochalasin D and latrunculin B on mechanical properties of cells. *J. Cell Sci.*, **114**(Pt. 5), 1025–1036.
33. Lu, L., Oswald, S. J., Ngu, H., Yin, F. C. (2008). Mechanical properties of actin stress fibers in living cells. *Biophys. J.*, **95**(12), 6060–6071.
34. Martens, J. C., Radmacher, M. (2008). Softening of the actin cytoskeleton by inhibition of myosin II. *Pflugers. Arch.*, **456**(1), 95–100.
35. Barakat, A. I. (2001). A model for shear stress-induced deformation of a flow sensor on the surface of vascular endothelial cells. *J. Theor. Biol.*, **210**, 221–236.
36. Schmitz, J., Benoit, M., Gottschalk, K. E. (2008). The viscoelasticity of membrane tethers and its importance for cell adhesion. *Biophys. J.*, **95**(3), 1448–1459.
37. del Alamo, J. C., Norwich, G. N., Li, Y. S., Lasheras, J. C., Chien, S. (2008). Anisotropic rheology and directional mechanotransduction in vascular endothelial cells. *Proc. Natl. Acad. Sci. U. S. A.*, **105**, 15411–15416.
38. Watanabe Nakayama, T., Machida, S., Afrin, R., Ikai, A. (2010). Microscoop for manipulation of micro-objects: Use of fabricated cantilever with atomic force microscope. *Small*, **6**, 2853-2857.

39. Hotulainen, P., Lappalainen, P. (2006). Stress fibers are generated by two distinct actin assembly mechanisms in motile cells. *J. Cell. Biol.*, **173**, 383–394.

40. Sawada, Y., Tamada, M., Dubin Thaler, B. J., Cherniavskaya, O., Sakai, R., Tanaka, S., Sheetz, M. P. (2006). Force sensing by mechanical extension of the Src family kinase substrate p130Cas. *Cell*, **127**, 1015–1026.

41. Hirata, H., Tatsumi, H., Sokabe, M. (2008). Mechanical forces facilitate actin polymerization at focal adhesions in a zyxin-dependent manner. *J. Cell. Sci.*, **121**(Pt. 17), 2795–2804.

42. Lele, T. P., Pendse, J., Kumar, S., Salanga, M., Karavitis, J., Ingber, D. E. (2006). Mechanical forces alter zyxin unbinding kinetics within focal adhesions of living cells. *J. Cell. Physiol.*, **207**(1), 187–194.

43. Guo, W. H., Wang, Y. L. (2007). Retrograde fluxes of focal adhesion proteins in response to cell migration and mechanical signals. *Mol. Biol. Cell*, **18**(11), 4519–4527.

44. Smith, M. A., Blankman, E., Gardel, M. L., Luettjohann, L., Waterman, C. M., Beckerle, M. C. (2010). A zyxin-mediated mechanism for actin stress fiber maintenance and repair. *Dev. Cell.*, **19**, 365–376.

45. Hansen, S. D., Mullins, R. D. (2010). VASP is a processive actin polymerase that requires monomeric actin for barbed end association. *J. Cell. Biol.*, **191**, 571–584.

46. Watanabe-Nakayama, T., Saito, M., Machida, S., Kishimoto, K., Afrin, R., Ikai, A. (2013). Requirement of LIM domains for the transient accumulation of paxillin at damaged stress fiberes. *Biol. Open.*, **2**, 667–674.

# Chapter 8

# Shape and Mechanical Cues Underlying Cellular Homeostasis in Soft Organs

**Jordi Alcaraz and Pere Roca-Cusachs**

*Facultat de Medicina, Universitat de Barcelona,*
*Casanova 143, Barcelona, 08036, Spain*

jalcaraz@ub.edu

Tissue cells are subjected to a variety of biochemical and biophysical stimuli from their local microenvironment. The ability of cells to sense these stimuli and orchestrate appropriate responses is essential for the integration of form and function underlying tissue homeostasis. Conversely, failure of any of these processes and the consequent loss of form and function integration are hallmarks of a large family of mechanobiology diseases including fibrosis and cancer. In this chapter, we discuss our previous studies aiming to understand how cells integrate form, mechanics, and function in the context of mammary and lung tissues. Special attention is given to our findings highlighting the prominent role of integrin receptors and their signaling factors, matrix metalloproteinases and transforming growth factor beta in the mechanical homeostasis of soft tissues. In addition, we describe studies linking cell shape, epigenetics, and chromatin organization. Based on the

*Cells, Forces, and the Microenvironment*
Edited by Charles M. Cuerrier and Andrew E. Pelling

ISBN 978-981-4613-36-1 (Hardcover), 978-981-4613-37-8 (eBook)
www.panstanford.com

latter and related findings, we argue that normal mechanical and morphological cues are protective against mechanobiology diseases.

## 8.1 Introduction: Biochemical Cues Cannot Do It Alone

In normal conditions, many organs in our body are very soft compared to the rigid structures of the musculo-skeletal system. Examples of such very soft organs include the liver, kidney, lung, brain, breast, and pancreas [1]. A common important property of these soft organs is their compartmentalization. Thus, the part of the organ that performs the tissue-specific function (i.e., the parenchyma) is generally separated from the parts of the organ (i.e., the stroma) that provide structural and "logistic" support to the parenchyma by a specialized extracellular matrix (ECM) known as basement membrane (BM). For example, the parenchyma in the mammary tissue includes the epithelial-rich alveolar/acinar structures responsible for secreting milk during lactation, whereas the stroma includes the fibroblast-rich connective tissue, blood capillaries and nerves [2]. In addition, tissues in each compartment exhibit an exquisite integration of form and function. Thus, mammary acini are organized in a three-dimensional (3D) thin layer of polarized epithelial cells that enclose a central lumen to facilitate the vectorial secretion of milk proteins, whereas the surrounding stroma is rich in fibrillar collagens and other ECM filaments that provide mechanical support to the whole organ. Intriguingly, a hallmark of many prevalent diseases affecting soft organs is a chronic loss of normal tissue elasticity in which tissues become too stiff—as in fibrosis, cancer, or sclerosis—with a concomitant loss of tissue compartmentalization and function.

Most previous studies have assumed that the maintenance of tissue architecture and its loss in diseased conditions arises from signaling triggered by soluble molecular cues such as cytokines, growth factors, and hormones. In addition, studies carried out in the last three decades have underscored the role of the insoluble ECM in controlling tissue architecture [3]. However, compelling evidence gathered in recent years by our group and others has indicated that any attempt to understand tissue organization in normal and diseased conditions is incomplete without considering the role of biophysical cues. In particular, we have examined how cell shape and tissue elasticity contribute to the regulation of basic physiological processes including cell proliferation, differentiation, and migration in the context of lung and mammary tissues. In addition, we have investigated how cues from cell shape and tissue elasticity become awry in—and contribute to the progression of—pathological conditions, with special emphasis in fibrosis and cancer.

## 8.2 Overview of Tools to Study Cell Shape and Mechanics

### 8.2.1 Tools to Provide ECM Signaling to Cultured Cells: 2D versus 3D Cultures

A major experimental challenge when dealing with tissue cells in culture is to obtain results that are physiologically relevant. For this purpose, a suitable strategy is based on providing essential biochemical signals found in vivo to cells in culture. The sources of biochemical signals in the context of the tissue microenvironment include the ECM, neighbor cells, and soluble signals that are released locally or transmitted from other organs, such as hormones [3]. Of note, most cell culture studies are carried out by seeding cells on two-dimensional (2D) glass or plastic substrata. In these conditions, cells can be exposed to signals from neighbor cells and exogenous soluble factors, whereas exogenous ECM signals are often missing. To overcome this limitation, we have used three different approaches depending on the specific experiment. First, we have used three-dimensional (3D) cultures based on growing cells embedded in gels of native ECM components such as reconstituted basement membrane/Matrigel (rBM) or type I collagen (COLI) (Fig. 8.1) [4,5]. Second, we have overlaid cells cultured in 2D with ECM components diluted in the culture medium [4]. Alternatively, we have used 2D substrata coated with a thin layer of ECM components [4,6].

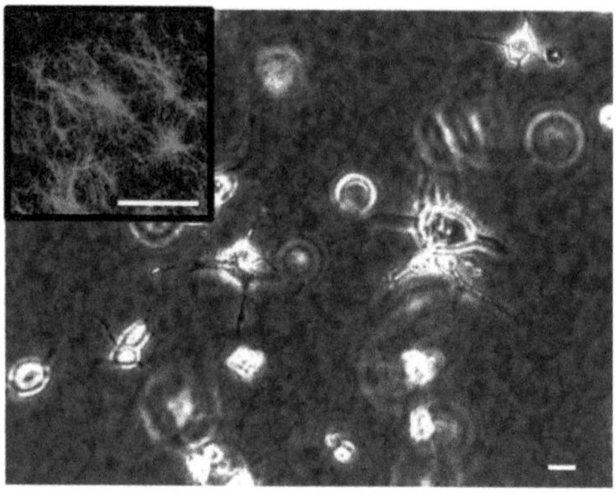

**Figure 8.1** Phase contrast image of a 3D culture of A549 lung carcinoma cells cultured embedded in a COLI gel. The insert shows details of the COLI fibers imaged by confocal reflection microscopy. Scale bars, 20 μm. Image courtesy of J. Alcaraz.

### 8.2.2 Tools to Study Cell Shape

To study the effects of cell shape in cellular functions, we have used two different methods that enable controlling cell spreading by limiting the extent of adhesion between cells and the underlying substratum. The first and simplest method was pioneered by Folkman and Moscona in 1978 [7], and is based on coating 2D substrata with a layer of poly (2-hydroxyethyl methacrylate) (polyHEMA), a chemically inert material that is nonadhesive to proteins and cells, thereby forcing cells to remain round [4]. The second and most sophisticated method is based on micropatterning by microcontact printing [6,8]. For this purpose, micrometric pits of specific shapes are etched on silicon or photoresist substrates (the "master") using photolithography or E-beam lithography. Then, a replica of the structures is obtained by polymerizing a soft polymer on the master like poly(dimethylsiloxane) (PDMS). This PDMS replica or "stamp," which contains raised features instead of pits, is then coated with ECM proteins and placed in contact with the target surface, which is typically a bacteriological petri dish [9]. Because only the raised features contact the target surface, the ECM protein is effectively "stamped" following the shapes of the original pits. After blocking the non-stamped regions with non-adhesive solution, cells will be only able to attach on the stamped areas [6].

In addition to the above mentioned standard approaches to control cell adhesion, we used a more advanced approach to pattern cells in soft substrata with tunable elasticity like polyacrylamide (PAA) gels instead of petri dishes [10]. In this case, the PDMS stamp is replicated from raised structures in the master, rather than pits. Using a spinner device, the stamp is made so thin that its height is lower than the master structures. The end result is a PDMS membrane with holes corresponding to the original raised features in the master. After placing the membrane in contact with the soft gel and coating ECM protein through it, only the areas with holes will be coated and will therefore support cell adhesion (Fig. 8.2) [6].

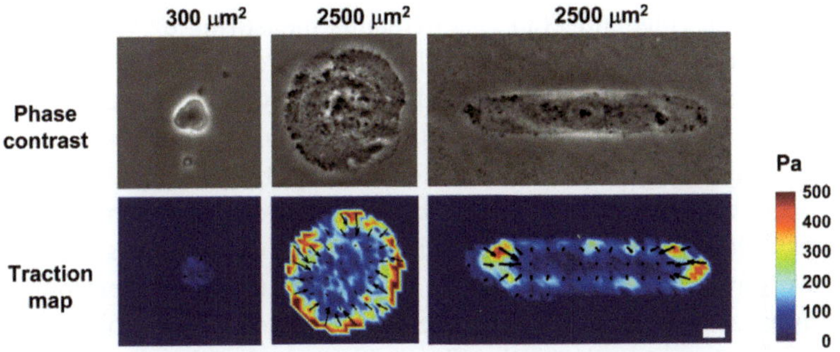


**Figure 8.2** Top: Images of single endothelial cells cultured on a PAA gel with micropatterned adhesive islands with distinct sizes and shapes. Bottom: Corresponding maps of traction forces measured by traction microscopy. Note that the traction forces are globally lower in the elongated than in the circular patterns. Scale bar, 10 μm. Reprinted from [6] with permission.

### 8.2.3 Tools to Study Mechanical Cues

A variety of experimental approaches from different fields including biomaterials, nanotechnology and bioengineering have been applied to study the mechanical properties and mechanoregulatory events in the context of cells, ECM and cell–ECM interactions (reviewed in [11]). Of paramount interest when studying mechanical cues is to be able to define them in the first place. For this purpose, we have benefited from the versatility of atomic force microscopy (AFM) to probe the mechanical properties of cells, ECM gels, or small tissues at the nanoscale and in their physiologic aqueous environment [4–6,12]. By measuring the sample resistance to loading forces applied with standard AFM tips in either quasi-static or oscillatory conditions, we have been able to assess Young's modulus ($E$) and the complex shear modulus ($G^*$), respectively, which are indicative of either the elastic ($E$) or viscoelastic behavior of the sample. Specifically, $G^*$ measurements are commonly analyzed in terms of $G^* = G' + iG''$, where $G''$ is the (elastic) storage modulus indicative of sample's solid-like resistance to deformation, and $G''$ is the (viscous) loss modulus indicative of the fluid-like resistance to flow [6]. Since biological samples are more solid-like than viscous in most conditions, assessing $E$ is enough in many occasions. However, the viscoelastic nature of cells and ECM gels demands measuring $G^*$ when an in-depth mechanical and structural analysis is required.

Mammary, pulmonary and other very soft tissues exhibit $E \sim 0.1$–$10$ kPa, whereas fibrotic and tumorigenic processes may render them up to ~20–30-fold stiffer [13,14]. Accordingly, it is desirable to study how cells behave when exposed to a normal- or pathological-like mechanical microenvironment. For this purpose, we have identified two useful and complementary experimental approaches based on two types of biomaterials. The oldest and simplest approach is the floating/attached gel assay [11]. In this assay, 3D cultures of cells in hydrogels attached to their stiff container (i.e., walls of a multiwell plate or similar) are compared with those using hydrogels floating in the culture media (i.e., unattached). In these conditions, we found a threefold difference in $E$ when using COLI gels [4,15]. This assay is advantageous in that it is 3D. However, the maximum $E$ attainable is ~1 kPa, which remains below the pathologic range. Alternatively, we have used the PAA gel assay in which cells are seeded on the surface of PAA gels coated with a thin layer of ECM [4,6]. This 2D assay was first introduced by Pelham and Wang in [16], and is advantageous in that PAA gels can be prepared to exhibit any $E$ value within the whole physiopathological range (~0.1–100 kPa).

A third group of important techniques for studying cell mechanics and mechanoregulation events are those that probe the forces between cells and their local ECM. A variety of set-ups have been introduced in recent years for this purpose [11]. We found four of them particularly useful in probing different mechanical aspects involved in cell–ECM interactions. Each of these four techniques is unique in its resolution and type of mechanical stress being applied or measured, and include AFM with cantilevers provided with flat-ended tips [17], micropillar arrays [18], traction microscopy (TM) [6], and magnetic tweezers (MT) [18,19]. Details on their principle of operation can be found elsewhere [11]. In brief, they are based on either the bending of microfabricated mechanical sensors (AFM, micropillars), manipulating magnetic micrometer-sized beads with magnetic fields (magnetic tweezers), or in tracking fluorescent nanoparticles embedded in an elastic PAA gel (traction microscopy) (Fig. 8.2). All these techniques have in common that they enable cells to attach to ECM molecules in liquid, to form focal adhesions or their precursors with either controlled or uncontrolled size, and to probe different mechanical aspects of local cell–ECM adhesions including adhesion strength, traction forces, and force transmission. Of note, they enable quantifying cell mechanoresponses to different ECM molecules, soluble biochemical stimuli, or force regimes.

## 8.3 Cell Shape: Regulatory Role and Mechanisms

It has been long recognized that each cell type is characterized by a specific morphology, which is expected to support its specific functions. Yet, our understanding of how cell shape controls cell functions is still in its infancy. We have provided new insights on this old biological question by examining how cell shape controls two important cellular processes: proliferation and differentiation. In addition, we have shed some light on the potential molecular and biophysical mechanisms underlying cell shape signaling in these processes.

### 8.3.1 Cell Shape and Proliferation

In their pioneer work, Folkman and Moscona used the polyHEMA assay to show for the first time that cell spreading and DNA synthesis were tightly coupled in nontransformed cells [7]. This work clearly illustrated that biochemical cues (i.e., growth factors in this case) alone are not sufficient to predict a biological response (proliferation), and provided a proof-of-principle that biochemical signaling could be strongly regulated by the biophysical status of the cell (spreading). Subsequent work in this field suggested that the biophysical signal linking spreading and proliferation was the internal mechanical tension (prestress) built within the actin cytoskeleton (CSK) as cells spread out [20]. However, controversial findings prompted us to search for alternative biophysical cues [6]. For this purpose, we examined single lung endothelial cells after modulating independently three cell shape-related variables with micropatterned adhesive islands: spreading, pointedness (i.e., presence of corners) and elongation (ratio between cell length and width). In these conditions, cells were exposed to mitogenic biochemical cues by adding serum to the culture media, and their proliferative status was assessed by measuring DNA synthesis. In addition, we examined two types of biophysical variables: those related to either cellular contractility (assessed by AFM and TM) or to nuclear size and organization. The latter were assessed by measuring nuclear volume and the apparent chromatin condensation with optical methods [6]. A linear increase of DNA synthesis with spreading was observed independently of all other geometrical variables. Likewise, nuclear volume of cells in G1 of the cell cycle consistently increased with spreading regardless pointedness and elongation, whereas apparent chromatin condensation decreased in the same conditions. In contrast, cell contractility was strongly dependent on both spreading and elongation (Fig. 8.2).

Our results revealed that nuclear volume of cells in G1 is predictive of the probability of a single cell to enter into S-phase, whereas cellular contractility is not. Moreover, they supported that spreading is linked biophysically to proliferation through an increase in nuclear size and the concomitant decrease of chromatin condensation, which may render DNA more accessible to the replication machinery (see Section 8.2.2 for further discussion in this topic). However, how cells control the size of their nucleus remains an open question.

### 8.3.2 Cell Shape and Differentiation

Although the notion that cells in tissues integrate form and function is widely accepted, the mechanistic role of cell shape in tissue-specific functions still remains largely ignored in most cell biology studies. As pointed out elegantly by pioneering studies from Mina J. Bissell, a possible explanation for this lack of appreciation could be that most cell biology studies have been conducted in 2D cultures and, in these conditions, cells typically exhibit an altered morphology and experience a partial loss of differentiation [21]. Of note, studies carried out with mammary epithelial cells (MECs) demonstrated that ECM integrates form and function in vivo and in 3D cultures. Specifically, it was found that the BM component laminin-111 (LM1) was necessary to induce polarity and acinar morphogenesis in vivo [22] and, together with lactogenic hormones, it was required to express the most abundant milk protein β-casein in MECs in culture [23]. Further studies in culture revealed that exposing MECs to LM1 induced cell rounding prior to β-casein expression. Interestingly, it was reported that pre-rounding MECs with polyHEMA in the absence of LM1 was sufficient to induce the expression of the milk protein lactoferrin, and poised cells to express β-casein very rapidly (<8 h) in response to LM1 and hormones (reviewed in [11]). We expanded these previous findings by showing that despite the specific increase in the expression of milk proteins, global gene expression was reduced in 3D cultures with LM1-rich gels. Moreover, we observed that this reduced transcriptional activity was concomitant with an increased expression of both heterochromatin markers and apparent chromatin condensation, and with a decrease in cell and nuclear size (Fig. 8.1) [8]. In addition we showed that rounding MECs in the absence of exogenous ECM with either micropatterning techniques or by disrupting the actin CSK phenocopied ECM-dependent heterochromatinization [13]. Therefore, all these findings reveal a link between ECM signaling, cell shape and alterations in both global and tissue-specific gene expression that is mediated partially through changes in chromatin organization.

### 8.3.3 Cell Shape, Chromatin Condensation, and Epigenetics

Although the biophysical cues linking cell shape with cellular functions are becoming elucidated, the molecular mechanisms underlying these links remain poorly understood. However, our work in this field has provided some clues that may help define some of these mechanisms. First, it is worth reminding that chromatin in interphase cells is embedded in an extremely crowded environment [24]. Previous studies have shown that macromolecular crowding leads to volume exclusion effects, which in turn may increase the tendency of macromolecular association [25]. Conversely, a reduction in macromolecular crowding may act otherwise. Thus, it is conceivable that cells have adapted to this universal biophysical organizing principle to link changes in nuclear size and chromatin condensation with alterations in the crowding conditions within the nucleus. In support of this interpretation we observed an inverse relationship between nuclear volume and apparent chromatin condensation in single endothelial cells [6]. Since the level of chromatin condensation is a central regulator of many biological processes, our work underlines the need to direct future studies towards understanding how nuclear volume is regulated and maintained.

Another clue on the mechanisms underlying the links between cell shape and function came from the observation that LM1 induced a global decrease in both acetylated histone 3 (acH3) and histone 4 (acH4), and in nuclear size in MECs. Importantly, we also observed that pre-rounding cells in the absence of LM1 was sufficient to induce similar epigenetic and morphological changes, along with a global increase in apparent chromatin condensation [8]. Thus, in agreement with the observations made in endothelial cells, we found a marked association between nuclear size and chromatin condensation in MECs. Interestingly, we also found that the latter biophysical properties of the nucleus were modulated in concert with epigenetic markers (i.e., histone acetylation) in response to cell shape changes. However, the causal relationship between size/macromolecular crowding in the nucleus and epigenetic alterations in the chromatin remain to be elucidated.

## 8.4 Tissue Mechanics: Mechanical Aspects and Molecular Regulators

It is an everyday-life fact that each tissue and organ in our body exhibits well-defined mechanical properties to carry out their specific functions. For instance, bones exhibit large stiffness to sustain our weight and support our movements, whereas lungs are very soft and elastic to enable the cyclic changes in lung volume required for breathing [1]. A consequence of this observation is that our bodies must have homeostatic mechanisms to preserve tissue and organ mechanics, thereby preventing bones from becoming soft like a lung, or lungs from becoming stiff like a bone. Surprisingly, our current understanding of these homeostatic mechanisms remains extremely limited. Nonetheless, a large body of work has identified three critical structures in controlling tissue mechanics, namely (from outside-in the cell): ECM, integrin ECM receptors, and the CSK. Likewise, there is ample evidence that the connections between these structures, although dynamic, enable transmitting forces bidirectionally between cells and their microenvironment [26].

### 8.4.1 ECM Mechanics and Turnover

The ECM is the natural scaffold of our tissues and organs. It is a complex 3D filamentous network of entangled proteins and polysaccharides whose biochemical composition and structural organization exhibits marked tissue specificity. ECM components can be roughly classified into four main categories, each one with distinctive mechanical features (see [27] for a detailed review). Fibrous glycoprotein components like collagens or elastin are the most abundant and control tissue strength and resilience. Amorphous components like glygosaminoglycans and proteoglycans are highly hydrated, which enable forming cushion-like structures that resist compression. Adaptor/adhesive proteins like fibronectin or laminins facilitate cell attachment or movement within the ECM, and are essential in transmitting forces outside-in and inside-out cells. Finally, cross-linkers such as lysyl-oxidase or transglutaminase can mechanically stabilize the ECM by increasing its strength and rendering it more resistant to degradation.

We have learned important lessons on ECM mechanics by performing AFM measurements on cell-free ECM gels. Special attention has been given to COLI, since it is the most abundant ECM component and is largely responsible for tissue strength in most tissues, including mammary and lung [28]. Specifically, we have dissected how ECM density, composition, and degradation alters the microelasticity of the gel. Major lessons obtained in acellular gels are summarized as follows:

- *The applicability of ECM gels as tissue surrogates in terms of mechanical properties is limited to very soft tissues*

  All $E$ values of COLI gels obtained in different conditions fall in the range 0.1–1 kPa, which indicate that COLI gels mimic somewhat the mechanical microenvironment of very soft tissues, including normal mammary tissue and partially activated mammary stroma [14], but not that of lung [13] or other stiffer tissues.

- *ECM enzymatic degradation softens gels very rapidly*

  Enzymatic COLI degradation by collagenases induced a rapid (within minutes) marked reduction in $E$ that increased with time (Fig. 8.3a) [5]. Given the abundance of COLI in vivo, these findings suggest that any imbalance between COLI deposition and degradation emerges as an efficient strategy to alter the local mechanical strength of the tissue.

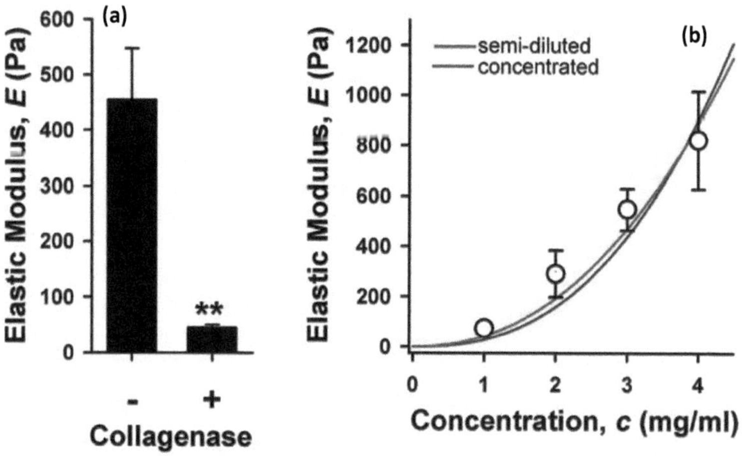


**Figure 8.3** AFM measurements of $E$ of COLI gels after collagenase degradation (a) and as a function of COLI concentration (b). Data in B were fitted to semiflexible polymer models corresponding to semi-diluted (red) or concentrated (blue) gels. Reprinted from [5] with permission.

- *Increasing ECM density stiffens gels in a nonlinear fashion*

  We found that $E$ of cell-free COLI gels increases with COLI concentration ($c$) as $E \sim c^2$ within the range 1–4 mg/ml [5], in agreement with current theoretical models of semiflexible polymers in which the fibril length is larger than the pore or mesh size (Fig. 8.3B).

- *The relationship between ECM composition and microelasticity is highly nonlinear*

  To assess how ECM composition affects its mechanics, we examined how increasing COLI content in LM1 gels affected $E$ of the gel [4]. We found that gels remained soft ($E \sim 0.1$ kPa) for up to ~90% COLI content, and abruptly stiffen for larger COLI concentrations, illustrating the nonlinear relation between ECM composition and ECM microelasticity.

- *ECM gels exhibit power-law viscoelasticity*

  Dynamic measurements carried out by AFM by our group and with other techniques elsewhere have revealed that COLI gels are viscoelastic, with an (elastic) $G'$ slightly higher than the (viscous) $G''$ for frequencies up to ~10 Hz. These results indicate that ECM gels are solid-like materials at low frequencies. In addition, both $G'$ and $G''$ increased with frequency ($f$) following a power-law of very weak exponent (i.e., $G' \sim G'' \sim f^{\alpha}$, with a power-law exponent $\alpha < 1$) [29,30]. Similar mechanical features have been reported in different types of soft glassy materials that are ageing. However, understanding the physics of soft power-law viscoelastic materials like ECM gels is still an open question and a matter of active research [31]. Interestingly, the power-law viscoelasticity observed in ECM gels appears to be a hallmark of ECM in vivo, according to recent AFM measurements on decellularized lung sections [32].

### 8.4.2 Integrin ECM Receptors: Role in Force Transmission and Mechanoresponse

The molecular machinery that connects cells to the ECM determines the mechanical behavior of cells and their response to mechanical stimuli. Among the molecules involved, integrins and the proteins that link them to the CSK play a crucial role [17,33]. Using a variety of biomechanical tools, we have unraveled different mechanical and molecular hallmarks of integrins and integrin-binding proteins, with special emphasis on fibronectin receptors, which are summarized below.

- *Integrin activation occurs within seconds*

  Previous studies have shown that local attachment to the ECM in cultured cells begins with a dot-like initial adhesion that progressively matures into ~1 μm$^2$ focal complex rich in integrins, actin, vinculin and few other focal adhesion proteins in a timescale of seconds to minutes, and eventually forms larger and more tightly bound focal adhesions. To study the dynamics of adhesive sites in detail, we microfabricated flat-ended AFM (FE-AFM) tips with a ~1 μm$^2$ cross section, and coated them with a synthetic peptide containing the tri-aminoacid Arg-Gly-Asp (RGD) sequence, which is an integrin-specific binding domain present in fibronectin and other ECM components (Fig. 8.4). We found that applying a sustained compressive loading to the surface of either lung epithelial cells or fibroblasts for ~30 s was sufficient to strengthen cell–RGD attachment more than fourfold compared to control tips [33]. This strengthening was associated with a rapid increase in local cell rigidity or reinforcement, which required actin polymerization, tyrosine phosphatases and $Ca^{2+}$ signaling [6].

**Figure 8.4** Scanning electron microscopy images of flat-ended AFM tips at different magnifications. Tips were nanofabricated with a cross section ~1 μm$^2$ and coated with ECM proteins to probe cell mechanoresponses in real time to local exogenous forces mediated through few integrins. Available from [17] under Creative Commons Attribution-Noncommercial licence.

- *Cell–ECM force transmission is asymmetric*

  To examine cell mechanoresponse to compression and tension, we assess local cell resistance to either compressive loading or tensional loading applied with RGD-coated FE-AFM tips by measuring the corresponding *E* values during compression and tension [9]. We observed that cell mechanical resistance to compression was four- to sixfold higher than in tension, depending on the cell type. This mechanical asymmetry suggests that different cellular structures are involved in resisting compression and tension. Moreover, they may provide a built-in mechanism for sensing force directionality.

- *Cell–ECM adhesion strength is cell-type specific*

  Using RGD-coated FE-AFM tips, we determined that cell–RGD adhesion strength was strongly dependent on the cell-type. Specifically, we found that focal adhesion precursors in lung fibroblasts developed adhesive forces up to ~1 nN, whereas lung epithelial cells developed half that value (Fig. 8.5). Such large force generating capabilities observed in fibroblasts may facilitate the ECM remodeling functions that are a hallmark of these cells.

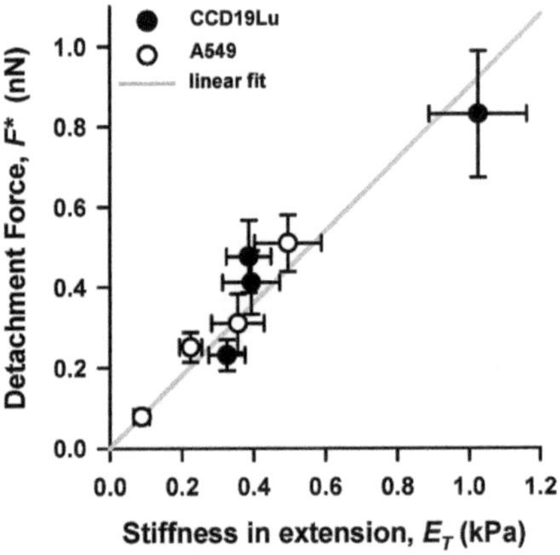


**Figure 8.5** Local cell adhesion strength increases with cell stiffness. The detachment force between a FE-AFM tip and either lung fibroblasts (CCD-19Lu) or epithelial cells (A549) was measured with AFM and plotted as a function of the corresponding Young's modulus measured with the same tip. Available from [17] under Creative Commons Attribution-Noncommercial licence.

- *Division of labor between two fibronectin receptors* ($\alpha_5\beta_1$ *and* $\alpha_v\beta_3$) *in terms of adhesion strength and mechanosensing*

  Using magnetic tweezers, we showed that the main fibronectin integrin receptor $\alpha_5\beta_1$ provides adhesion strength to fibronectin matrices. Indeed, cells in which $\alpha_5\beta_1$ had been blocked could no longer hold on to fibronectin-coated magnetic beads after force application [19]. This is consistent with the finding that $\alpha_5\beta_1$–fibronectin bonds increase their stability upon force application (catch bond behavior) [34]. In contrast, $\alpha_v\beta_3$ (another fibronectin-binding integrin) did not contribute to adhesion strength. However, whereas fibronectin-coated beads attached to control cells progressively reduced the amplitude of their oscillations in response to pulsatile forces, beads attached to cells with inhibited $\alpha_v\beta_3$ did not. Thus, $\alpha_v\beta_3$ is necessary for the mechanosensing process that normally leads to adhesion reinforcement in response to force. The mechanism underlying this function is unclear, although it is conceivable that precisely because they do not hold high forces, $\alpha_v\beta_3$–fibronectin bonds are constantly unbinding and rebinding, inducing downstream signaling such as activation of Src-family kinases [35].

- *Intracellular molecular machinery underlying cell–ECM force transmission by fibronectin receptors ($\alpha_5\beta_1$ and $\alpha_v\beta_3$)*

  A division of mechanical tasks occurs at the level of the adaptor proteins that link integrins to the actin CSK. Among the former, only a very small number can simultaneously bind integrins and actin, being thus of special interest (Fig. 8.6) [33]. Like $\alpha_v\beta_3$, talin is required for force detection and adhesion reinforcement [19]. This is consistent with our earlier finding that force application to talin unfolds previously buried binding sites to vinculin, providing a mechanism for force detection [36]. However, the very flexible nature of talin molecules [33] suggests that while talin stretch may provide a mechanosensing mechanism, it may not be very effective at transmitting forces. Indeed, we recently used a micropillar assay to show that another direct integrin–actin linker, α-actinin, is the main molecule transmitting forces between the cytoskeleton and the ECM [18]. Interestingly, using an in vitro binding assay we also found that talin competes with α-actinin for the binding to integrin $\beta_3$, but cooperates for the binding to integrin $\beta_1$. Thus, talin and α-actinin interact with each other to bind integrins, and detect and transmit forces.

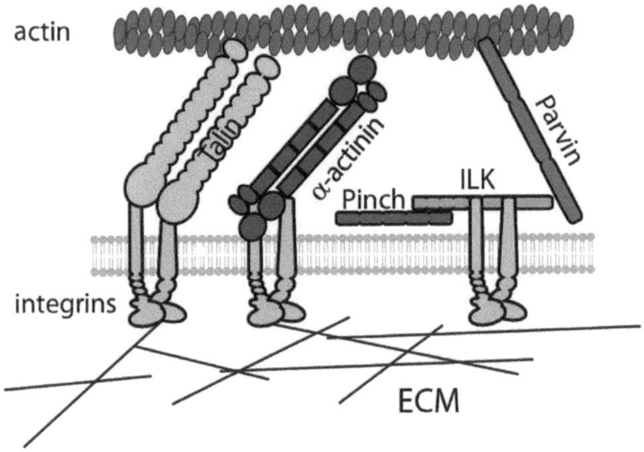


**Figure 8.6** Molecules like talin or α-actinin are able to directly link integrins to actin, thus connecting the cytoskeleton to the ECM. Other complexes (like the ILK–Pinch–Parvin complex) also provide an actin-integrin linkage which involves more than one molecule. Image courtesy of P. Roca-Cusachs. More details in [33].

## 8.4.3 Mechanical Hallmarks of Cells

We have exploited the versatility of AFM to unravel several hallmarks of cell mechanics, which collectively underscore the prominent mechanical role of the actin CSK. These hallmarks are summarized as follow:

- *Cell mechanics is cell-type specific*

  The simplest mechanical parameter than can be assessed with AFM is the local Young's modulus in response to compressive loading forces ($E_C$), which is indicative of cell resistance to compression. Table 8.1 shows a summary of $E_c$ values obtained in different cell types. For a given cell, $E_c$ can vary up to eightfold depending on the location, and tends to be lower in the nuclear region than in the perinuclear part [6]. Likewise, for a given location $E_c$ can vary ~2–8-fold depending on the biochemical stimuli and overall culture conditions. Yet an emerging property from Table 8.1 is that $E_C$ is cell-type specific, and allows roughly classifying cells in decreasing order of $E_c$ as endothelial/fibroblast > epithelial > neutrophils.

**Table 8.1** Young's elastic modulus measured in compression with AFM in different cell types cultured in 2D

| Cell type | Cell source | Tissue of origin | $E$(kPa) | Ref. |
|---|---|---|---|---|
| Epithelial | BEAS-2B | Human lung | 1.5 | [12] |
| | A549 | Human lung | 1.6 | [12] |
| | | Adenocarcinoma | 0.5–1.5 | [17] |
| | ScP2, EpH4 | Mouse mammary Tissue | 1.3 | [4] |
| Neutrophils | Primary | Rat blood | 0.2–1.5 | [37] |
| Endothelial | HMVEC-L | Human lung Microvascular | 3–25 | [6] |
| Fibroblast | CCD-19Lu | Human lung | 3–7 | [17] |

- *Cells exhibit power-law viscoelasticity akin of ECM gels*

Cells exhibit striking mechanical similarities with ECM gels in that their viscoelastic moduli exhibit a power-law dependence with $f$ with a weak exponent at low frequencies (i.e., $G' \sim G'' \sim f^{-\alpha}$, $\alpha < 1$) [12,37]. In addition, $G''$ exhibits a stronger dependence at higher frequencies that is consistent with $G'' \sim f^{-\alpha} + f^{3/4}$, in which the latter $f^{3/4}$ is due to entropic elasticity as predicted in some theoretical models [31]. Accordingly, as in cell-free ECM samples, cells behave as solid-like materials in the physiological (low) frequency range (i.e., $G' > G''$) (Fig. 8.7). The mechanical similarity between cells and ECM strongly suggests that the physical nature of their internal structure is also similar. This is not entirely surprising, since the structure of both cells and the ECM are largely based on filamentous proteinaceous networks: the cellular CSK and the ECM network itself. Moreover, this similarity suggests a viscoelastic matching between the cell and its microenvironment, which may be required to elicit either synchronous deformations in response to forces within the tissue, an efficient cell–ECM force transmission, or both.

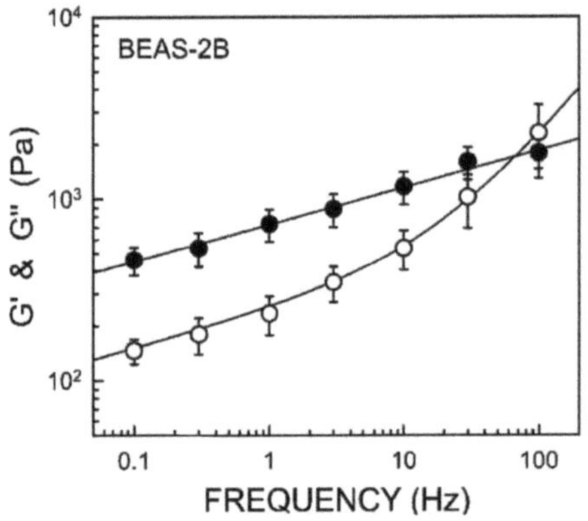


**Figure 8.7** AFM measurements of $G'$ (black) and $G''$ (white circles) of single lung epithelial cells (BEAS-2B cell line) cultured in 2D. Data were obtained by driving the vertical position of the cells with low-amplitude oscillations and measuring the corresponding force and indentation Reprinted from [12] with permission.

- *Prominent role of the actin CSK and myosin II in cell mechanics* Although both $E_c$ and $G'$ values are cell-type specific, a large body of evidence gathered by our group and others underscores the prominent role of the actin CSK in all the mechanical aspects of cells and cell–ECM interactions. Most of this evidence has been obtained owing to the availability of effective inhibitors of actin polymerization such as cythocalasin-D or latrunculin B. Inhibiting actin polymerization reduces $E_c$ more than twofold in epithelial cells, fibroblasts, endothelial cells and neutrophils cultured in 2D [4,6,17,37]. Likewise, an intact actin CSK was required to support cell–ECM adhesion strength in epithelial cells and fibroblasts [17]. In contrast, rounded epithelial, endothelial cells, and neutrophils exhibited little or no mechanical effects in response to actin polymerization inhibitors, which can be explained by the low F-actin content in these conditions. In addition to F-actin, we observed a strong positive correlation between phosphorylated non-muscle myosin II and $E_c$ values in epithelial cells [4]. Likewise, inhibiting myosin II ATPase activity with blebbistatin or its upstream effector Rho kinase (ROCK) with Y-27632 in epithelial cells cultured in 2D reduced $E_c$ values similarly to actin polymerization inhibitors. These findings support that large $E_c$ and adhesion strength require large F-actin and myosin II activity in virtually any adherent cell, whereas down-regulating either of them is sufficient to compromise the mechanical stability of cells and cell–ECM interactions. In contrast, we found that the mechanical effects of microtubule polymerization inhibitors were strongly cell-type dependent [17].

## 8.5 Integration of Form, Mechanics and Function: Emerging Molecular Players

Tissue cells are subjected to a variety of biochemical and biophysical stimuli from their local microenvironment. The ability of cells to sense these stimuli and orchestrate appropriate responses is essential for the integration of form and function that supports tissue homeostasis. The molecular mechanisms underlying these processes remain poorly understood. Here we review the lessons learned on the critical extra- and intra-cellular molecular features underlying differentiation, invasion, and repair in soft tissues.

Integrin ECM receptors, matrix metalloproteinase (MMPs) ECM degrading enzymes, and TGF-β1 are master regulators of tissue mechanical homeostasis.

### 8.5.1 $\alpha_3\beta_1$ Integrin and Differentiation

Previous studies with MECs in culture had identified the minimum set of biochemical and biophysical cues required for β-casein expression, which is a marker of tissue-specific function [23]. These cues included extrinsic microenvironmental factors (LM1 and the lactogenic hormone prolactin) and intrinsic signals elicited by a round cell shape. In vivo, cell morphology is largely regulated by cell–ECM and cell–cell mechanochemical interactions. Since cell–cell signaling is not required for β-casein expression, we focused on cell–ECM interactions. We used the floating gel and the PAA gel assays described in Section 8.2.3 to examine the effects of extracellular rigidity on MEC differentiation. We found that β-casein expression was maximum when cells were cultured in soft gels with $E$ mimicking mammary tissue elasticity. In these conditions, single MECs exhibited biomimetic $E_c$ values—i.e., $E_c$ comparable to that measured on mammary tissue extracts—as revealed by AFM. Using functional blocking antibodies and genetic approaches we identified $\alpha_3\beta_1$ integrin as the LM1-specific receptor responsible for the integration of extracellular mechanical and biochemical cues underlying such maximum β-casein expression. In addition, inhibiting this integrin rendered $E_c$ of single MECs beyond the physiological range [4]. These results revealed that functional differentiation in MECs requires a normal-like extracellular rigidity and is concomitant with a unique mechanical state of the cell. The latter intrinsic feature was associated with intermediate F-actin levels and low myosin II activity, in agreement with previous observations that Rho signaling needs to be downregulated for acinar morphogenesis of MECs [14] and that integrin-linked kinase (ILK) activity and Rac1 signaling are involved in β-casein expression [38]. It is conceivable that such biomimetic cellular elasticity is indicative of a unique level of signaling conveyed by β1 integrin and Rac1/Rho, since these molecules are all important regulators of the actin–myosin cytoskeleton and are also necessary for β-casein expression. These results illustrate how extrinsic (ECM composition and mechanics, hormones) and intrinsic (integrins, cell shape, CSK organization and mechanics) factors are integrated at the single-cell level in normal physiologic conditions.

### 8.5.2 MMP14 and Collective Cell Invasion

Collective cell invasion (CCI) through interstitial collagenous ECM is a crucial process in both developmental and repair programs. In normal conditions, CCI is highly regulated to enable the transient alteration of tissue architecture without compromising its global function. The collagenous ECM is thought to act as a structural and mechanical barrier against CCI, particularly at large collagen densities. Analyzing mammary branching morphogenesis in 3D cultures in COLI gels revealed 2 strategies that can independently minimize ECM mechanical resistance during CCI. First, MECs adopt a narrow tube-like geometry during invasion, which is optimal in reducing the mechanical opposition from the ECM according to theoretical models of contact elasticity. Second, the rigidity of the collagenous ECM is reduced at invasive fronts due to its degradation by matrix metalloproteinases (MMPs), as indicated by AFM. Genetic approaches further revealed that the membrane-bound MMP14 specifically mediates the degradation of COLI at invasive fronts [5]. These observations underscore the prominent role of MMP14 in compromising the mechanical integrity of the collagenous ECM to facilitate CCI, and reveal a geometric strategy to further reduce ECM constraints against invasion. Subsequent studies revealed that MMP14 proteolytic activity is necessary for invasion through dense (≥3 mg/ml) but not dilute (≤1 mg/ml) COLI gels. In contrast, signals downstream of the transmembrane/cytoplasmic domain of MMP14 to MAPK activation were required for CCI in all gel densities. Intriguingly, the latter signaling required MMP14 direct association with β1 integrin [15]. These findings illustrate a tissue mechanical homeostatic mechanism, since MMP14 degrades the surrounding collagenous ECM only as required by high COLI density during CCI.

### 8.5.3 TGF-β1 and Repair

Upon injury, the normal structural and mechanical integrity of the tissue becomes compromised. In normal physiological conditions, tissue cells can repair even severe injury by forming scar tissue, a provisional structure generated by ECM deposition, fibroblast proliferation and transient activation, angiogenesis, and immune cell influx [39]. Activated fibroblasts/myofibroblasts are the key cells involved in scar tissue formation, producing proteases that degrade the damaged tissue, and synthesizing COLI and other ECM components required to form the wound provisional matrix. At later stages, myofibroblasts also mediate ECM contraction through the formation of stress fibers rich in alpha smooth muscle actin (α-SMA) to further expedite the repair process [40,41]. Completion of injury repair is followed by degradation of the provisional ECM and apoptosis of the myofibroblasts. Among the soluble factors that accumulate at injury sites, transforming growth factor beta 1 (TGF-β1) has been identified as a potent fibroblast activator. Thus, previous studies of TGF-β1-induced fibroblast activation in culture revealed that α-SMA expression is associated with increased cell contractility as revealed by traction microscopy measurements [42], and that extracellular hardening can promote fibroblast activation further [43]. We conducted AFM experiments to show that TGF-β1 increases $E_c$ of cultured lung fibroblasts nearly twofold concomitantly with a marked increase in α-SMA expression (Fig. 8.8; J. Alcaraz, unpublished). These experiments reveal the reciprocal interactions between extrinsic and intrinsic factors in rendering stiffer tissues during repair, since (extrinsic) TGF-β1 enhances fibroblast contractility through α-SMA expression (intrinsic), which induces extracellular hardening (extrinsic), which activates fibroblasts further.

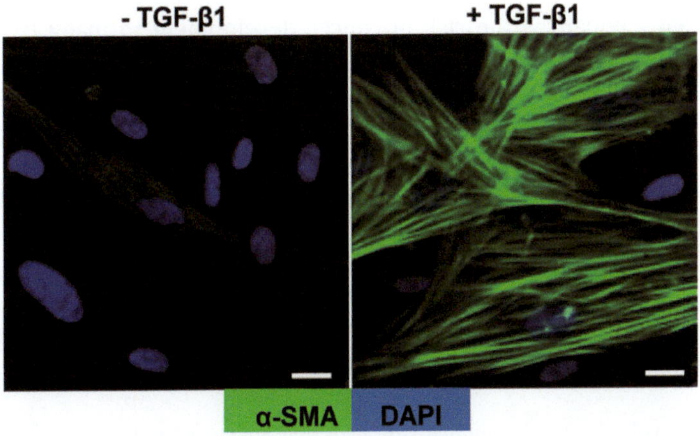


**Figure 8.8** Fluorescence images of the nuclei and α-SMA content of lung fibroblasts cultured in 2D with or without the potent activator TGF-β1. Scale bars, 15 μm. Images courtesy of J. Alcaraz.

## 8.6 Aberrant Mechanical and Morphological Cues in Fibrosis and Cancer

Tissue stiffness may be transiently compromised during injury, although physiologic repair mechanisms bring back both normal tissue elasticity and function in a timely manner. In contrast, bones become permanently soft and fragile in patients with osteoporosis or arthritis. Conversely, lungs exhibit sustained tissue stiffening at sites of tumor growth or fibrotic lesions. These are not isolated examples, but rather illustrations of the general observation that normal tissue stiffness becomes chronically compromised in a large number of human diseases in which tissues become either too stiff or too soft [44].

We have focused on two very prevalent mechanobiology diseases in which tissues become too stiff: fibrosis and cancer. Tissue hardening in fibrosis and cancer has been associated with abnormal wound healing in which myofibroblasts evade apoptosis and develop dysfunctional repair mechanisms that lead to perpetuated scarring or fibrosis, and ultimately loss of tissue function [45,46]. Indeed, a relationship between fibrosis and tumor formation was suspected ever since the early identification of tumors near scar tissue. Subsequent work revealed that fibrosis greatly enhances the risk of cancer development in many tissues and organs including the lung [46], and led to the interpretation of fibrotic and some tumor tissues as "wounds that never heal" [45]. Accordingly, it is unsurprising that abnormal tissue hardening is a common feature of both fibrogenesis and tumorigenesis. The relative contributions of the different microenvironmental alterations that occur during dysfunctional repair to tissue hardening is likely to depend on the particular disease and organ, and overall remain poorly understood. Here we provide illustrative examples of how mechanical and morphological cues become awry in fibrosis and cancer in soft tissues.

A solid body of evidence supports that sustained tissue hardening is necessary for fibrogenesis and tumor progression, whereas maintenance of normal tissue elasticity is protective against these mechanoregulatory diseases.

## 8.6.1 Aberrant Features in Lung Fibrosis

Among the different lung fibrotic diseases, idiophatic pulmonary fibrosis (IPF) is the most common and lethal, with a mean survival of 5 years for only 20–30% of the patients after diagnosis. In addition, many patients with IPF develop lung cancer [47]. Below we describe selected extrinsic and intrinsic alterations in lung fibrosis that are indicative of a loss of mechanical homeostasis and are associated with the progression of the disease.

- *Abnormal tissue hardening in IPF*

  This hardening is remarkable because lung elasticity is rather conserved in mammals, with an average $E \sim 5$ kPa [13,48].

In contrast, fibrotic tissue may be macroscopically two- to threefold stiffer, and microscopically up to 30-fold stiffer than the median normal lung, as revealed by recent high-resolution tissue measurements with the nanotechnique AFM [23]. These data indicate that the mechanical microenvironment in IPF is closer to muscle or even bone rather than to normal lung [1].

- *Lung stiffness is a prognostic factor of IPF*

  Lung stiffness is one of the few independent factors predicting survival in IPF patients [49]. This fact underlines that normal lung elasticity is not only important for breathing, but that abnormal lung elasticity may be a major driving force in IPF progression.

- *Abnormal myofibroblast accumulation in IPF*

  Histologically, IPF is associated with the pattern of usual interstitial pneumonia, which is characterized by the appearance of active areas of fibrosis (referred to as fibroblast foci) densely populated with proliferating fibroblasts and myofibroblasts (Fig. 8.9a) in the background of abundant deposition of collagen and other scar ECM components [47]. Although the mechanisms underlying such fibroblast accumulation remain ill defined, a recent elegant study conducted by Liu and coworkers [13] reported reduced apoptosis and increased proliferation in lung fibroblasts cultured in stiff but not soft PAA gels.

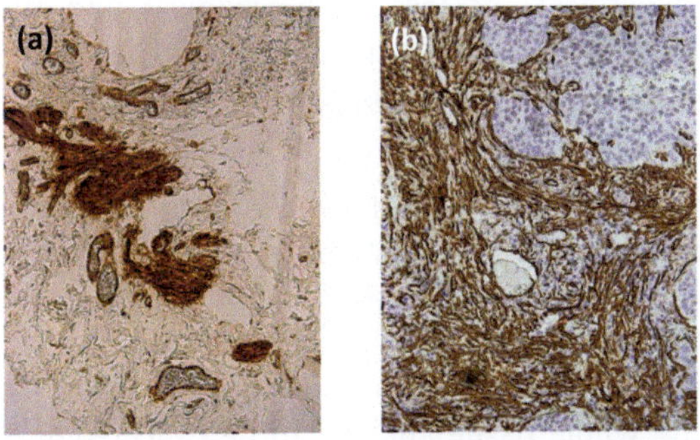


**Figure 8.9** Bright-field images of α-SMA (reddish color) distribution in histologic samples obtained from patients with idiopathic pulmonary fibrosis (a) or lung adenocarcinoma (b). Images were obtained with a 20× objective. Courtesy of A. Giménez and M. Puig.

- *Dysfunctional mechanochemical signaling in IPF*

  IPF is characterized by the up-regulation of oxygen radicals and other fibrogenic soluble factors such as TGF-β1 and angiotensin-II [47]. Concomitantly, a down-regulation of potentially anti-fibrotic factors such as prostaglandin E2 (PGE2) and elastin has been reported [13]. Of note, studies in culture have revealed that matrix stiffening is necessary for TGF-β1-induced fibroblast differentiation into α-SMA-positive myofibroblasts [50], whereas it decreases the expression of the potentially anti-fibrotic pathway COX-2/PGE2 [13]. These findings provide further support of a positive feedback between extrinsic and intrinsic mechanochemical alterations as a major driving mechanism in lung fibrosis.

### 8.6.2 Aberrant Features in Cancer

The stroma of many solid tumors including breast and non-small cell lung cancer (NSCLC) is typically desmoplastic, which is characterized by the overabundance of myofibroblasts, dense deposition of COLI and other fibrotic ECM components, new blood vessel formation and immune cell infiltration [45,46]. Data gathered from different cancer types have revealed that the desmoplastic stroma supports tumor cell growth, invasion, metastasis and even resistance to treatment [51]. Below we provide selected extrinsic and intrinsic mechanochemical alterations in cancer in soft organs that are illustrative of loss of mechanical and morphological homeostasis and are associated with tumor progression.

- *Accumulation of myofibroblasts in breast, lung, and other desmoplastic tumors*

  Myofibroblasts are the most abundant cell type within the stroma of desmoplastic tumors, and their accumulation is a hallmark of most solid tumors including breast and major NSCLC subtypes (Fig. 8.9b). However, the percentage of desmoplastic stroma (and their myofibroblast content) within the whole tissue appears to be cancer-type specific. The mechanical characteristics of tumor-associated myofibroblasts remained ill defined, although they are expected to be comparable to activated fibroblasts from non-malignant tissue.

- *Tissue rigidity increases in mammary tumors and destroys functional and morphological mammary epithelial cell differentiation in culture*

  Most previous studies of the mechanical alterations in tumors have been carried out in the context of breast cancer. These studies revealed that mammary tumors may be ~15–25-fold stiffer than normal mammary tissue both in humans and rodents, as revealed by AFM [14,52]. This tissue hardening has been associated with the overabundance of myofibroblasts and the deposition of fibrillar collagens organized into parallel bundles [28]. Since the latter effects appear to be common in all desmoplastic tumors, similar tissue stiffening is expected in lung cancer and other cancer types. Indeed, these mechanical alterations often enable surgeons to detect tumors by direct palpation [1]. Such dense collagenous stroma has been identified as indicative of poor prognosis and increased invasion [28]. Moreover, it limits the access of soluble therapeutic factors to cancer cells by increasing diffusion hindrance [29,53]. In culture studies we showed that increasing matrix rigidity up to tumor-like values was sufficient to inhibit β-casein expression in MECs [4]. Likewise, in a seminal work carried out by Paszek and coworkers it was found that matrix stiffening prevented 3D cellular organization into acinar-like structures, and enhanced the malignant phenotype of breast cancer cells [14]. Collectively, these findings support that tissue hardening contributes to tumor progression, whereas normal tissue elasticity is protective against it [4].

- *Dysfunctional mechanochemical signaling in cancer*

  Virtually all the relevant molecular players and pathways involved in tissue mechanical homeostasis are deregulated in cancer. With respect to those intrinsic proteins described in the previous section, it has been reported that $\alpha_3\beta_1$ is downregulated in breast, lung and other cancer types [54], and that MMP14 plays a crucial role in cancer cell invasion [15]. In terms of extrinsic factors, a loss of basement membrane integrity has been reported in breast cancer in vivo [22] and in MECs cultured in stiff substrata [14], and deregulation of TGF-β1 signaling is ubiquitous in many cancer types. In terms of cell mechanics, it has been recently shown that a fraction of breast cancer cells become abnormally soft, which may provide them with enhanced invasive capacities [52].

- *Cancer cells exhibit an abnormally large nuclei in NSCLC*

  A hallmark of many cancer cells is altered chromatin structure and organization, which elicits often aberrant nuclear sizes and shapes. In NSCLC, abnormally large nuclei are frequently observed (Fig. 8.10). Based on the studies summarized in Section 8.3, it is tempting to speculate that such large nuclei may contribute to the enhanced proliferative activity by rendering chromatin less condensed, thereby becoming more accessible to the replication machinery. However, testing this hypothesis awaits future investigations.

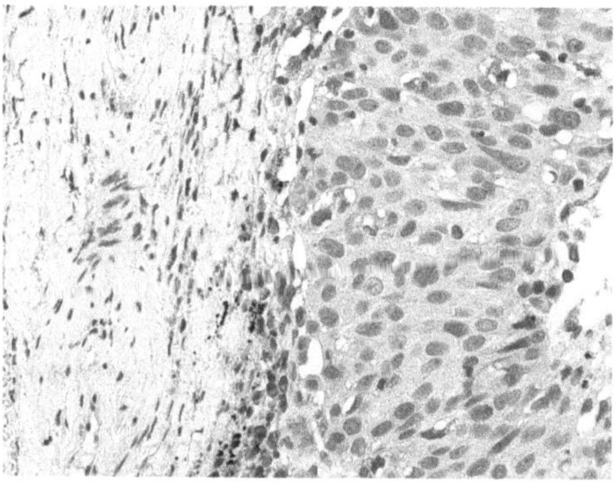

**Figure 8.10** Bright-field image of an hematoxylin-eosin staining of a lung adenocarcinoma tissue sample. Nuclei are stained in blue. Note the large and round nuclei of carcinoma cells in the left side of the image compared to the small and elongated nuclei of stromal cells shown in the right half side. Image courtesy of A. Velásquez.

## 8.7 Translational Implications

The common dependence of fibrosis, cancer and other mechanobiology diseases on abnormal tissue mechanics and/or morphologies raises the possibility that therapeutic strategies aiming to rescue normal tissue mechanoregulation could be applied to these diseases. In support of this hypothesis, an inhibitor of lysyl oxidase-like-2 (LOXL2)—a collagen cross-linker that increases tissue mechanical strength overexpressed in tumors and fibrotic lesions—effectively prevented the progression of cancer and fibrosis in lung, liver and breast in animal models [1,55]. Likewise, an inhibitor of cell-dependent force generation through the actomyosin cytoskeleton was effective against the progression of skin cancer in mice [56]. The recent finding that rescuing normal interstitial fluid pressure in an in vivo model of pancreatic cancer enhanced the delivery of standard chemotherapeutics by preventing the collapse of the microvasculature [53] further illustrates the clinical benefits of targeting abnormal tissue mechanics.

## References

1. Butcher, D. T., Alliston, T., and Weaver, V. M. A tense situation: Forcing tumour progression. *Nat. Rev. Cancer*, 2009. **9**(2), 108–122.
2. Bissell, M. J., and Radisky, D. Putting tumours in context. *Nat. Rev. Cancer*, 2001. **1**(1), 46–54.
3. Nelson, C. M., and Bissell, M. J. Of extracellular matrix, scaffolds, and signaling: Tissue architecture regulates development, homeostasis, and cancer. *Annu. Rev. Cell Dev. Biol.*, 2006. **22**, 287–309.
4. Alcaraz, J., et al. Laminin and biomimetic extracellular elasticity enhance functional differentiation in mammary epithelia. *EMBO J.*, 2008. **27**(21), 2829–38.
5. Alcaraz, J., et al. Collective epithelial cell invasion overcomes mechanical barriers of collagenous extracellular matrix by a narrow tube-like geometry and MMP14-dependent local softening. *Integr. Biol. (Camb.)*, 2011. **3**(12), 1153–1166.
6. Roca-Cusachs, P., et al. Micropatterning of single endothelial cell shape reveals a tight coupling between nuclear volume in G1 and proliferation. *Biophys. J.*, 2008. **94**(12), 4984–4995.
7. Folkman, J., and Moscona, A. Role of cell shape in growth control. *Nature*, 1978. **273**(5661), 345–349.
8. Le Beyec, J., et al. Cell shape regulates global histone acetylation in human mammary epithelial cells. *Exp. Cell Res.*, 2007. **313**(14), 3066–3075.
9. Tan, J. L., et al. Simple approach to micropattern cells on common culture substrates by tuning substrate wettability. *Tissue Eng.*, 2004. **10**(5–6), 865–872.

10. Wang, N., et al. Micropatterning tractional forces in living cells. *Cell Motil. Cytoskeleton*, 2002. **52**(2), 97–106.
11. Alcaraz, J., Nelson, C. M., and Bissell, M. J. Biomechanical approaches for studying integration of tissue structure and function in mammary epithelia. *J. Mammary Gland Biol. Neoplasia*, 2004. **9**(4), 361–374.
12. Alcaraz, J., et al. Microrheology of human lung epithelial cells measured by atomic force microscopy. *Biophys. J.*, 2003. **84**(3), 2071–2079.
13. Liu, F., et al. Feedback amplification of fibrosis through matrix stiffening and COX-2 suppression. *J. Cell Biol.*, 2010. **190**(4), 693–706.
14. Paszek, M. J., et al. Tensional homeostasis and the malignant phenotype. *Cancer Cell*, 2005. **8**(3), 241–254.
15. Mori, H., et al. Transmembrane/cytoplasmic, rather than catalytic, domains of Mmp14 signal to MAPK activation and mammary branching morphogenesis via binding to integrin beta 1. *Development*, 2013. **140**(2), 343–352.
16. Pelham, R. J., Jr., and Wang, Y. Cell locomotion and focal adhesions are regulated by substrate flexibility. *Proc. Natl. Acad. Sci. U. S. A.*, 1997. **94**(25), 13661–13665.
17. Acerbi, I., et al. Integrin-specific mechanoresponses to compression and extension probed by cylindrical flat-ended AFM tips in lung cells. *PLoS One*, 2012. **7**(2), e32261.
18. Roca-Cusachs, P., et al. Integrin-dependent force transmission to the extracellular matrix by alpha-actinin triggers adhesion maturation. *Proc. Natl. Acad. Sci. U. S. A.*, 2013. **110**(15), E1361–E370.
19. Roca-Cusachs, P., et al. Clustering of alpha(5)beta(1) integrins determines adhesion strength whereas alpha(v)beta(3) and talin enable mechanotransduction. *Proc. Natl. Acad. Sci. U. S. A.*, 2009. **106**(38), 16245–16250.
20. Wang, N., and Ingber, D. E. Control of cytoskeletal mechanics by extracellular matrix, cell shape, and mechanical tension. *Biophys. J.*, 1994. **66**(6), 2181–2189.
21. Bissell, M. J., Hall, H. G., and Parry, G. How does the extracellular matrix direct gene expression? *J. Theor. Biol.*, 1982. **99**(1), 31–68.
22. Gudjonsson, T., et al. Normal and tumor-derived myoepithelial cells differ in their ability to interact with luminal breast epithelial cells for polarity and basement membrane deposition. *J. Cell Sci.*, 2002. **115**(Pt 1), 39–50.

23. Streuli, C. H., Bailey, N., and Bissell, M. J. Control of mammary epithelial differentiation: Basement membrane induces tissue-specific gene expression in the absence of cell–cell interaction and morphological polarity. *J. Cell Biol.*, 1991. **115**(5), 1383–1395.

24. Medalia, O., et al. Macromolecular architecture in eukaryotic cells visualized by cryoelectron tomography. *Science*, 2002. **298**(5596), 1209–1213.

25. Ellis, R. J. Macromolecular crowding: Obvious but underappreciated. *Trends Biochem. Sci.*, 2001. **26**(10), 597–604.

26. Maniotis, A. J., Chen, C. S., and Ingber, D. E. Demonstration of mechanical connections between integrins, cytoskeletal filaments, and nucleoplasm that stabilize nuclear structure. *Proc. Natl. Acad. Sci. U. S. A.*, 1997. **94**(3), 849–854.

27. Frantz, C., Stewart, K. M., and Weaver, V. M. The extracellular matrix at a glance. *J. Cell Sci.*, 2010. **123**(24), 4195–4200.

28. Egeblad, M., Rasch, M. G., and Weaver, V. M. Dynamic interplay between the collagen scaffold and tumor evolution. *Curr. Opin. Cell Biol.*, 2010. **22**(5), 697–706.

29. Galgoczy, R. et al., A spectrophotometer-based diffusivity assay reveals that diffusionhindrance of small molecules in extracellular matrix gels used in 3D cultures is dominated by viscous effects. *Colloids and Surfaces B: Biointerfaces*, 2014. **120**:200–207

30. Raeber, G. P., Lutolf, M. P., and Hubbell, J. A. Molecularly engineered PEG hydrogels: A novel model system for proteolytically mediated cell migration. *Biophys. J.*, 2005. **89**(2), 1374–1388.

31. Deng, L., et al. Fast and slow dynamics of the cytoskeleton. *Nat. Mater.*, 2006. **5**(8), 636–640.

32. Luque, T., et al. Local micromechanical properties of decellularized lung scaffolds measured with atomic force microscopy. *Acta Biomater.*, 2013. **9**(6), 6852–6859.

33. Roca-Cusachs, P., Iskratsch, T., and Sheetz, M. P. Finding the weakest link: Exploring integrin-mediated mechanical molecular pathways. *J. Cell Sci.*, 2012. **125**(13), 3025–3038.

34. Kong, F., et al. Demonstration of catch bonds between an integrin and its ligand. *J. Cell Biol.*, 2009. **185**(7), 1275–1284.

35. Arias-Salgado, E. G., et al. Specification of the direction of adhesive signaling by the integrin beta cytoplasmic domain. *J. Biol. Chem.*, 2005. **280**(33), 29699–29707.

36. del Rio, A., et al. Stretching single talin rod molecules activates vinculin binding. *Science*, 2009. **323**(5914), 638–641.

37. Roca-Cusachs, P., et al. Rheology of passive and adhesion-activated neutrophils probed by atomic force microscopy. *Biophys. J.*, 2006. **91**(9), 3508–3518.

38. Akhtar, N., and Streuli, C. H. Rac1 links integrin-mediated adhesion to the control of lactational differentiation in mammary epithelia. *J. Cell Biol.*, 2006. **173**(5), 781–793.

39. Singer, A. J., and Clark, R. A. Cutaneous wound healing. *N. Engl. J. Med.*, 1999. **341**(10), 738–746.

40. Hinz, B., Masters and servants of the force: The role of matrix adhesions in myofibroblast force perception and transmission. *Eur. J. Cell Biol.*, 2006. **85**(3–4), 175–181.

41. Tomasek, J. J., et al. Myofibroblasts and mechano-regulation of connective tissue remodelling. *Nat. Rev. Mol. Cell Biol.*, 2002. **3**(5), 349–363.

42. Chen, J., et al. Alpha-smooth muscle actin expression enhances cell traction force. *Cell Motil. Cytoskeleton*, 2007. **64**(4), 248–257.

43. Zhao, X. H., et al. Force activates smooth muscle alpha-actin promoter activity through the Rho signaling pathway. *J. Cell Sci.*, 2007. **120**(Pt 10), 1801–1809.

44. Ingber, D. E., Mechanobiology and diseases of mechanotransduction. *Ann. Med.*, 2003. **35**(8), 564–577.

45. Dvorak, H. F., Flier, J., and Frank, H. Tumors: Wounds that do not heal: Similarities between tumor stroma generation and wound-healing. *N. Eng. J. Med.*, 1986. **315**(26), 1650–1659.

46. Radisky, D. C., Kenny, P. A., and Bissell, M. J. Fibrosis and cancer: Do myofibroblasts come also from epithelial cells via EMT? *J. Cell Biochem.*, 2007. **101**(4), 830–839.

47. King, T. E., Jr., Pardo, A., and Selman, M. Idiopathic pulmonary fibrosis. *Lancet*, 2011. **378**(9807), 1949–1961.

48. Ebihara, T., et al. Changes in extracellular matrix and tissue viscoelasticity in bleomycin-induced lung fibrosis. Temporal aspects. *Am. J. Respir. Crit. Care Med.*, 2000. **162**(4 Pt 1), 1569–1576.

49. King, T. E., Jr., et al. Predicting survival in idiopathic pulmonary fibrosis: Scoring system and survival model. *Am. J. Respir. Crit. Care Med.*, 2001. **164**(7), 1171–1181.

50. Arora, P. D., Narani, N., and McCulloch, C. A. The compliance of collagen gels regulates transforming growth factor-beta induction of alpha-smooth muscle actin in fibroblasts. *Am. J. Pathol.*, 1999. **154**(3), 871–882.

51. Beacham, D. A., and Cukierman, E. Stromagenesis: The changing face of fibroblastic microenvironments during tumor progression. *Semin. Cancer Biol.*, 2005. **15**(5), 329–341.

52. Plodinec, M., et al. The nanomechanical signature of breast cancer. *Nat. Nanotechnol.*, 2012. **7**(11), 757–765.

53. Provenzano, P. P., et al. Enzymatic targeting of the stroma ablates physical barriers to treatment of pancreatic ductal adenocarcinoma. *Cancer Cell*, 2012. **21**(3), 418–429.

54. Tsuji, T., Physiological and pathological roles of alpha 3 beta 1 integrin. *J. Membr. Biol.*, 2004. **200**(3), 115–132.

55. Barry-Hamilton, V., et al. Allosteric inhibition of lysyl oxidase-like-2 impedes the development of a pathologic microenvironment. *Nat. Med.*, 2010. **16**(9), 1009–1017.

56. Samuel, M. S., et al. Actomyosin-mediated cellular tension drives increased tissue stiffness and beta-catenin activation to induce epidermal hyperplasia and tumor growth. *Cancer Cell*, 2011. **19**(6), 776–791.

# Chapter 9

# Muscle Contraction and Sarcomere Length Non-Uniformities

**Walter Herzog**

*Faculties of Kinesiology, Engineering, Medicine and Veterinary Medicine, The University of Calgary, Calgary, Canada*

wherzog@ucalgary.ca

## 9.1 Background

Muscle contraction has fascinated scientists for centuries [1] but had eluded satisfactory explanation until the microstructure of muscle assembly became apparent, and the role of contractile filaments had been elucidated in the 1950s. Prior to the 1930s, muscle contraction was almost universally thought to be linked to the production of lactic acid, but this idea was rejected when Lundsgaard [3] discovered that perfectly good contractions could be achieved when a muscle was poisoned with iodoacetic acid, which prevented lactic acid production. The role of lactic acid was then replaced by the breakdown of phosphorylcreatine (PCr), which was thought to trigger the folding of long strands of filaments in some unexplained way, thereby producing muscle shortening and contraction. However, with the discovery that PCr merely caused the re-phosphorylation of ADP to ATP, PCr hydrolysis

*Cells, Forces, and the Microenvironment*
Edited by Charles M. Cuerrier and Andrew E. Pelling

ISBN 978-981-4613-36-1 (Hardcover), 978-981-4613-37-8 (eBook)
www.panstanford.com

became a secondary reaction and protein folding was thought to be caused by the hydrolysis of ATP to ADP and inorganic phosphate [5].

A revolution in muscle mechanics occurred when it was discovered that contraction and force production was not caused by the folding/shortening of long filaments, but was caused by the relative sliding of two sets of filaments relative to each other [6,7]. These two sets of filaments were the thick (myosin) and thin (actin) filaments, which are composed primarily of myosin and actin proteins, respectively. These filaments made up the two-filament structure of sarcomeres, which is typically shown in textbooks of muscle physiology (Fig. 9.1). Sarcomeres are bound by the Z-bands at either end and are considered the smallest independent structural units for skeletal and cardiac muscle contraction, a concept we will critically examine below. Andrew Huxley [8] then proposed a first theory of how the relative sliding of thin past thick filaments was supposed to occur. This is now referred to as the cross-bridge theory, and although many details of this theory have been changed in the past half century, the basic assumptions of the initial theory have been retained to this day, including the two-contractile-filament structure of the sarcomere.

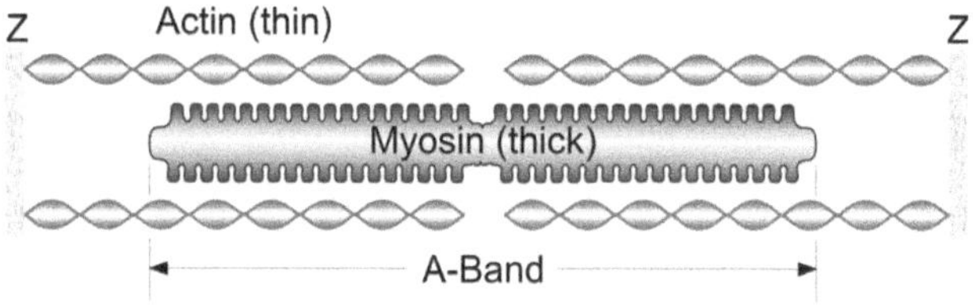


**Figure 9.1** Schematic two-filament structure of a sarcomere. Actin (thin) and myosin (thick) filaments make up the contractile proteins of the sarcomere and are thought to be the sole force regulatory proteins in skeletal muscle.

## 9.2 The Cross-Bridge Theory

The idea for the cross-bridge theory of muscle contraction had been introduced first in the early 1950s by Hugh Huxley [9] but was explained in detail, and formulated mathematically in its first inception by Andrew Huxley [8]. Huxley assumed that the thick (myosin) filaments contained side-pieces (cross-bridges) that interacted cyclically with the thin (actin) filaments and pulled the thin past the thick filaments. These cross-bridges were assumed to be uniformly arranged on the thick filaments and were supposed to interact with uniformly distributed, specialized attachment sites on the thin filaments. Cross-bridges were attached to the filament backbone by a linear elastic element, and cross-bridges were thought to move about their equilibrium position by thermal agitation (Brownian motion). Attachment of cross-bridges to actin was assumed to occur spontaneously and in accordance with rate constants based on the position of the cross-bridge equilibrium position relative to its nearest eligible attachment site on actin (Fig. 9.2a). One attachment/detachment cycle was associated with

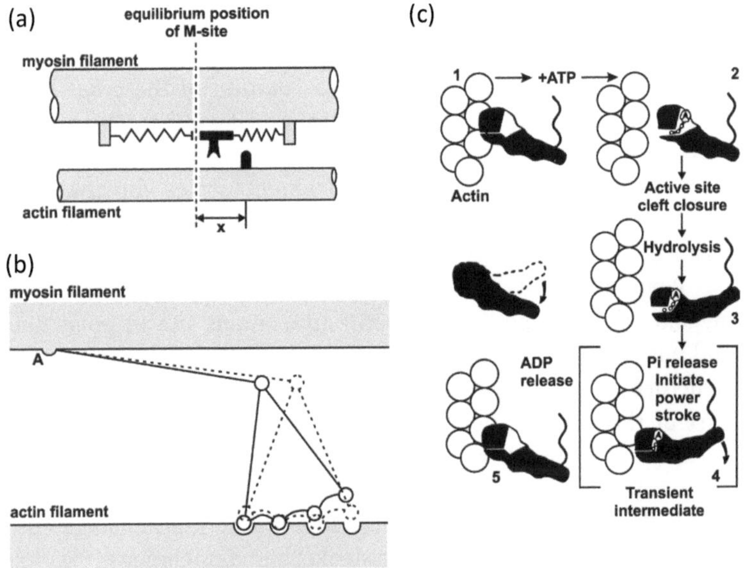


**Figure 9.2** Schematic illustration of three iterations of the cross-bridge theory. (a) Huxley's (1957) initial proposal of a two-state cross-bridge model; (b) Huxley and Simmons' (1971) rotating cross-bridge model with multiple attachment sites; and (c) Rayment et al.'s (1993) multi-state cross-bridge model with associated ATP hydrolysis states matching the mechanical attachment and detachment configurations of the cross-bridge cycle. Adapted from Huxley, 1957, Huxley and Simmons, 1971, and Rayment et al., 1993; reproduced with permission.

the hydrolysis of a single molecule of ATP. The rate function for attachment and detachment of cross-bridges to the thin filaments were derived by Huxley such that they reproduced what he considered the best available data on muscle contraction at that time, Hill's force–velocity relationship [10].

Refinements of the cross-bridge theory occurred over the past half century, but the basic idea has not been challenged seriously, despite some obvious shortcomings of the predictions of the theory, as we shall discuss below. Likely the most profound change in the cross-bridge theory was proposed by H. Huxley [11] when he suggested that cross-bridge and actin interactions could not occur in a stereotypical manner because of the changes in spacing between thin and thick filaments with changing sarcomere lengths. Huxley (1969) proposed that cross-bridges were tethered to the thick filament backbone through molecular joints that allowed the cross-bridges to move away from the thick filament and apply force on the thin filament through a rotation of the cross-bridge rather than through a linear pulling action (Fig. 9.2b). This rotation was thought to occur in steps of decreasing potential energies, was associated with multiple attachment states for the cross-bridges on the thin filament, and was expressed in mathematical terms by A. Huxley two years after H. Huxley's initial proposal [12].

Rayment et al. [13] described the atomic structure of the myosin cross-bridge and the actin attachment site in great detail and with an accuracy of approximately 4 Å (0.4 nm). Based on this structural detail, they proposed a cross-bridge action whereby one part of the cross-bridge head was fixed in a given configuration on the thin filament, while the remaining part of the cross-bridge head rotated and produced muscle contraction. In their theory, ATP hydrolysis was tightly associated with the mechanical steps of cross-bridge attachment, power stroke, and detachment (Fig. 9.2c). Although more refined cross-bridge models have been proposed in the meantime, Rayment's theory represents the general consensus of the scientific community and is the model of choice (or close to it) for most researchers in the field.

## 9.3 Limitations of the Cross-Bridge Theory

Although the cross-bridge theory makes good predictions for experimental results obtained for isometric (constant muscle length) and concentric (muscle shortening) contractions, forces, stiffness, and energetics for eccentric (muscle stretching) contractions are not well predicted. Huxley (1957) acknowledged this shortcoming of his theory when predicting forces and energy requirements during eccentric contractions that were unrealistically high. Although suggestions have been made to remedy this limitation of cross-bridge models, no structurally feasible analog to the cross-bridge model or associated mathematical framework has received more than just isolated support. In his classic book *Reflections on Muscle*, A. Huxley [14] remarks on the shortcomings of the cross-bridge theory in eccentric contractions. He suggested that eccentric contractions are quite different from concentric and isometric contractions, that there might be as of yet undiscovered structural elements that provide stability to eccentric contractions and prevent muscle damage, and that there will be surprises in store for those who try to understand the molecular workings of actively lengthening muscle. All these predictions made more than 30 years ago have come true and will be discussed below, although some of the detailed workings of eccentric contractions still remain a mystery.

However, many scientists ignored Huxley's (1980) predictions that eccentric contractions were inherently different from isometric/concentric contractions and that special features might account for the observed properties of actively lengthening muscles. These scientists continued to use the cross-bridge theory for eccentric contractions but added theories of unstable sarcomere length and force behavior to explain eccentrically induced muscle injuries, muscle creep, and the permanent force enhancement observed following actively lengthening. I would like to describe, discuss, and evaluate the basic assumptions of sarcomere instability and non-uniformity theories and scrutinize the experimental evidence upon which these theories are based. For such an evaluation, it will first be necessary to understand the force–length relationship of skeletal muscle contraction and its role within the cross-bridge theory.

## 9.4 The Force–Length Relationship

It has been known for a long time that the force a muscle can exert depends on its length (e.g., [15]). However, the reason for these length-dependent changes in force was only identified following the discovery of the sliding filament and cross-bridge theories. Gordon et al. [16] described the sarcomere force–length relationship based on experiments on isolated frog fibers. The relationship had an ascending and descending limb (Fig. 9.3) connected by a plateau region. Although the ascending limb portion of the relationship could not be explained directly with cross-bridge thinking, the plateau and descending limb of the force–length relationship perfectly matched the predictions of the cross-bridge theory. Specifically, the maximal, isometric, steady-state force of a sarcomere was predicted to be linearly related to the amount of thick and thin filament overlap, which in turn was directly related to sarcomere length. When overlap between myofilaments decreased, so did the proportion of cross-bridges that could be formed between actin and myosin, and thus the corresponding force. The experimental relationship found in single fibers matched this expectation, showing a linear decrease in force with sarcomere lengths (i.e., loss of myofilament overlap—Fig. 9.3).

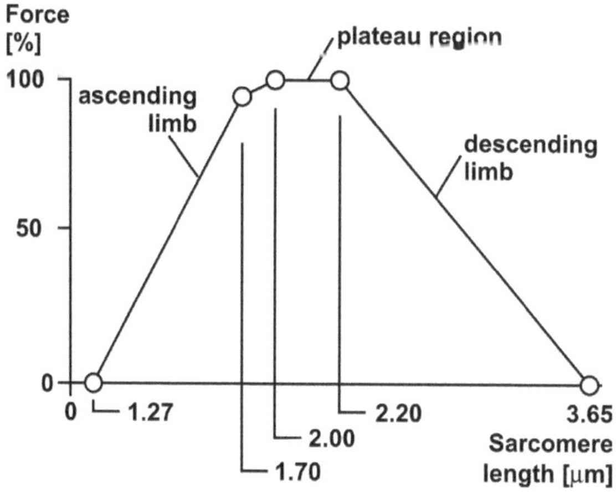


**Figure 9.3** Sarcomere force–length relationship as first described by Gordon et al. (1966). The plateau and descending limb regions are perfectly explained based on the cross-bridge theory. However, the descending limb, with its negative slope, was often interpreted (incorrectly) as a region of negative muscle stiffness, and thus was likened to a "softening" engineering material that was thought inherently unstable [2]. Adapted from Gordon et al., 1966; reproduced with permission.

The descending limb of the force–length relationship has a negative slope; that is, the maximal, isometric, steady-state force decreases with increasing sarcomere (fiber or muscle) length. On the surface of it, this seems to be equivalent to a softening material in engineering, which has a negative stiffness (slope of the force–length relationship) and thus exhibits unstable behavior. Indeed, Hill [17] was among the first to describe that segments of muscles on the descending limb of the force–length relationship cannot be stable, and length changes would occur. This idea was taken up by many scientists who tacitly assumed that muscle had negative stiffness, would be unstable, and would tear itself apart when stretched onto the descending limb of the force length relationship (e.g., [2,18]). Thus, the idea of instability, and the associated development of sarcomere length non-uniformities on the descending limb of the force–length relationship, was born. This instability and sarcomere length non-uniformity was thought to be the primary cause for many properties of skeletal muscle, including the so-called creep behavior [16], residual force enhancement [19] and stretch-induced damage and loss of force [20,21]. Before commenting on these properties in more detail, the concept of instability will be discussed below.

## 9.5 Instability on the Descending Limb of the Force–Length Relationship

As shown in Fig. 9.3, the descending limb of the force–length relationship has a negative slope. This negative slope has been associated with reflecting a negative stiffness of muscle in that part of the force–length relationship. If this was the case, and a muscle that is actively stretched would indeed become progressively weaker, this property would correspond to a softening material and it would be unstable [22]. However, before examining this issue further, a more philosophical question needs to be answered: Why would a universal motor for muscle contraction evolve in such a way that it is unstable over approximately 60% of its working range? It does seem counter-intuitive that nature would produce such blunder, and this might explain Huxley's (1980) foresight when he suggested that we really do not understand eccentric contractions on the descending limb of the force–length relationship, and that there are surprises in store.

Nevertheless, let's have a closer look at the question of instability. It seems that one of the great misconceptions in muscle mechanics is the idea that because the slope of the descending limb of the force–length relationship is negative (a muscle or sarcomere becomes weaker with increasing length), its stiffness is negative too, and consequently the muscle behaves like a softening material. However, this assumption is incorrect. When an active muscle is stretched on the descending limb of the force–length relationship its force increases, thereby exhibiting positive stiffness, or an increase in force with increasing muscle length (e.g., [23]). Furthermore, following active elongation, a muscle's force is substantially higher than the corresponding isometric force at that same length. This result, referred to as residual force enhancement, has been observed on all structural levels of muscle, such as, in vivo human skeletal muscles activated voluntarily or electrically [24], isolated muscle preparations [25], single fibers [23,26], single myofibrils [4], single sarcomeres [27], and isolated half sarcomeres [4]. The misunderstanding of the descending limb of the force–length relationship has arisen because the static force–length property of muscles had been assumed to reflect its dynamic properties [17,18].

## 9.6 Sarcomere Length Non-Uniformity

Another misconception associated with instability on the descending limb of the force–length relationship is the notion that sarcomeres become non-uniform in length and force and thus cannot exist and be stable next to each other on the descending limb of the force–length relationship. Although it is well established that sarcomeres in muscles [28], fibers [29] and myofibrils [4] are non-uniform in length, when stretched onto the descending limb of the force–length relationship the anticipated "quick pulling apart" of sarcomeres [2], sometimes referred to as "popping" of sarcomeres [30] or "overextension" [31] does not seem to occur. Myofibrils activated on the descending limb of the force–length relationship, or pulled actively onto that limb, exhibit perfectly stable lengths of sarcomeres, even though they are by no means uniform in length [32] (Fig. 9.4). In fact, sarcomeres (and half sarcomeres) in fibers and myofibrils seem more stable [23] and more uniform [4] when actively stretched onto the descending limb of the force–length relationship.

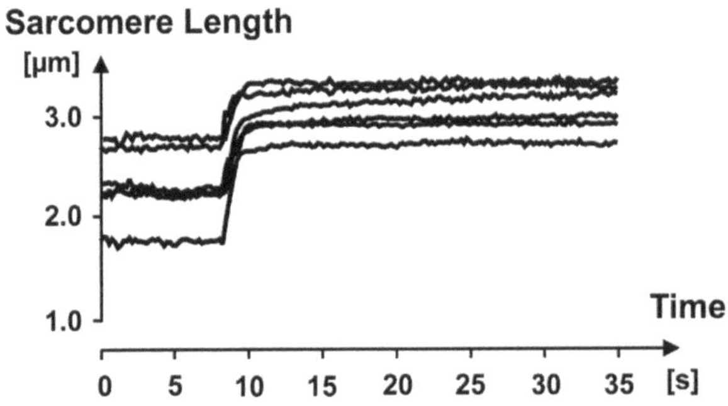


**Figure 9.4** Sarcomere lengths vs. time trace for all sarcomeres of a single, activated myofibril stretched onto the descending limb of the force-length relationship. Note the non-uniformity in length prior to stretching, the stretching of all sarcomeres when the myofibril is stretched, and the stability (constancy) of sarcomere lengths following stretching when all sarcomeres are at a length corresponding to the descending limb of the force–length relationship. Rabbit psoas myofibril at 21°C.

## 9.7 Muscle Injury

Sarcomere instability, and the associated development of sarcomere length non-uniformities, with sarcomeres rapidly pulled beyond actin-myosin filament overlap (popped, overextended), has been thought to be the primary cause for the loss of force following damaging eccentric muscle contractions. This thinking is based on histological evidence of sarcomere disruption in actively stretched muscles compared to control muscles that were not stretched, but were activated isometrically or concentrically [21,33,34]. These studies were performed in entire muscle or single fiber preparations where sarcomeres are well embedded in series and in parallel with other sarcomeres through structural proteins, such as desmin, titin, and nebulin. This arrangement of literally thousands of sarcomeres gives rise to a mathematically redundant system in which muscle force cannot be associated with a single sarcomere, or even a group of sarcomeres. Thus, this type of evidence is statistical (more sarcomere disruption is associated with greater loss in muscle force) [21] and thus cannot be used to imply causality. However, in a single myofibril preparation, where each and all sarcomeres are strictly arranged in series, force measured at the end of a myofibril can be directly related to the force in each of the sarcomeres and its associated length. When performing damaging eccentric contractions in single myofibrils, Panchangam [31] made the surprising observation that myofibrils with overextended sarcomeres had smaller losses in force compared to myofibrils without overextended sarcomeres, suggesting, if anything at all, that sarcomere overextension protected a muscle from stretch-induced force loss rather than causing it. This first-ever study on damage in isolated myofibrils will need careful examination through further experiments, but it does question the idea that sarcomere overextension is equivalent to "muscle damage" and that sarcomere overextension is the primary cause for loss of muscle force following damaging eccentric muscle contractions.

## 9.8 Residual Force Enhancement and Sarcomere Length Non-Uniformity

In the following, we will discuss the force enhancement property of skeletal muscles in some detail. Residual force enhancement, or force enhancement for short, is insofar an intriguing property of skeletal muscle contraction as it cannot be explained with current cross-bridge models [35]. Therefore, understanding the molecular basis of residual force enhancement may help in refining, or reformulating, how muscles contract. It is even more intriguing, that despite systematic evidence of residual force enhancement prior to the first formulation of the cross-bridge theory [25], it was ignored at that initial stage [8] and in future developments of the theory (e.g., [11–13]). Finally, and this might be the most intriguing part of the story, force enhancement has been explained for the past several decades with sarcomere instability and associated development of sarcomere length non-uniformities, without any direct evidence [18,19,30,36,37].

Residual force enhancement has been observed consistently since the early 1950s [25]. It is formally defined as the increase in steady-state, isometric force following active lengthening of a muscle compared to the corresponding isometric force not preceded by stretch. Force enhancement has been observed on all structural levels, from in vivo muscles [24] to mechanically isolated sarcomeres [27] (Fig. 9.5). Force enhancement is known to increase with increasing stretch magnitude [23,25], at least until a threshold is reached [38], does not depend on the speed of stretch [39,40], is greater on the descending than the ascending limb of the force–length relationship [36], can be abolished instantaneously by deactivating a muscle [25,36], and is not associated with a substantial change in stiffness [41]. In 2002, we also discovered that force enhancement is associated with a passive structural element and proposed that force enhancement may have two independent components; an active and a passive component [42].

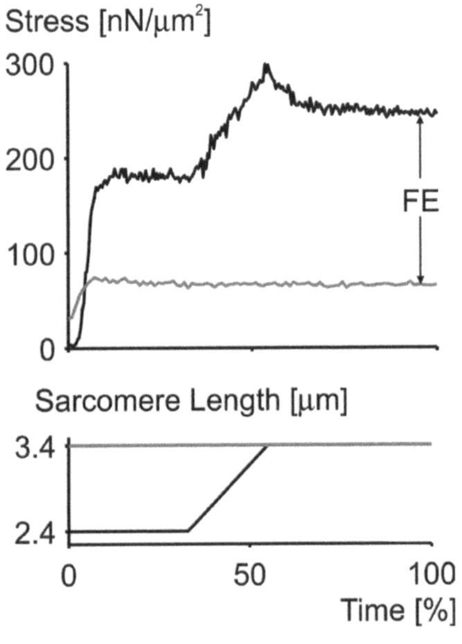


**Figure 9.5** Force (stress) and sarcomere length as a function of time for a single, mechanically isolated sarcomere stretched from 2.4 to 3.4 µm. Note the substantial residual force enhancement (FE) and also the enhancement of force beyond the initial isometric force at optimal sarcomere length (2.4 µm).

According to the sarcomere instability and non-uniformity theory, force enhancement can only occur in parts of the force–length relationship where sarcomere lengths are unstable; the descending limb [2,17,36], and it contains the following basic assumptions:

(i) Purely isometric contractions occur with essentially uniform sarcomeres

(ii) When actively lengthened onto the descending limb of the force–length relationship, sarcomeres become unstable and develop length non-uniformities with a small number of sarcomeres becoming overstretched (popped).

(iii) The overstretched sarcomeres can only produce passive force and this passive force matches the active forces of the non-popped sarcomeres

(iv) Non-popped sarcomeres are merely stretched by a small amount (or not at all), and thus remain at about their pre-stretch length

(v) The shorter length of the non-popped sarcomeres provides greater actin-myosin filament overlap than the overlap of the essentially uniform sarcomeres of the isometric reference contraction, and this increased overlap provides the enhanced force (Fig. 9.6).

The primary predictions of the sarcomere length instability theory are as follows:

(i) Force enhancement cannot occur on the stable, ascending limb of the force length relationship as sarcomere length non-uniformities would not develop there.

(ii) Force in the enhanced state cannot exceed the maximal isometric forces achieved at optimal sarcomere lengths (on the plateau of the force–length relationship).

(iii) Sarcomeres in the enhanced state have two distinct lengths: the popped sarcomeres are at a length where actin-myosin filament overlap has ceased to exist, and the non-popped sarcomeres are at approximately their initial pre-stretch length (Fig. 9.6).

(iv) Isometric reference contractions are associated with approximately uniform sarcomere lengths.

(v) The passive forces of the overextended group of sarcomeres match the active forces of the non-overextended sarcomeres.

(vi) Force enhancement cannot occur in a single sarcomere, since, by definition a single sarcomere cannot be non-uniform with other sarcomeres (although the special case of half-sarcomere non-uniformity will be discussed below).

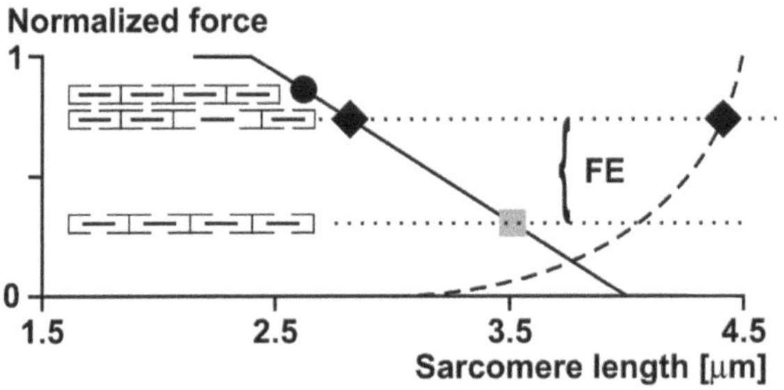


**Figure 9.6** Schematic illustration of residual force enhancement (FE) according to the sarcomere length non-uniformity theory. The solid line represents the plateau region and the descending limb of the force–length relationship. The dashed line represents the passive force-extension curve of the sarcomeres. Finally, the dotted lines represent the expected reference force (lower line) for uniform sarcomeres in an isometric contraction and the enhanced force following stretch (upper line) when sarcomeres are non-uniform. The black circle and grey square indicate two purely isometric contractions with essentially uniform sarcomeres on the descending limb of the force–length relationship (solid line). The two black diamonds indicate the two sarcomere lengths when the myofibril is stretched actively from the short to the long lengths. Because of the presumed instability of sarcomere forces and lengths on the descending limb of the force–length relationship, sarcomeres are assumed to be divided into two groups of distinct lengths, as shown. The resulting non-uniformity in sarcomere lengths then gives the enhanced force compared to the isometric force because of the increased overlap for some of the sarcomeres and the corresponding passive force for the "popped" sarcomeres.

All predictions associated with this theory have been shown to be violated.

(i) Force enhancement has been shown to occur on the ascending limb of the force–length relationship in whole muscle (e.g., [25,36]), single fibers [43], and sarcomeres and half sarcomeres [4,44].

(ii) Forces in the enhanced state have been shown to exceed the purely maximal isometric reference forces on the plateau of the force–length relationship in muscles [45], fibers [43,46], myofibrils, and single sarcomeres [27,47].

(iii) There is no evidence of two distinct groups of sarcomere lengths following active stretching and force enhancement [48] (Fig. 9.4).

(iv) Isometric contractions have been associated with substantial sarcomere length non-uniformities in muscles [28], single fibers [29] and isolated myofibrils [4,32].

(v) Forces in isolated sarcomeres cannot be measured in fiber or muscle preparations, so this prediction cannot be tested easily. However, there have been no reports of two distinct populations of sarcomere lengths following active muscle stretching; therefore this prediction becomes obsolete.

(vi) We demonstrated in a first-ever study on mechanically isolated sarcomeres that force enhancement is a property of single sarcomeres and that forces in the enhanced state can exceed the corresponding purely isometric reference forces at optimal sarcomere lengths by as much as 50% [27].

## 9.9 The Smallest Contractile Unit of Muscle Contraction

Textbooks of muscle physiology will state that the sarcomere is the smallest independent structural unit of muscle contraction. Therefore, demonstrating that force enhancement occurs in a single sarcomere [27] should have theoretically eliminated the possibility for sarcomere length instability and non-uniformity to play a role in this finding. However, Telley [49] argued that not the sarcomere, but the half-sarcomere, should be considered the smallest independent structural unit for muscle force production. Theoretical work on half sarcomere dynamics then confirmed that force enhancement was perfectly possible using just a single sarcomere preparation, if one allows for the individual halves of the sarcomere to adopt different lengths [50]. The possibility of half-sarcomere non-uniformities playing a major role in residual force enhancement was reignited in a topical review by Edman [19], and associated letters to the editor [51,52].

The role of half-sarcomere dynamics and non-uniformity became even more confused when it was argued recently that *"stretch-induced non uniformities of half-sarcomere lengths ... significantly increase the level of force enhancement"* [53]. These authors showed results of half-sarcomere non-uniformities in isolated sarcomeres and myofibrils from rabbit psoas muscles following active lengthening and for purely isometric reference contractions. The confusion arises from two results in their article that seem to defy their conclusion statement above. The first result deals with the half sarcomere non-uniformities obtained in isometric contractions and after active lengthening. There do not seem to be stretch-induced non uniformities of half-sarcomere lengths. If anything at all, half sarcomere length non uniformities seemed reduced after stretch compared to the isometric reference contractions, in agreement with previous findings in the same preparation [4] and in single fibers [23].

The second result deals with the amount of force enhancement as a function of half sarcomere non uniformity, as quantified by A-band shifts (Fig. 9.7). The regression line obtained by Rassier and Pavlov (2012) can be approximated by the following equation:

$$\text{Force enhancement } [\%] = \text{A-band displacement } [\text{nm}] + 8$$

Therefore, force enhancement is about 8% in the absence of any A-band shift (or half sarcomere length non-uniformity) and is about 20% for the greatest measured A-band shift of 12 nm measured in a single sarcomere. However, an A-band shift of 12 nm corresponds to an increase in myofilament overlap of about 1.5% in a half sarcomere, which corresponds to an increase in force of 1.5%. However, the experimentally measured force enhancement was 20%, thus only a tiny amount of the observed force enhancement could be explained theoretically with the increased myofilament overlap due to half sarcomere length non-uniformity. In summary, the lack of increased half sarcomere length non-uniformity with active muscle lengthening compared to isometric reference contractions, and the negligible change in active force arising from the observed half sarcomere length non uniformities suggest that (half-) sarcomere length non uniformities likely play no role, or at best a marginal role, in residual force enhancement.

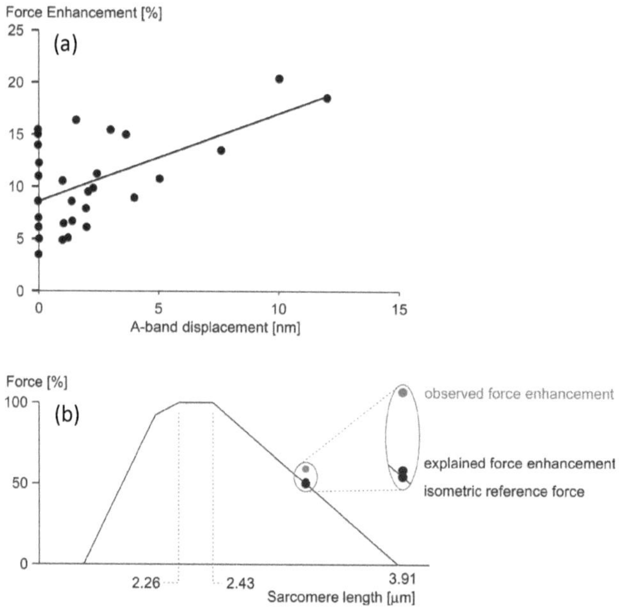


**Figure 9.7** (a) Force enhancement as a function of A-band shifts in a single sarcomere. The maximal A-band shift observed (about 12 nm) would produce an increase in myofilament overlap and associated increase in force of approximately 1.5%, while the measured increase in force (force enhancement) was about 20% (b). Therefore, even if one assumed that half sarcomere lengths are perfectly uniform for purely isometric contractions (which they are not [4]), the A-band shift would only explain a tiny amount of the measured force enhancement, as indicated in (b). Blue dot = isometric reference force; purple dot = force enhancement that can be explained by the maximal A-band shift observed experimentally in (a); brown dot = observed force enhancement in (a). Figure a from Pavlov and Rassier, 2012; reproduced with permission.

What then is the smallest independent unit of muscle contraction? Since serially aligned sarcomeres are perfectly stable on the descending limb of the force–length relationship, despite great differences in lengths (Fig. 9.4), it appears that sarcomeres are connected across the z-lines and re-enforced across their centers. Therefore, the smallest independent unit of muscle contraction becomes "several" sarcomeres: not one sarcomere and not a half sarcomere. If we accept this notion, then the idea of sarcomere length non uniformities becomes irrelevant as one sarcomere in a myofibril is linked to its neighbors by distinct structures crossing the z-line. Suggestions of such linkages across the z-line include actin and titin [54], and linkages across the vulnerable center of sarcomeres can easily be envisaged by passive reinforcement of the central A-band region by titin binding to actin [40,55] (Fig. 9.8).

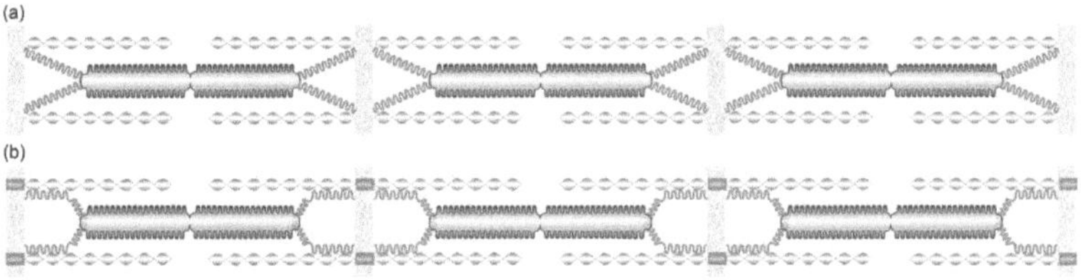


**Figure 9.8** Schematic illustration of serial sarcomeres that are independent force generators (a), and serial sarcomeres that are interconnected force generators (b). In the interconnected force generator scheme, it is assumed that sarcomeres are connected to each other across the z-line by structural proteins (e.g., actin) that transmit force from one sarcomere to its neighbors. Furthermore, half sarcomere instability in this scheme is prevented by titin binding to actin, thereby reinforcing the "vulnerable" myosin filament from being moved far from its central sarcomere location.

## 9.10 Conclusions

It is an accepted fact that sarcomeres are non-uniform in lengths in muscles, single fibers, and isolated myofibrils. Based on Hill's (1953) instability concept, some scientists have suggested that sarcomere length non-uniformities occur primarily (exclusively) on the descending limb of the force–length relationship and develop primarily (exclusively) when a muscle or fiber is actively lengthened. This suggestion is likely too simplistic as there is ample evidence in muscles, fibers and myofibrils that sarcomeres are non-uniform in length even on the ascending limb of the force–length relationship, in the passive state and for isometric contractions [4,28,29,48].

Sarcomere length non-uniformities have also been thought to be the cause for loss of force following damaging eccentric exercise, for the creep behavior of muscle, and for the residual force enhancement and force depression [16,18,19,30,36,37] observed in dynamically contracting muscles. Most of the research on these topics has been performed on muscle or single-fiber preparations, which have the limitation that lengths of single sarcomeres cannot be measured, that sarcomeres are arranged in a complex and mechanically redundant manner, and that forces in muscles and fibers cannot be related to the contractile state of individual sarcomeres. All these limitations can be overcome when using single myofibrils, which are made up of serially arranged sarcomeres whose lengths and force can be uniquely and accurately related at any instant in time [56]. Research on single myofibrils, single sarcomeres, and mechanically isolated half-sarcomeres suggests the following:

- Sarcomeres and half sarcomeres are naturally non-uniform in length in passive preparations, in isometric contractions and in actively lengthened preparations.
- Sarcomere length non-uniformities do not increase with active lengthening beyond that observed for isometric contractions.
- Sarcomeres and half sarcomeres are perfectly stable on the descending limb of the force–length relationship.
- Loss of force in eccentrically damaged muscle is reduced in preparations with overextended (popped) sarcomeres compared to preparations without overextended sarcomeres, suggesting that overextension may be a mechanism to limit loss of muscle force, rather than causing it.
- Force enhancement occurs in the absence of sarcomere or half sarcomere length non-uniformities.
- Force enhancement is associated with (half) sarcomere length non-uniformities that are similar or smaller than those observed in purely isometric reference contractions.
- Even in the absence of non-uniformities in isometric reference contractions, (half) sarcomere non-uniformities and associated increases in myofilament overlap, can only explain a minute fraction of the total force enhancement observed experimentally.

From these observations, we suggest that sarcomere length non-uniformities are a normal occurrence of muscle contraction with no particular effect on instability, muscle injury, force depression, or force enhancement properties in skeletal muscles.

## References

1. Needham, D. M. *Machina Carnis*. Cambridge: Cambridge University Press; 1971.
2. Zahalak, G. I. Can muscle fibers be stable on the descending limbs of their sarcomere length-tension relations? *J. Biomech.*, 1997; **30**, 1179–1182.
3. Lundsgaard, E. Untersuchungen über Muskelkontraktionen ohne Milchsäurebildung. *Biochem. Z.*, 1930; **217**, 162–177.
4. Joumaa, V., Leonard, T. R., Herzog, W. Residual force enhancement in myofibrils and sarcomeres. *Proc. R. Soc. B*, 2008; **275**, 1411–1419.
5. Huxley, A. F. Cross-bridge action: Present views, prospects, and unknowns. In: *Skeletal Muscle Mechanics: From Mechanisms to Function*. (Herzog, W., ed.), West Sussex, UK: John Wiley & Sons, Ltd., 2000. pp. 7–31.
6. Huxley, A. F., Niedergerke, R. Structural changes in muscle during contraction. Interference microscopy of living muscle fibres. *Nature*, 1954; **173**, 971–973.
7. Huxley, H. E., Hanson, J. Changes in cross-striations of muscle during contraction and stretch and their structural implications. *Nature*, 1954; **173**, 973–976.
8. Huxley, A. F. Muscle structure and theories of contraction. *Prog. Biophys. Biophys. Chem.*, 1957; **7**, 255–318.
9. Huxley, H. E. Electron microscope studies of the organization of the filaments in striated muscle. *Biochim. Biophys. Acta*, 1953; **12**, 387–394.
10. Hill, A. V. The heat of shortening and the dynamic constants of muscle. *Proc. R. Soc. Lond.*, 1938; **126**, 136–195.
11. Huxley, H. E. The mechanism of muscular contraction. *Science*, 1969; **164**, 1356–1366.
12. Huxley, A. F., Simmons, R. M. Proposed mechanism of force generation in striated muscle. *Nature*, 1971; **233**, 533–538.

13. Rayment, I., Holden, H. M., Whittaker, M., Yohn, C. B., Lorenz, M., Holmes, K. C., et al. Structure of the actin-myosin complex and its implications for muscle contraction. *Science*, 1993; **261**, 58–65.

14. Huxley, A. F. *Reflections on Muscle*. Liverpool: Liverpool University Press; 1980.

15. Blix, M. Die Laenge und die Spannung des Muskels. *Skand. Arch. Physiol.*, 1891; **3**, 295–318.

16. Gordon, A. M., Huxley, A. F., Julian, F. J. The variation in isometric tension with sarcomere length in vertebrate muscle fibres. *J. Physiol. (Lond)*, 1966; **184**, 170–192.

17. Hill, A. V. The mechanics of active muscle. *Proc. R. Soc. Lond*, 1953; **41**, 104–117.

18. Morgan, D. L. New insights into the behavior of muscle during active lengthening. *Biophys. J.*, 1990; **57**, 209–221.

19. Edman, K. A. Residual force enhancement after stretch in striated muscle. A consequence of increased myofilament overlap? *J. Physiol.*, 2012; **590**(Pt 6), 1339–1345.

20. Armstrong, R. B., Ogilvie, R. W., Schwane, J. A. Eccentric exercise-induced injury to rat skeletal muscle. *J. Appl. Physiol.*, 1983; **54**(1), 80–93.

21. Proske, U., Morgan, D. L. Muscle damage from eccentric exercise: Mechanism, mechanical signs, adaptation and clinical applications. *J. Physiol. (Lond)*, 2001; **537**(Pt 2), 333–345.

22. Allinger, T. L., Epstein, M., Herzog, W. Stability of muscle fibers on the descending limb of the force-length relation. A theoretical consideration. *J. Biomech.*, 1996; **29**, 627–633.

23. Edman, K. A. P., Elzinga, G., Noble, M. I. M. Residual force enhancement after stretch of contracting frog single muscle fibers. *J. Gen. Physiol.*, 1982; **80**, 769–784.

24. Lee, H. D., Herzog, W. Force enhancement following muscle stretch of electrically and voluntarily activated human adductor pollicis. *J. Physiol. (Lond)*, 2002; **545**, 321–330.

25. Abbott, B. C., Aubert, X. M. The force exerted by active striated muscle during and after change of length. *J. Physiol. (Lond)*, 1952; **117**, 77–86.

26. Edman, K. A., Elzinga, G., Noble, M. I. Enhancement of mechanical performance by stretch during tetanic contractions of vertebrate skeletal muscle fibres. *J. Physiol. (Lond)*, 1978; **281**, 139–155.

27. Leonard, T. R., Duvall, M., Herzog, W. Force enhancement following stretch in a single sarcomere. *Am. J. Physiol. Cell Physiol.*, 2010; **299**(6), C1398–C1401.

28. Llewellyn, M. E., Barretto. R. P. J., Delp, S. L., Schnitzer, M. J. Minimally invasive high-speed imaging of sarcomere contractile dynamics in mice and humans. *Nature*, 2008; **454,** 784–788.

29. Huxley, A. F., Peachey, L. D. The maximum length for contraction in vertebrate striated muscle. *J. Physiol. (Lond)*, 1961; **156,** 150–165.

30. Morgan, D. L. An explanation for residual increased tension in striated muscle after stretch during contraction. *Exp. Physio.*, 1994; **79,** 831–838.

31. Panchangam, A., Herzog, W. Sarcomere overextension reduces stretch-induced tension loss in myofibrils of rabbit psoas. *J. Biomech.*, 2011; **44**(11), 2144–2149.

32. Rassier, D.E., Herzog, W., Pollack, G.H. Stretch-induced force enhancement and stability of skeletal muscle myofibrils. *Adv. Exp. Med. Biol.*, 2003; **538,** 501–515.

33. Butterfield, T. A. Eccentric exercise in vivo: Strain-induced muscle damage and adaptation in a stable system. *Exerc. Sport Sci. Rev.*, 2010; **38**(2), 51–60.

34. Lieber, R. L., Woodburn, T. M., Fridén, J. Muscle damage induced by eccentric contractions of 25% strain. *J. Appl. Physiol.*, 1991; **70,** 2498–2507.

35. Walcott, S., Herzog, W. Modeling residual force enhancement with generic cross-bridge models. *Math. Biosci.*, 2008; **216,** 172–186.

36. Morgan, D. L., Whitehead, N. P., Wise, A. K., Gregory, J. E., Proske, U. Tension changes in the cat soleus muscle following slow stretch or shortening of the contracting muscle. *J. Physiol. (Lond)*, 2000; **522**(pt 3), 503–513.

37. Edman, K. A., Tsuchiya, T. Strain of passive elements during force enhancement by stretch in frog muscle fibres. *J. Physiol. (Lond)*, 1996; **490**(pt 1), 191–205.

38. Bullimore, S. R., Leonard, T. R., Rassier, D. E., Herzog, W. History-dependence of isometric muscle force: Effect of prior stretch or shortening amplitude. *J. Biomech.*, 2007; **40,** 1518–1524.

39. Herzog, W., Duvall, M., Leonard, T. R. Molecular mechanisms of muscle force regulation: a role for titin? *Exerc. Sport Sci. Rev.*, 2012; **40**(1), 50–57.

40. Herzog, W., Leonard, T., Joumaa, V., Duvall, M., Panchangam, A. The three filament model of skeletal muscle stability and force production. *Mol. Cell Biomech.*, 2012; **9**(3), 175–191.

41. Sugi, H., Tsuchiya, T. Stiffness changes during enhancement and deficit of isometric force by slow length changes in frog skeletal muscle fibres. *J. Physiol. (Lond)*, 1988; **407,** 215–229.

42. Herzog, W., Leonard, T. R. Force enhancement following stretching of skeletal muscle: A new mechanism. *J. Exp. Biol.*, 2002; **205,** 1275–1283.

43. Peterson, D., Rassier, D. E., Herzog, W. Force enhancement in single skeletal muscle fibres on the ascending limb of the force–length relationship. *J. Exp. Biol.*, 2004; **207,** 2787–2791.

44. Pun, C., Syed, A., Rassier, D. E. History-dependent properties of skeletal muscle myofibrils contracting along the ascending limb of the force–length relationship. *Proc. Biol. Sci.*, 2010; **277**(1680), 475–484.

45. Schachar, R., Herzog, W., Leonard, T. R. The effects of muscle stretching and shortening on isometric forces on the descending limb of the force–length relationship. *J. Biomech.*, 2004; **37**(6), 917–926.

46. Rassier, D. E., Herzog, W., Wakeling, J. M., Syme, D. Stretch-induced, steady-state force enhancement in single skeletal muscle fibers exceeds the isometric force at optimal fibre length. *J. Biomech.*, 2003; **36,** 1309–1316.

47. Leonard, T. R., Herzog, W. Regulation of muscle force in the absence of actin-myosin based cross-bridge interaction. *Am. J. Physiol. Cell Physiol.*, 2010; **299,** C14–C20.

48. Rassier, D. E., Herzog, W., Pollack, G. H. Dynamics of individual sarcomeres during and after stretch in activated single myofibrils. *Proc R Soc Lond B* 2003; **270,** 1735–1740.

49. Telley, I. A., Stehle, R., Ranatunga, K. W., Pfitzer, G., Stussi, E., Denoth, J. Dynamic behaviour of half-sarcomeres during and after stretch in activated rabbit psoas myofibrils: Sarcomere asymmetry but no "sarcomere popping". *J. Physiol. (Lond)*, 2006; **573**(Pt 1), 173–185.

50. Campbell, K. S. Interactions between connected half-sarcomeres produce emergent mechanical behavior in a mathematical model of muscle. *PLoS Comput. Biol.*, 2009; **5**(11), e1000560.

51. Herzog, W., Leonard, T. R. Residual force enhancement: The neglected property of striated muscle contraction. *J. Physiol.*, 2013; **591**(Pt 8), 2221.

52. Edman, K. A. Reply from K. A. P. Edman. *J. Physiol.*, 2013; **591**(Pt 8), 2223.

53. Rassier, D. E., Pavlov, I. Force produced by isolated sarcomeres and half-sarcomeres after an imposed stretch. *Am. J. Physiol. Cell Physiol.*, 2012; **302**(1), C240–C248.

54. Higuchi, H., Umazume, Y. Localization of the parallel elastic components in frog skinned muscle fibers studied by the dissociation of the A- and I-bands. *Biophys. J.*, 1985; **48**(1), 137–147.

55. Nishikawa, K. C., Monroy, J. A., Uyeno, T. E., Yeo, S. H., Pai, D. K., Lindstedt, S. L. Is titin a "winding filament"? A new twist on muscle contraction. *Proc. Biol. Sci.*, 2012; **279**(1730), 981–990.

56. Herzog, W., Lee, E. J., Rassier, D. E. Residual force enhancement in skeletal muscle. *J. Physiol. (Lond)*, 2006; **574**(3), 635–642.

Chapter 10

# Microfluidic Modeling of Cancer Metastasis

**Zeta Tak For Yu,**[a,b,*] **Koh Meng Aw Yong,**[a,b*] **and Jianping Fu**[a,b,c,*]
[a]*Integrated Biosystems and Biomechanics Laboratory,*
[b]*Department of Mechanical Engineering,*
[c]*Department of Biomedical Engineering, University of Michigan,*
*Ann Arbor, MI 48109-2125, USA*
jpfu@umich.edu

## 10.1 Introduction

The majority of solid tumor cancers such as breast, colon, and prostate can be treated successfully through surgical resection of the primary tumor with more than 90% of the patients enjoying a long-term survival rate. However, survival rates decrease once invasion of cancer cells into surrounding local tissue, such as lymph nodes, occurs. The survival rates further decrease if the cancer is found in distal sites in the body such as lung, liver, and bone [1]. Once this happens, the cancer becomes metastatic.

The metastatic process involves multiple stages. Cancer cells have to first degrade the extracellular environment and invade through the matrix surrounding the primary tumor. They further need to intravasate and survive in the circulatory system as

*These authors contributed equally to this work.

*Cells, Forces, and the Microenvironment*
Edited by Charles M. Cuerrier and Andrew E. Pelling

ISBN 978-981-4613-36-1 (Hardcover), 978-981-4613-37-8 (eBook)
www.panstanford.com

circulating tumor cells (CTCs). Once they encounter a suitable environment for colonization such as the liver, lung, and bone, they extravasate from the circulation and form metastases (Fig. 10.1). The diagnosis of metastases today depend largely on imaging technology, for example, x-ray, magnetic resonance imaging (MRI), and positron emission tomography (PET), to name a few [2]. These imaging techniques, however, are limited by their specificity and sensitivity of detection.

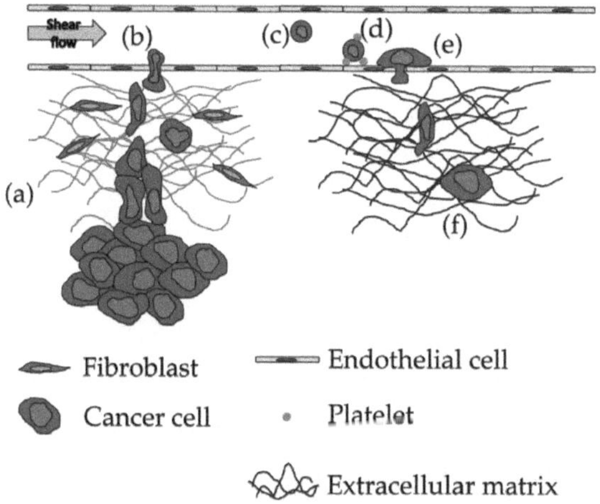


**Figure 10.1** Schematic of the metastatic cascade. (a) Invasion of cancer cells through the extracellular matrix. (b) Intravasation of cancer cells into the circulatory system. (c) Circulating tumor cells (CTCs). (d) Platelet aggregation promotes adhesion of CTCs to endothelium. (e) Extravasation of cancer cells into distal sites. (f) Cancer cells colonize distal sites, forming metastases.

Microfluidics, the manipulation of fluid at the microscale, has emerged as a new and useful technology for biological research and clinical study [3]. Microfluidic devices and systems are commonly made through soft-lithography of polydimethylsiloxane (PDMS), a polymer that offers optical transparency, gas permeability, ease-of-use and biocompatibility [4]. PDMS-based microfluidic devices have demonstrated numerous benefits for biological and biomedical applications, such as cost effectiveness, high-throughput automated operation, high spatiotemporal resolution [5], low consumption of biochemical reagents [6], integration capability with three-dimensional (3D) extracellular matrix (ECM) scaffold

[7] and coculture system [8], proteomic analysis at a single-cell resolution [9], to just name a few.

A significant challenge facing cancer research is determining if and when the cancer in the patient will start to invade. Successful understanding and modeling of the metastatic process will enable more effective strategies to combat it. In this chapter, we will briefly describe some of the crucial steps of metastasis: invasion, intravasation into the circulatory system and finally colonization of distal sites. We will further describe the conventional methods used to study these steps in the metastatic cascade, their limitations and how the recent development in microfluidic modeling of cancer is facilitating a better understanding of the metastatic process.

## 10.2 The Metastatic Process

### 10.2.1 Invasion

Invasion is one of the first steps in metastasis. Cancer cells acquire genetic mutations or changes to its epigenetic landscape that trigger the invasion process. These changes to the genetic or epigenetic status in cancer cells can have a direct effect on the intrinsic ability of the cancer cell to invade as well as indirectly on the surrounding stromal cells to produce ECM proteins and growth factors such as TGF-β that promote cancer cell invasion [10]. Once cancer cells have acquired the necessary mutations, they can break from the primary tumor site invading as single cells or collectively and via a variety of different methods. One widely studied process during cancer invasion is the epithelial–mesenchymal transition (EMT), where cancer cells start to express genes that enable them to transition from an epithelial origin to appear and behave more like a mesenchymal cell. As a result of this phenotypic transition, cancer cells secrete enzymes such as matrix metalloproteases that remodel the surrounding ECM and facilitate invasion [10]. Cancer cells can also maneuver their way through the matrix without degrading it; during this process, they adopt an amoeboid phenotype and squeeze through the matrix instead [11].

The physical properties of both cancer cells and ECM can change during cancer progression. While these changes are an effect of acquired genetic or epigenetic abnormalities, they play an important role in affecting cancer progression, including invasion. For one, stiffening of the ECM is commonly observed in cancer and is associated with promoting tumor initiation and invasion [12–14]. Interestingly, greater cell deformability is a general trait of metastatic cancer cells [15–17]. Other physical properties of the ECM relevant to cancer invasion include interstitial pressure. Solid tumors typically have increased interstitial pressure. While it is thought increased interstitial pressure hinders chemotherapy by posing as an obstacle to effective drug delivery, there is also recent data suggesting that interstitial pressure can regulate invasion [18,19].

### 10.2.2 Circulating Tumor Cells

Cancer cells tend to invade along a chemical gradient, typically toward the circulatory system where a richer source of nutrient exists. Intravasation into the circulation system can occur through either newly formed or existing capillaries [20,21]. Once in the circulatory system, CTCs must survive a variety of environmental factors such as anoikis (cell death induced by loss of attachment) and shear flow exerted by the circulation, as well as evade the immune system [22,23]. Furthermore, coagulation with platelets can occur on CTCs that may result in the cancer cells being trapped and dying in micro-capillaries. At the same time, there is also evidence suggesting that such interactions between CTCs and platelets may actually facilitate the metastatic process [24]. Despite the many obstacles facing the survival of CTCs and how different they behave in the circulation, their presence in patients has been suggested to be a good prognostic indicator of patient survival in several types of cancer and is an important area of study [25,26].

### 10.2.3 Colonization of Distal Sites

To successfully colonize a distal site in the body, CTCs must first find a suitable location that is conducive for proliferation. The most common sites of metastasis, apart from the lymph nodes are the liver, lung and bone [27]. This is facilitated by the presence of chemogradients that attract cancer cells to these sites [28]. Before invading and colonizing secondary sites, CTCs must first extravasate from the circulation. As mentioned previously, CTCs come into contact with coagulation factors while in circulation and may interact with platelets. This CTC–platelet aggregate may facilitate binding with selectins found on the endothelium that help arrest CTCs [24]. The tumor cells can then extravasate through remodeling endothelial cell–cell junctions [29].

Once tumor cells have successfully infiltrated the secondary site, it must be able to survive within this foreign environment. It is believed that the ability of cancer cells to adapt to a new environment is limited and a majority of tumor cells die before establishing a colony [30,31]. During this initial stage of colonization, tumors cells may remain in a state of dormancy for several years. They remain as micro lesions known as micrometastases and cannot be detected or treated with conventional methods [32,33]. Metastatic lesions may survive as solitary cells or as a small mass through passive diffusion of nutrients. To form larger lesions, cancer cells have to ensure sufficient nutrients are present to support their growth. They do so by inducing angiogenesis, the formation of new blood vessels [34], creating a tumor vasculature that allows for increased tumor burden to be supported. The ECM in the secondary site most likely plays an important role in determining whether a metastatic lesion proliferates or remains dormant. Under the right conditions, interactions between the cancer cell and local ECM can induce dormancy [35]. Conversely, under favorable conditions, these metastatic lesions may start to proliferate [36].

### 10.2.4 Conventional Technologies to Study Cancer Metastasis

The traditional method of characterizing invasion potential of cancer cells is done using the Boyden chamber invasion assay, which is performed ex vivo and involves a transwell insert with a porous membrane coated with ECM proteins on it. Cancer cells are seeded onto the insert and placed into another well containing a chemoattractant. In the presence of a chemogradient, cancer cells will actively invade through the ECM protein coated porous membrane of the transwell to the bottom well. The number of cells that successfully invade into the bottom well is used as a measure of invasion potential. While the Boyden chamber invasion assay is efficient at determining how invasive a cancer cell is, visualizing the kinetics of the invasion process is difficult. Further, it is not possible to study invasion in the presence of the microenvironmental parameters discussed above such as interstitial pressure using the Boyden chamber.

An alternative to the Boyden chamber is through the use of a 3D matrix composed of extracellular proteins, such as collagen. Typically, cancer cells are grown within a collagen gel and their invasive properties can be observed by following their migration through the matrix. The EMT process has been well studied using this method [11]. However, a drawback is that it is hard to establish invasion in the presence of a chemogradient within a collagen gel that has polymerized in a tissue culture dish.

Histological analysis of the tumor can also be used to study cancer invasion. The tumor and surrounding tissue are removed through biopsy, fixed and embedded, typically in paraffin. After embedding, slices of the fixed tumor or tissue are obtained to perform histological analysis. A major drawback of this procedure is that cells are not viable, making studying the dynamic invasion process or downstream analyses involving live cells impossible.

Intravital tumor imaging is an alternative way of studying live cancer cell invasion kinetics in an in vivo setting. This method typically involves generating a xenograft tumor in an animal. The animal is restrained and kept alive while the tumor exposed for monitoring through the use of a microscope. The mode of detection used in intravital tumor imaging is typically based on fluorescence or bioluminescence. The advantage of this method is that the tumor is grown in an in vivo setting similar to its native environment and the dynamic invasion process can be studied over time. However, this technique is costly to perform and difficult to execute [2].

There are several platforms currently available that are used for capture and studying CTCs from blood specimens [37]. They mainly involve antibody-based approaches targeting markers of epithelial cells to separate CTCs from non-tumor cells of hematological origin [26,38]. A drawback to using such method is that CTCs are poorly characterized and known to be heterogeneous. Further, during the metastatic process, cancer cells may undergo EMT and lose their epithelial markers and express a totally different set of genes [39].

Currently, detecting metastases is done mainly using imaging technologies, which require imaging agents to be taken up by cancer cells before detection. The challenges lie in designing molecules that will be specifically taken up by cancer cells and at the same time compatible with the mode of detection. An example of the most successful and widely used imaging agent is 2-deoxy-2-(18F)fluoro-D-glucose (FDG), a glucose analogue. As improvements to both imaging technologies as well as imaging agents continue, we can expect increased sensitivity in future imaging tests as well as better specificity toward tumor cells.

The current tools used have played an instrumental role in furthering the understanding of metastasis. However, each system has their limitations that prevent a deeper understanding of metastasis as a multistep dynamic process. However, with advancement in technology, these limitations are slowly removed. In recent years, the field of microfluidics especially has been stepping up as a major player in helping to improve our understanding of metastasis.

## 10.3 Microfluidics for Studying Cancer Metastasis

While the use of microfluidics in studying cancer metastasis is extensive, it is worth noting that two typical chip configurations, as shown in Fig. 10.2, have been most commonly used. The first configuration contains three microchannels in which the central and the side channels are connected by an array of blocks forming fluidic constrictions or barriers [40,41]. Owing to dominant surface tension force, the block barrier serves as a cage to confine the biological matrix solution in the central channel and prevent the solution from flowing into the side microchannels. As the matrix can be embedded and gelated with cells, drugs, or other biological ingredients for cell immobilization and localization, this configuration is suitable to study cancer invasion, intravasation, and extravasation, and the EMT (see detailed discussion in Section 10.2). In the second configuration, two main microchannels are connected with interconnecting microchannels, which are typically as thin as 3 μm to minimize convection flow between the two microchannels [42,43]. This configuration is often used to study directional cell migration such as chemotaxis and cell deformation through the interconnecting microchannels under a chemical concentration gradient.

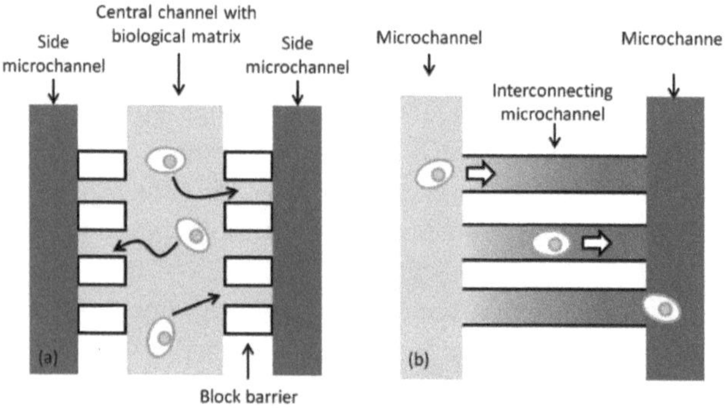


**Figure 10.2** Typical microfluidic chip configurations for modeling metastasis. (a) A central microchannel has a cage-like structure to confine the shape of the biological matrix. (b) Interconnecting microchannels confine cell migration between two main microchannels.

### 10.3.1 Microfluidic Modeling of Epithelial–Mesenchymal Transition

Combining the capability of generating 3D microenvironment and requiring only small amounts of chemicals, a recent microfluidic study by Aref et al. using lung adenocarcinoma A549 spheroids, cancer cells that can revert from an intermediate mesenchymal-like phenotype to an epithelial-like phenotype, demonstrated that microfluidics could offer a power approach for therapeutic drug screening for EMT [44]. In this study, A549 spheroids suspended in collagen I hydrogel were first seeded in the central microchannel using an array of fluidic constrictions and surface tension effect as shown in Fig. 10.2a. Human umbilical vein endothelial cells (HUVECs) were subsequently loaded to the two side channels to form a HUVEC monolayer along the gel surface. Except for one of the 13 drugs tested using this microfluidic platform, significantly lower the half maximal inhibitory concentration (IC50) doses necessary to inhibit EMT was observed for spheroid maintained in the 3D microenvironment as compared to carcinoma cells in isolation in conventional two-dimensional (2D) microwell systems, underscoring the significant difference of cancer cells in response to drug treatments between 2D and 3D, and between monoculture and co-culture systems.

### 10.3.2 Microfluidics for Capture and Informative Analysis of Circulating Tumor Cells

Microfluidic chips have demonstrated reliable capture of CTCs from whole blood of cancer patients. One of the pioneering devices, reported by Nagrath et al., involved flowing patient blood onto a microfabricated flow chamber containing an array of pillars conjugated with antibodies against surface markers of cancer cells [45]. Since then, numerous creative concepts using microfluidics have been proposed to improve capture efficiency and purity of CTCs directly from blood specimens by taking advantage of the differences in biophysical and surface properties between cancer and non-cancer cells. One example involved capturing CTCs using nanoscale rough surfaces etched on glass slides. This technique is advantageous as it does not require antibodies to capture the CTCs and makes use of the intrinsic preference of CTCs for adhesion on rough surfaces over smooth ones [46]. Another example involved the generation of a spiral shaped microfluidic channel. As blood was continuously passed through the microfluidic channel, differential centrifugal forces were exerted on the blood cells and CTCs that allowed the cells to separate based on size [47]. A third example combined microfluidic chaotic mixing using herringbone structures with silicon nanopillar surfaces coated with anti-EpCAM antibodies (Fig. 10.3) [48]. Owing to enhanced cell–surface interactions as well as increased cell capture surface area, superb capture efficiency for CTCs was reported by Wang et al. (>95%).

Microfluidics has also been applied as model systems to study adhesion of CTCs [49]. For example, Zhen et al. proposed a simplified biophysical model to study the effects of cell receptor and surface ligand density on dynamic states of adhesion of CTCs on a microfluidic channel functionalized with capture antibodies. Their biophysical model was based on a receptor-coated sphere moving above a solid surface immobilized with capture ligands. The mathematical analysis and modeling for capture of CTCs were based on calculation and numerical simulation of Langevin equation and an empirical formula with receptor–ligand bonds modeled as linear springs separated by a gap. The authors also examined two breast cancer cell lines, MDA-MB-231 and BT-20, both expressing EpCAM, in microfluidic channels coated with anti-EpCAM or anti-N-cadherin antibodies. Besides three dynamic states (firm adhesion, rolling adhesion, and free motion) CTCs going through as verified by experiments, simulation and analysis, Zhen et al. were able to estimate the cell–surface gap and spring constant properly. Importantly, all measured and simulated results could be generalized as an exponential correlation between the CTC capture ratio and the normalized flow rate.

**Figure 10.3** Functionalized nanostructured substrate combined with a microfluidic chaotic mixer to capture circulating tumor cells with high efficiency. Adapted from [48]. Reprinted with permission from John Wiley and Sons.

Similarly, Song et al. performed an interesting comparative experiment on tumor cell adhesion modulated by endothelium [50]. After a confluent monolayer of human dermal microvascular endothelial cells (HDMECs) was cultured on top of a semi-porous polyester membrane sandwiched between top and bottom PDMS microchannels, the authors found that twice as many breast cancer cells could adhere to the endothelium when HDMECs were treated with CXCL12 basally compared to apically. Such result suggests that the orientation or polarity of the endothelium can be critical in regulating vascular transport and arrest and retention of CTCs.

In addition to studying the physical properties of CTCs, microfluidics has also contributed toward a better understanding of CTC biology. By coupling the use of a herringbone microfluidic chip with antibodies into a single platform, it was possible to isolate and characterize CTCs from breast cancer patients [51]. From this study, the authors found that EMT markers were enriched within the isolated CTCs as compared to cancer cells within primary tumor, reinforcing the clinical importance of EMT as a key player in the metastatic process. In a separate study by Ameri et al., the authors generated a xenograft model of human breast cancer in mice and used a magnetic-based microfluidic device to isolate CTCs generated from these xenograft models. Briefly, magnetic beads containing antibodies recognizing EpCAM were added to blood harvested from mice and the labeled blood was passed through the microfluidic device that allowed automated recovery of CTCs. The authors were further able to demonstrate that isolated CTCs behaved more aggressively than the cells from the primary tumor in response to hypoxia and established a relationship between hypoxia and CTCs [52].

The use of microfluidics has also helped improve on monitoring clinical progression of cancer. In particular, Maheswaran et al. isolated circulating tumor cells from lung cancer patients using a microfluidic device and analyzed these CTCs for EGFR mutations. The authors found mutated EGFR in CTCs isolated from patients that underwent tyrosine kinase inhibitor therapy. These mutations conferred resistance to tyrosine kinase inhibitors. Furthermore, they studied CTCs isolated from pre-treatment patients and observed a negative correlation between pre-existing mutated EGFR in CTCs from pre-treatment patients and survival. This work demonstrates the capability of using microfluidics in improving clinical prognosis and perhaps even predict therapy outcome [53].

### 10.3.3 Microfluidics to Study Cancer Cell Migration

By incorporating a microfluidic gradient generator to produce flows with laterally uniform, linear, polynomial, or complex concentration gradients of soluble molecules including epidermal growth factor (EGF), anti-EGF and CXCL12 ligands, researchers studied human metastatic breast cancer cell line MDA-MB-231 in terms of cell motility, speed, and directionality [54–56]. Results illustrated that chemotaxis of metastatic cancer cells could depend on the shape of chemical gradient profile as well as the chemical concentration range.

Integrating microfluidics with modern microscopy technologies allows for real-time observation of tumor cell migration in geometrically confined environment, which is difficult in conventional assays. Two research groups applied microfluidic channel structures shown in Fig. 10.2b to emulate migration of brain cancer stem cells through interstitial spaces and that of breast cancer cells by the influence of nuclear deformation through endothelial-lined capillaries [57,58]. In both cases, the interconnecting microfluidic channels with their sizes ranging from 3–5 μm were assembled by either reversible or irreversible PDMS bonding against glass slides. The microfluidic chips were further coated with poly-L-lysine or fibronectin, to enhance cell attachment. Time-lapse imaging was carried out to capture real-time dynamics of cell migration along the interconnecting channels for a period of 2 days.

With a more sophisticated fabrication to embed micro-valves in microfluidic chips, researchers were able to use phase-contrast microscopic time-lapse images to detail effects of cell–cell interactions on cell migration through paracrine signaling [57]. A simple microfluidic patterning technique was also recently reported by Wang et al. that could facilitate screening of potential anti-migratory agents, beneficial for drug discovery compared to conventional wound-healing assay [59]. Specifically, after three cancer cell lines with different metastatic potentials were individually plated and confined inside microchannels of a PDMS stamp, the stamp was removed and free movements of monolayers of cancer cells were imaged over time. Migration rate of cancer cells under the treatment of two anticancer drugs, curcumin and apigenin, was successfully evaluated using this microfluidic patterning technique.

### 10.3.4 Microfluidic Tools to Study Interactions of Cancer Invasion, Intravasation, and Extravasation with the Microenvironment

Several research groups [7,40,41,60] studied invasion of tumor cells through endothelium and/or ECM using the two microfluidic configurations shown in Fig. 10.2. The essential elements in such microfluidic cancer metastasis models are the ECM, commonly made of collagen, Matrigel and agarose, as well as the endothelial monolayer, commonly using HUVECs. By exploiting the dominant effect of surface tension, ECM prepolymers can be localized and polymerized inside interconnecting microchannels or micro-cages in microfluidic chips. These gelated ECM media are porous to allow processes like chemotaxis and immunostaining by diffusion or convection of chemoattractants and biomolecules.

Using microfluidic cancer metastasis models with gelated ECM media, researchers successfully demonstrated sustained maintenance of concentration profiles of soluble factors for a prolonged period of time by simply connecting a microchannel made in ECM with a small source and a large sink [61]. Periodically adding factors and replenishing the sink by an operator, the microchip could generate pseudo-steady linear as well as non-linear concentration gradients up to 10 days. Such microfluidic tools were successfully used to study invasion of metastatic rat mammary adenocarcinoma cells (MtLN3) into surrounding matrix.

Recently, Shin et al. reported a microfluidic chip containing serially connected chambers and external screw valves that could be used to study both cancer intravasation and extravasation simultaneously [62]. In the intravasation chamber, colon cancer cells, either metastatic LOVO or non-metastatic SW480, were embedded in polymerized Matrigel. The extravasation chamber was coated in sequence with poly-L-lysine, fibronectin and HUVECs. The two chambers were flown with media with shear stress of a physiological range (1–5 dyne·cm$^{-2}$). By counting the number of cancer cells escaped from the intravasation chamber and arrested onto the extravasation chamber, the ability of cancer cells to intravasate and extravasate under different drug treatment conditions were characterized by the authors.

### 10.3.5 Microfluidic Study of Cancer Cell Deformation

Mobility of cancer cells is connected to the physical and mechanical properties of the cells and the surrounding microenvironment such as cell size and deformability, ECM porosity and deformability, and blood vessel size and pressure. Recently, different microfluidic cell deformability assays have been successfully developed to allow single cancer cells to flow or migrate through confining structures such as microscale orifices and channels. These cell deformability assays have been proven useful to study (i) how cancer cells traverse through blood vessel during metastasis [63,64], (ii) difference in cell deformability between benign and malignant cancer phenotypes [65], and (iii) mechanical effects on behaviors of cancer cells through sub-nucleus physical confinement [42]. Experimental parameters such as cell entry time, transit velocity, elongation index, motility, viability, proliferation, have been commonly analyzed in such microfluidic cell deformability assays to quantify how cancer cell migration and motion are dictated by their intrinsic deformability property. In addition to microfluidic confining structures, optofluidics tools, such as optical tweezers, have been recently integrated with microfluidics to examine functional correlations between intrinsic deformability property of cancer cells and their metastatic potential [17].

### 10.3.6 Microfluidic Modeling of Angiogenesis

Instead of fabricating typical PDMS microfluidic chips, a research group constructed a plate structure formed by aggregating poly(lactic-co-glycolic acid) (PLGA) particles to resemble and study tumor angiogenesis. The plate attached with HUVECs was laid out within the hydrogel matrix, and cancer cells were either placed in the center or spread evenly to mimic the initial phase of a tumor before vascularization or a highly vascularized tumor respectively. Their engineered tumors showed greater drug resistance compared to cancer cells cultured in a traditional 3D setting [5].

### 10.3.7 Microfluidics for Cancer Imaging

Integrated microfluidic radioassays for glycolysis analysis in small tumor cell populations were recently developed by Vu et al. to detect and image very low activity levels of beta emitting isotope [66]. This microfluidic radioassay has achieved highly sensitive imaging of a radioactive tracer 18F-FDG uptake in small mouse melanoma cell populations down to a single-cell level (Fig. 10.4). Further, by precisely controlling dynamic operations under in situ imaging and subsequent data modeling, this microfluidic radioassay was shown to be capable of obtaining kinetic rate constants of 18F-FDG metabolism. This microfluidic radioassay system suggested an exciting new way to quantitatively study transport and reaction of biomolecules within cancer cells at the culture scale.

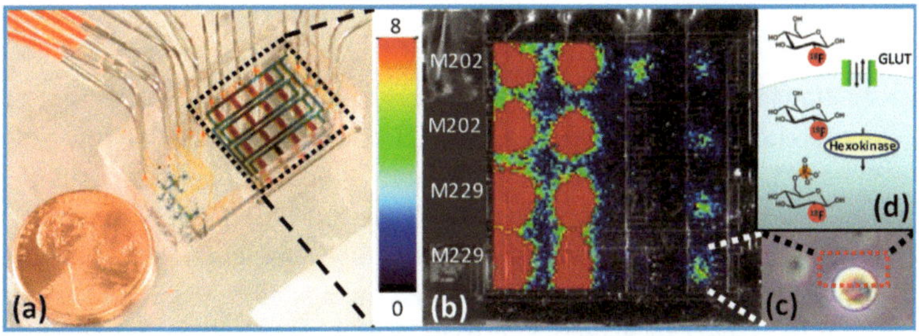


**Figure 10.4** FDG uptake as a way to image cancer metabolism. (a) Schematic of the integrated microfluidic radioassay and the corresponding (b) Radioassay image. (c) Micrograph showing a single cell in a chamber. (d) Glycolysis kinetics studied using the integrated microfluidic radioassay. Adapted from [66]. Reprinted with permission from the Society of Nuclear Medicine and Molecular Imaging, Inc.

Micro image cytometry (MIC) technology, a system composed of a microfluidic cell array chip, image acquisition and cytometry analysis, was recently developed by Sun et al. to study cancer cells [67]. Coupled with systems pathology analysis, the MIC technology developed by Sun et al. was shown to be capable of quantitative, single-cell proteomic analysis of multiple signaling molecules using only about 1,000 single cells. Using MIC, simultaneous measurements of four critical signaling proteins (EGFR, PTEN, phospho-Akt and phospho-S6) relevant to the oncogenic PI3K/Akt/mTOR signaling pathway had been achieved in individual cancer cells by Sun et al., with their results showing meaningful correlations between measurements of minute patient samples and clinical prognosis.

## 10.4 Conclusion and Perspective

In this chapter we have reviewed and highlighted the biological significance of using microfluidics to study and model cancer metastasis. While not a comprehensive review of all available microfluidic devices, the various microfluidic models discussed here have suggested microfluidics as a promising and powerful research tool for new and in-depth understanding of cancer metastasis as compared to traditional assays.

## References


1. Howlader, N., Noone, A. M., Krapcho, M., Neyman, N., Aminou, R., Altekruse, S. F., Kosary, C. L., Ruhl, J., Tatalovich, Z., Cho, H., Mariotto, A., Eisner, M. P., Lewis, D. R., Chen, H. S., Feuer, E. J., and Cronin, K. A. (2012). SEER Cancer Statistics Review, 1975–2009 (Vintage 2009 Populations), National Cancer Institute. Bethesda, MD, based on November 2011 SEER data submission, posted to the SEER web site. http://seer.cancer.gov/csr/1975_2009_pops09/.
2. Condeelis, J., and Weissleder, R. (2010). In vivo imaging in cancer, Cold Spring Harb. *Perspect. Biol.*, **2**, a003848.
3. El-Ali, J., Sorger, P. K., and Jensen, K. F. (2006). Cells on chips, *Nature*, **442**, 403–411.
4. Sia, S. K., and Whitesides, G. M. (2003). Microfluidic devices fabricated in poly(dimethylsiloxane) for biological studies, *Electrophoresis*, **24**, 3563–3576.
5. Lee, W., and Park, J. (2012). The design of a heterocellular 3D architecture and its application to monitoring the behavior of cancer cells in response to the spatial distribution of endothelial cells, *Adv. Mater.*, **24**, 5339–5344.
6. Yu, Z. T., Kamei, K., Takahashi, H., Shu, C. J., Wang, X., He, G. W., Silverman, R., Radu, C. G., Witte, O. N., Lee, K. B., and Tseng, H. R. (2009). Integrated microfluidic devices for combinatorial cell-based assay, Biomed. *Microdevices*, **11**, 547–555.
7. Haessler, U., Teo, J. C., Foretay, D., Renaud, P., and Swartz, M. A. (2012). Migration dynamics of breast cancer cells in a tunable 3D interstitial flow chamber, *Integr. Biol. (Camb.)*, **4**, 401–409.
8. Kamei, K., Guo, S., Yu, Z. T., Takahashi, H., Gschweng, E., Suh, C., Wang, X., Tang, J., McLaughlin, J., Witte, O. N., Lee, K. B., and Tseng, H. R. (2009). An integrated microfluidic culture device for quantitative analysis of human embryonic stem cells, *Lab Chip*, **9**, 555–563.
9. Fan, R., Vermesh, O., Srivastava, A., Yen, B. K., Qin, L., Ahmad, H., Kwong, G. A., Liu, C. C., Gould, J., Hood, L., and Heath, J. R. (2008). Integrated barcode chips for rapid, multiplexed analysis of proteins in microliter quantities of blood, *Nat. Biotechnol.*, **26**, 1373–1378.
10. Xu, J., Lamouille, S., and Derynck, R. (2009). TGF-beta-induced epithelial to mesenchymal transition, *Cell Res.*, **19**, 156–172.
11. Wolf, K., Mazo, I., Leung, H., Engelke, K., von Andrian, U. H., Deryugina, E. I., Strongin, A. Y., Brocker, E. B., and Friedl, P. (2003). Compensation

mechanism in tumor cell migration: mesenchymal-amoeboid transition after blocking of pericellular proteolysis, *J. Cell Biol.*, **160**, 267–277.

12. Provenzano, P. P., Inman, D. R., Eliceiri, K. W., Knittel, J. G., Yan, L., Rueden, C. T., White, J. G., and Keely, P. J. (2008). Collagen density promotes mammary tumor initiation and progression, *BMC Med.*, 6, 11-7015-7016-7011.
13. Levental, K. R., Yu, H., Kass, L., Lakins, J. N., Egeblad, M., Erler, J. T., Fong, S. F., Csiszar, K., Giaccia, A., Weninger, W., Yamauchi, M., Gasser, D. L., and Weaver, V. M. (2009). Matrix crosslinking forces tumor progression by enhancing integrin signaling, *Cell*, **139**, 891–906.
14. Pathak, A., and Kumar, S. (2012). Independent regulation of tumor cell migration by matrix stiffness and confinement, *Proc. Natl. Acad. Sci. U. S. A.*, **109**, 10334–10339.
15. Swaminathan, V., Mythreye, K., O'Brien, E. T., Berchuck, A., Blobe, G. C., and Superfine, R. (2011). Mechanical stiffness grades metastatic potential in patient tumor cells and in cancer cell lines, *Cancer Res.*, **71**, 5075–5080.
16. Remmerbach, T. W., Wottawah, F., Dietrich, J., Lincoln, B., Wittekind, C., and Guck, J. (2009). Oral cancer diagnosis by mechanical phenotyping, *Cancer Res.*, **69**, 1728–1732.
17. Guck, J., Schinkinger, S., Lincoln, B., Wottawah, F., Ebert, S., Romeyke, M., Lenz, D., Erickson, H. M., Ananthakrishnan, R., Mitchell, D., Kas, J., Ulvick, S., and Bilby, C. (2005). Optical deformability as an inherent cell marker for testing malignant transformation and metastatic competence, *Biophys. J.*, **88**, 3689–3698.
18. Heldin, C. H., Rubin, K., Pietras, K., and Ostman, A. (2004). High interstitial fluid pressure—an obstacle in cancer therapy, *Nat. Rev. Cancer*, **4**, 806–813.
19. Tien, J., Truslow, J. G., and Nelson, C. M. (2012). Modulation of invasive phenotype by interstitial pressure-driven convection in aggregates of human breast cancer cells, *PloS One*, **7**, e45191.
20. Balkwill, F. (2003). Chemokine biology in cancer, *Semin. Immunol.*, **15**, 49–55.
21. Blood, C. H., and Zetter, B. R. (1990). Tumor interactions with the vasculature: angiogenesis and tumor metastasis, *Biochim. Biophys. Acta.*, **1032**, 89–118.
22. Simpson, C. D., Anyiwe, K., and Schimmer, A. D. (2008). Anoikis resistance and tumor metastasis, *Cancer Lett.*, **272**, 177–185.

23. Igney, F. H., and Krammer, P. H. (2002). Immune escape of tumors: apoptosis resistance and tumor counterattack, *J. Leukoc. Biol.*, **71**, 907–920.

24. Konstantopoulos, K., and Thomas, S. N. (2009). Cancer cells in transit: The vascular interactions of tumor cells, *Ann. Rev. Biomed. Eng.*, **11**, 177–202.

25. de Bono, J. S., Scher, H. I., Montgomery, R. B., Parker, C., Miller, M. C., Tissing, H., Doyle, G. V., Terstappen, L. W., Pienta, K. J., and Raghavan, D. (2008). Circulating tumor cells predict survival benefit from treatment in metastatic castration-resistant prostate cancer, *Clin. Cancer Res.*, **14**, 6302–6309.

26. Riethdorf, S., Fritsche, H., Muller, V., Rau, T., Schindlbeck, C., Rack, B., Janni, W., Coith, C., Beck, K., Janicke, F., Jackson, S., Gornet, T., Cristofanilli, M., and Pantel, K. (2007). Detection of circulating tumor cells in peripheral blood of patients with metastatic breast cancer: a validation study of the cell search system, *Clin. Cancer Res.*, **13**, 920–928.

27. Disibio, G., and French, S. W. (2008). Metastatic patterns of cancers: Results from a large autopsy study, *Arch. Pathol. Lab. Med.*, **132**, 931–939.

28. Kakinuma, T., and Hwang, S. T. (2006). Chemokines, chemokine receptors, and cancer metastasis, *J. Leukoc. Biol.*, **79**, 639–651.

29. Stoletov, K., Kato, H., Zardouzian, E., Kelber, J., Yang, J., Shattil, S., and Klemke, R. (2010). Visualizing extravasation dynamics of metastatic tumor cells, *J. Cell Sci.*, **123**, 2332–2341.

30. Luzzi, K. J., MacDonald, I. C., Schmidt, E. E., Kerkvliet, N., Morris, V. L., Chambers, A. F., and Groom, A. C. (1998). Multistep nature of metastatic inefficiency: Dormancy of solitary cells after successful extravasation and limited survival of early micrometastases, *Am. J. Pathol.*, **153**, 865–873.

31. Tsuji, K., Yamauchi, K., Yang, M., Jiang, P., Bouvet, M., Endo, H., Kanai, Y., Yamashita, K., Moossa, A. R., and Hoffman, R. M. (2006). Dual-color imaging of nuclear-cytoplasmic dynamics, viability, and proliferation of cancer cells in the portal vein area, *Cancer Res.*, **66**, 303–306.

32. Townson, J. L., Ramadan, S. S., Simedrea, C., Rutt, B. K., MacDonald, I. C., Foster, P. J., and Chambers, A. F. (2009). Three-dimensional imaging and quantification of both solitary cells and metastases in whole mouse liver by magnetic resonance imaging, *Cancer Res.*, **69**, 8326–8331.

33. Naumov, G. N., Townson, J. L., MacDonald, I. C., Wilson, S. M., Bramwell, V. H., Groom, A. C., and Chambers, A. F. (2003). Ineffectiveness of doxorubicin treatment on solitary dormant mammary carcinoma cells or late-developing metastases, *Breast Cancer Res. Treat.*, **82**, 199–206.
34. Folkman, J. (2002). Role of angiogenesis in tumor growth and metastasis, *Semin. Oncol.*, **29**, 15–18.
35. Weaver, V. M., Petersen, O. W., Wang, F., Larabell, C. A., Briand, P., Damsky, C., and Bissell, M. J. (1997). Reversion of the malignant phenotype of human breast cells in three-dimensional culture and in vivo by integrin blocking antibodies, *J. Cell Biol.*, **137**, 231–245.
36. Goodison, S., Kawai, K., Hihara, J., Jiang, P., Yang, M., Urquidi, V., Hoffman, R. M., and Tarin, D. (2003). Prolonged dormancy and site-specific growth potential of cancer cells spontaneously disseminated from nonmetastatic breast tumors as revealed by labeling with green fluorescent protein, *Clin. Cancer Res.*, **9**, 3808–3814.
37. Pantel, K., and Alix-Panabieres, C. (2010). Circulating tumour cells in cancer patients: challenges and perspectives, *Trends Mol. Med.*, **16**, 398–406.
38. Fehm, T., Hoffmann, O., Aktas, B., Becker, S., Solomayer, E. F., Wallwiener, D., Kimmig, R., and Kasimir-Bauer, S. (2009). Detection and characterization of circulating tumor cells in blood of primary breast cancer patients by RT-PCR and comparison to status of bone marrow disseminated cells, *Breast Cancer Res.*, **11**, R59.
39. Mani, S. A., Guo, W., Liao, M. J., Eaton, E. N., Ayyanan, A., Zhou, A. Y., Brooks, M., Reinhard, F., Zhang, C. C., Shipitsin, M., Campbell, L. L., Polyak, K., Brisken, C., Yang, J., and Weinberg, R. A. (2008). The epithelial-mesenchymal transition generates cells with properties of stem cells, *Cell*, **133**, 704–715.
40. Shin, Y., Kim, H., Han, S., Won, J., Jeong, H. E., Lee, E. S., Kamm, R. D., Kim, J. H., and Chung, S. (2013). Extracellular matrix heterogeneity regulates three-dimensional morphologies of breast adenocarcinoma cell invasion, *Adv. Healthc. Mater.*, **2**, 790–794.
41. Chaw, K. C., Manimaran, M., Tay, F. E., and Swaminathan, S. (2007). Matrigel coated polydimethylsiloxane based microfluidic devices for studying metastatic and non-metastatic cancer cell invasion and migration, *Biomed. Microdevices*, **9**, 597–602.
42. Mak, M., Reinhart-King, C. A., and Erickson, D. (2013). Elucidating mechanical transition effects of invading cancer cells with a subnucleus-scaled microfluidic serial dimensional modulation device, *Lab Chip*, **13**, 340–348.

43. Huang, Y., Agrawal, B., Clark, P. A., Williams, J. C., and Kuo, J. S. (2011). Evaluation of cancer stem cell migration using compartmentalizing microfluidic devices and live cell imaging, *JoVE*, **58**, e3297–e3297.

44. Aref, A. R., Huang, R. Y., Yu, W., Chua, K. N., Sun, W., Tu, T. Y., Bai, J., Sim, W. J., Zervantonakis, I. K., Thiery, J. P., and Kamm, R. D. (2013). Screening therapeutic EMT blocking agents in a three-dimensional microenvironment, *Integr. Biol. (Camb.)*, **5**, 381–389.

45. Nagrath, S., Sequist, L. V., Maheswaran, S., Bell, D. W., Irimia, D., Ulkus, L., Smith, M. R., Kwak, E. L., Digumarthy, S., Muzikansky, A., Ryan, P., Balis, U. J., Tompkins, R. G., Haber, D. A., and Toner, M. (2007). Isolation of rare circulating tumour cells in cancer patients by microchip technology, *Nature*, **450**, 1235–1239.

46. Chen, W., Weng, S., Zhang, F., Allen, S., Li, X., Bao, L., Lam, R. H., Macoska, J. A., Merajver, S. D., and Fu, J. (2013). Nanoroughened surfaces for efficient capture of circulating tumor cells without using capture antibodies, *ACS Nano*, **7**, 566–575.

47. Hou, H. W., Warkiani, M. E., Khoo, B. L., Li, Z. R., Soo, R. A., Tan, D. S., Lim, W. T., Han, J., Bhagat, A. A., and Lim, C. T. (2013). Isolation and retrieval of circulating tumor cells using centrifugal forces, *Sci. Rep.*, **3**, 1259.

48. Wang, S., Liu, K., Liu, J., Yu, Z. T., Xu, X., Zhao, L., Lee, T., Lee, E. K., Reiss, J., Lee, Y. K., Chung, L. W., Huang, J., Rettig, M., Seligson, D., Duraiswamy, K. N., Shen, C. K., and Tseng, H. R. (2011). Highly efficient capture of circulating tumor cells by using nanostructured silicon substrates with integrated chaotic micromixers, *Angew. Chem. Int. Ed. Engl.*, **50**, 3084–3088.

49. Zheng, X., Cheung, L. S., Schroeder, J. A., Jiang, L., and Zohar, Y. (2011). Cell receptor and surface ligand density effects on dynamic states of adhering circulating tumor cells, *Lab Chip*, **11**, 3431–3439.

50. Song, J. W., Cavnar, S. P., Walker, A. C., Luker, K. E., Gupta, M., Tung, Y. C., Luker, G. D., and Takayama, S. (2009). Microfluidic endothelium for studying the intravascular adhesion of metastatic breast cancer cells, *PLoS One*, **4**, e5756.

51. Yu, M., Bardia, A., Wittner, B. S., Stott, S. L., Smas, M. E., Ting, D. T., Isakoff, S. J., Ciciliano, J. C., Wells, M. N., Shah, A. M., Concannon, K. F., Donaldson, M. C., Sequist, L. V., Brachtel, E., Sgroi, D., Baselga, J., Ramaswamy, S., Toner, M., Haber, D. A., and Maheswaran, S. (2013). Circulating breast tumor cells exhibit dynamic changes in epithelial and mesenchymal composition, *Science*, **339**, 580–584.

52. Ameri, K., Luong, R., Zhang, H., Powell, A. A., Montgomery, K. D., Espinosa, I., Bouley, D. M., Harris, A. L., and Jeffrey, S. S. (2010). Circulating tumour cells demonstrate an altered response to hypoxia and an aggressive phenotype, *Br. J. Cancer*, **102**, 561–569.

53. Maheswaran, S., Sequist, L. V., Nagrath, S., Ulkus, L., Brannigan, B., Collura, C. V., Inserra, E., Diederichs, S., Iafrate, A. J., Bell, D. W., Digumarthy, S., Muzikansky, A., Irimia, D., Settleman, J., Tompkins, R. G., Lynch, T. J., Toner, M. and Haber, D. A. (2008). Detection of mutations in EGFR in circulating lung-cancer cells, *N. Engl. J. Med.*, **359**, 366–377.

54. Wang, S. J., Saadi, W., Lin, F., Minh-Canh Nguyen, C., and Li Jeon, N. (2004). Differential effects of EGF gradient profiles on MDA-MB-231 breast cancer cell chemotaxis, *Exp. Cell. Res.*, **300**, 180–189.

55. Saadi, W., Wang, S. J., Lin, F., and Jeon, N. L. (2006). A parallel-gradient microfluidic chamber for quantitative analysis of breast cancer cell chemotaxis, *Biomed. Microdevices*, **8**, 109–118.

56. Mosadegh, B., Saadi, W., Wang, S. J., and Jeon, N. L. (2008). Epidermal growth factor promotes breast cancer cell chemotaxis in CXCL12 gradients, *Biotechnol. Bioeng.*, **100**, 1205–1213.

57. Zheng, C., Zhao, L., Chen, G., Zhou, Y., Pang, Y., and Huang, Y. (2012). Quantitative study of the dynamic tumor-endothelial cell interactions through an integrated microfluidic coculture system, *Anal. Chem.*, **84**, 2088–2093.

58. Fu, Y., Chin, L. K., Bourouina, T., Liu, A. Q., and VanDongen, A. M. (2012). Nuclear deformation during breast cancer cell transmigration. *Lab Chip*, **12**, 3774–3778.

59. Wang, Y., Chen, Z. L., Xiao, L., Du, Z. Y., Han, X. X., Yu, X. D., and Lu, Y. L. (2012). Evaluating cell migration in vitro by the method based on cell patterning within microfluidic channels, *Electrophoresis*, **33**, 773–779.

60. Zhang, Q., Liu, T., and Qin, J. (2012). A microfluidic-based device for study of transendothelial invasion of tumor aggregates in realtime, *Lab Chip*, **12**, 2837–2842.

61. Abhyankar, V. V., Toepke, M. W., Cortesio, C. L., Lokuta, M. A., Huttenlocher, A., and Beebe, D. J. (2008). A platform for assessing chemotactic migration within a spatiotemporally defined 3D microenvironment, *Lab Chip*, **8**, 1507–1515.

62. Shin, M. K., Kim, S. K., and Jung, H. (2011). Integration of intra- and extravasation in one cell-based microfluidic chip for the study of cancer metastasis, *Lab Chip*, **11**, 3880–3887.

63. Chaw, K. C., Manimaran, M., Tay, E. H., and Swaminathan, S. (2007). Multi-step microfluidic device for studying cancer metastasis, *Lab Chip*, **7**, 1041–1047.

64. Chaw, K. C., Manimaran, M., Tay, F. E., and Swaminathan, S. (2006). A quantitative observation and imaging of single tumor cell migration and deformation using a multi-gap microfluidic device representing the blood vessel, *Microvasc. Res.*, **72**, 153–160.

65. Hou, H. W., Li, Q. S., Lee, G. Y., Kumar, A. P., Ong, C. N., and Lim, C. T. (2009). Deformability study of breast cancer cells using microfluidics, *Biomed. Microdevices*, **11**, 557–564.

66. Vu, N. T., Yu, Z. T., Comin-Anduix, B., Sondergaard, J. N., Silverman, R. W., Chang, C. Y., Ribas, A., Tseng, H. R., and Chatziioannou, A. F. (2011). A beta-camera integrated with a microfluidic chip for radioassays based on real-time imaging of glycolysis in small cell populations, *J. Nucl. Med.*, **52**, 815–821.

67. Sun, J., Masterman-Smith, M. D., Graham, N. A., Jiao, J., Mottahedeh, J., Laks, D. R., Ohashi, M., DeJesus, J., Kamei, K., Lee, K. B., Wang, H., Yu, Z. T., Lu, Y. T., Hou, S., Li, K., Liu, M., Zhang, N., Wang, S., Angenieux, B., Panosyan, E., Samuels, E. R., Park, J., Williams, D., Konkankit, V., Nathanson, D., van Dam, R. M., Phelps, M. E., Wu, H., Liau, L. M., Mischel, P. S., Lazareff, J. A., Kornblum, H. I., Yong, W. H., Graeber, T. G., and Tseng, H. R. (2010). A microfluidic platform for systems pathology: multiparameter single-cell signaling measurements of clinical brain tumor specimens, *Cancer Res.*, **70**, 6128–6138.

## Chapter 11

# Engineering Strategies to Recapitulate the Tumor Microenvironment

**Andrew Rape and Sanjay Kumar**

*Department of Bioengineering, University of California, Berkeley, 274A Stanley Hall #1762, Berkeley, CA 94720-1762, USA*

skumar@berkeley.edu

The tumor microenvironment is increasingly recognized as an important contributor to the processes of growth, invasion, and metastasis. Molecular and cellular targets within the microenvironment may thus offer new and fruitful targets for therapeutic intervention. However, progress in understanding and clinically leveraging this regulation has been limited by the complexity of the in vivo microenvironment, which does not easily allow clear dissection of specific regulatory effects. This in turn has created a strong need for in vitro engineered model systems that offer highly precise and independent control of a variety of extracellular parameters that can faithfully recapitulate the tumor microenvironment. In this chapter, we review recent progress in the development of such systems. We discuss microenvironmental signals that regulate tumor growth in vivo, focusing on the extracellular signals a cell receives in the most critical steps in

*Cells, Forces, and the Microenvironment*
Edited by Charles M. Cuerrier and Andrew E. Pelling

ISBN 978-981-4613-36-1 (Hardcover), 978-981-4613-37-8 (eBook)
www.panstanford.com

cancer progression, including tumor initiation, growth, and metastasis. We then review strategies that have been developed to recreate in vitro important aspects of the cancer microenvironment. Building upon and adapting the lessons learned from tissue engineering, we then prospectively explore two specific paradigms that may enable improved dissection of in vitro signals: decellularized matrices and synthetic matrices.

## 11.1 Introduction

Cancer remains one of the deadliest diseases in the United States, claiming more than 577,000 lives in 2012 alone. American men and women face lifetime probabilities of developing cancer of 45% and 38%, respectively [118]. Despite this high incidence and the pressing need for effective treatments, decades of intensive research have only produced a slow decline in mortality rates, suggesting an urgent need to revisit our traditional understanding of tumor progression and broaden the search for potential therapeutic targets.

An important goal in modern cancer therapeutics, including surgical resection, chemotherapy, and radiation therapy, is to selectively destroy or impair tumor cells while minimizing collateral damage to normal host cells. While radiation and systemic chemotherapy are certainly cytotoxic, their imperfect selectivity produces devastating side effects that frequently limit their deployment in the clinic. As a result, there has been an emphasis on improved characterization of tumor heterogeneity to identify the most important tumor-populating cells as well as the development of increasingly cancer-specific therapeutic approaches based on selective targeting of tumor cells based on their aberrant metabolism, receptor expression, or signal transduction.

In the quest to target tumor cells with increasing specificity, cell-intrinsic processes such as intracellular signaling cascades and cell proliferation have received much of the attention. However, mounting evidence indicates that the cell-extrinsic tumor microenvironment plays a critical role in controlling tumor behavior. Key components of the tumor microenvironment include theextracellular matrix (ECM), the solid-state biopolymeric scaffold that surrounds cells in tissue; stromal cells, which frequently regulate tumor growth through juxtacrine and paracrine signaling as well as ECM remodeling; and soluble factors such as growth factors and hormones. Just as tumors are frequently characterized by cell-intrinsic dysfunction, so too is the microenvironment often aberrant, which in turn suggests that the microenvironment may represent a relatively unexplored source of untapped therapeutic targets.

Because of the immense complexity of the microenvironment, it has been extremely challenging to determine how individual microenvironmental parameters contribute to tumor progression in vivo. For these reasons, there has been growing interest in engineering reductionist, in vitro model systems that mimic essential features of the in vivo microenvironment while allowing a level of reproducibility and control not achievable in vivo.

In addition to improving our mechanistic understanding of how the microenvironment regulates tumor progression, these reverse-engineered synthetic systems may offer new and unexpected opportunities for molecular screening and help to streamline the use of animal models for drug discovery and evaluation. This is analogous to the recent emergence of "organ-on-a-chip" platforms, which hold great promise as reductionist tissue-like units that could be integrated into microfluidic devices and used for high-throughput drug discovery [47]. Similarly, one could envision recapitulating key functional units of the tumor microenvironment on the microscale, incorporating tumor cells, and using the resulting cell-material platforms for discovery and screening. Such models may help bridge the gap between the culture dish and animal model and thereby improve the predictive power of these in vitro paradigms.

The goal of this chapter is to illustrate the ways the tumor microenvironment has been modeled in vitro and to describe how microenvironmental signals influence the many stages of cancer progression. We will begin with a brief overview of the roles of the ECM in controlling tumor progression, and then we will discuss in depth the many experimental systems used to model this complex environment in vitro. Finally, we will conclude with a discussion of how some next-generation microenvironmental platforms developed in the field of tissue engineering can be used and adapted for modeling the tumor microenvironment.

## 11.2 Tumor-Microenvironment Interactions

The extracellular environment plays a critical role in virtually all aspects of tumor progression, invasion, and metastasis, including aberrant cell growth, enlargement of the primary tumor, tumor angiogenesis, migration of tumor cells away from the primary tumor, survival in the bloodstream and lymphatics, and finally metastasis to distal tissues. In this section, we will briefly review key microenvironmental factors that potentially influence tumor cell behavior at each stage of malignant progression, and how these factors are believed to influence this progression. Our overview is intended to provide brief background for the rest of this chapter, which will focus the reverse-engineering and modeling of these interactions in vitro. We would refer the reader to one of the many excellent reviews on the subject for a more comprehensive background [12,21,46,70,115].

### 11.2.1 Primary Tumor

During the initial stages of cancer development, a tumor consists of a few cells, which are typically undetectable with routine clinical technology. The origin of these cells remains controversial and may vary from tumor to tumor—for example, various lines of evidence support models in which tumors arise from tissue cells that have de-differentiated, from tissue stem cells with deregulated cell cycle control, or both. Whatever their source, tumor cells share the property of dysfunctional and inappropriately regulated growth [91,134]. As with most normal cells, the growth and survival of cells in solid tumors often depends on the engagement of the ECM [13]. The ECM is a complex macromolecular network that includes proteins such as collagen, fibronectin, and laminin as well as proteoglycans, polysaccharides, and other biopolymers [151]. In addition to dictating tissue structure and mechanics, the ECM provides adherent cells with important biochemical and biophysical inputs that control cellular growth; for example, altered presentation or removal of these inputs can stimulate proliferation or trigger apoptosis. Integrins are the best-characterized system of receptors through which cells adhere to the ECM [48]. Upon heterotypic dimerization and binding to a cognate peptide motif in an ECM protein, these transmembrane proteins cluster and induce assembly of intracellular adhesive complexes at the cell–ECM interface that mechanically couple the ECM to the cytoskeleton and can regulate the activity of a variety of mitogenic signaling pathways, including those mediated by ERK, Src, and FAK. Importantly, the expression level and complement of both ECM proteins and integrins are significantly altered in many tumors, which may serve to enhance cell growth, motility, and other behaviors relevant to tumor progression [64,104,114]. In fact, these alterations have been explored as potential targets for therapeutic intervention, such as monoclonal antibodies targeted against integrins [36] and tumor-enriched ECM proteins [97].

While integrins have long been understood to signal through classical receptor-mediated mechanisms, it has become clear over the past two decades that mechanical context in which integrins are engaged can strongly influence the resulting signaling. For example, it is now widely acknowledged that ECM stiffness can strongly regulate a wide variety of fundamental cell behaviors and that integrins play a critical, proximal role in stiffness-sensing. For example, increased matrix stiffness can strongly enhance cell division both in vitro and in vivo [55,67,132]. Tumors are often clinically identified by their increased stiffness [8,53], leading to the notion that this increase in stiffness may not only be a byproduct of tumor development, but that it may play a functional role in this progression [63,86,93].

As deregulated cell division continues, the tumor reaches a critical size, approximately 2–3 mm, beyond which the replication of tumor cells near the center of the tumor is limited by the diffusion of key nutrients through the tumor [26,28,139]. To overcome this critical bottleneck, the tumor secretes soluble factors, including vascular endothelial growth factor (VEGF) [23] and basic fibroblast growth factor (bFGF) [15,31,45,135], which serve to recruit endothelial cells from the host vasculature to form new vessels to supply the tumor. This process, known as tumor angiogenesis, allows for the delivery of oxygen and key metabolic precursors, facilitates the removal of metabolic waste products, and enables the tumor to grow substantially larger than the millimeter-scale size limits imposed by diffusion. More recently, it has become clear that tumor cells themselves may contribute directly to this process and thereby circumvent therapies that target host endothelial recruitment. For example, many tumor initiating cells may differentiate into endothelial cells that compete with host endothelial cells to populate angiogenic vessels [11,89,100,121,123].

Tumor cells often account for a comparatively modest fraction of the overall tumor mass [50], with the bulk of the tumor made up by fibrotic material and other deposited ECM, tumor-associated stromal cells such as immune cells, fibroblasts [82,141], and angiogenic endothelial cells. These tumor-associated cells often differ significantly from their counterparts found in normal tissue and indeed may contribute significantly to the progression of the tumor. Histopathological studies consistently show that the ratio of myofibroblasts to fibroblasts is enhanced at the edge of tumor, reminiscent of a wound healing response [7,107]. Often thought of as "activated fibroblasts," which may arise from the transformation of normal fibroblasts, these cells promote tumor growth through enhanced deposition of ECM and paracrine signaling through the release of growth factors and cytokines, such as stromal-cell derived factor 1 (SDF-1) [20,82].

Unregulated tumor growth, incomplete and disorganized angiogenesis, and an underdeveloped lymphatic system in the tumor mass leads to interstitial fluid pressures in the growing tumor that are often 50 times higher than that of normal tissue [43,66,119]. This interstitial pressure can lead directly to enhanced proliferation within the tumor, thus contributing to a positive feedback cycle in which enhanced pressures lead to enhanced growth, which further increases interstitial pressure [44,128]. In addition, increased pressure within the tumor can lead directly or indirectly to further angiogenesis through the regulations of VEGF or hypoxia-induced angiogenesis [42,75,102].

## 11.2.2 Metastasis

The majority of cancer deaths results from metastasis, which is characterized by the extrusion of tumor cells from the primary tumor, invasion of tumor cells into tissue, entry into the bloodstream and/or lymphatic system, transendothelial penetration of distal tissues, and establishment of a secondary tumor in that distal tissue [71,140].

In addition to its role in contributing to tumor growth, increased interstitial pressure may facilitate the migration of individual cells or small clusters of cells to break free from the tumor and migrate through the surrounding environment and, eventually, into the bloodstream. This increased intra-tumoral pressure may be transmitted to the surrounding healthy tissue, which can lead to thinning and/or rupture of that tissue's vascular basement membranes, facilitating the escape of tumor cells from the primary tumor [49,86].

If a tumor cell successfully escapes the primary tissue and enters the bloodstream, it is exposed to shear stresses in the vasculature, which can surprisingly enhance the ability of the cell to adhere to distal tissue endothelia, intravasate through the endothelial layer, and establish a secondary tumor [120]. Importantly, most metastatic tumors exhibit strong "tissue tropism," an observation first made by Paget more than a century ago [25,99]. This observation, frequently articulated as the "seed and soil" hypothesis, posits that the cells from specific tumors are calibrated to metastasize to and colonize in specific organs [84]. These and other observations have fueled the hypothesis that specific organs are especially hospitable for the establishment and metastasis of specific tumors by virtue of the physiological milieu they offer the tumor cells. In the years since Paget's observation, the biological basis for this tissue tropism has become better understood, with some lines of evidence pointing to the production of specific growth factors by the host organ [24] or the biophysical parameters of the host organ [57].

## 11.3 Engineering the Cancer Microenvironment

As illustrated above, extracellular signals in the tumor microenvironment collude with cell-intrinsic lesions to drive malignant progression. In vivo, cells must simultaneously integrate many extracellular signals at once, making it challenging to experimentally dissect how each signal individually contributes to phenotype. As discussed earlier, this need has spurred efforts to "reverse engineer" aspects of the in vivo environment in the form of simple, reproducible, and experimentally tunable paradigms in which individual extracellular cues relevant to tumor progression may be systematically manipulated.

### 11.3.1 Natural Matrices

As noted previously, tumor cells must successfully migrate through tissue ECM as they dissociate from the primary tumor and locally invade the host tissue. Initial efforts to create culture paradigms to investigate and dissect this process have emphasized reconstitution of full-length (native) ECM proteins in three-dimensional hydrogels. Given that collagen is the most abundant ECM protein in connective tissues and the most abundant protein in the body as a whole, attempts at recreating a simple, environment-mimetic system featured the encapsulation of cells within reconstituted collagen I scaffolds [65]. In this paradigm, collagen I is isolated from animal connective tissue, e.g., rat tail or bovine skin, and concentrated in an acidic environment. To create a gel from this precursor, the solution is neutralized and warmed to 37°C, which induces fiber assembly and gelation. Though simple, this paradigm has yielded powerful insight into basic mechanisms of tumor cell motility, including the importance of matrix metalloproteases (MMPs) in the digestion of ECM prior to cell migration [109–111].

One of the major drawbacks to using a simple collagen I gel is that it lacks many of the potentially bioactive components of the basement membrane in vivo, including the glycoprotein laminin, which has been shown to specifically promote attachment of certain tumor and cell types [125], and the glycoaminoglycans that normally decorate collagen fibers in vivo. This in turn has spurred efforts to develop reconstituted in vitro matrix preparations that retain these important properties. However, the relative scarcity of basement membrane in typical tissue sources precludes its high-yield isolation. A key step in overcoming this limitation was the recognition that murine Engelbreth–Holm–Swarm (EHS) tumors produce large quantities of laminin- and collagen IV-rich ECM, which compose a substantial portion of many basement membranes. By decellularizing the EHS tumor and solubilizing the remaining ECM, it was possible to isolate large quantities of basement membrane for reconstitution studies. This material, which is now marketed as Matrigel™, forms a gel when warmed to 37°C [56]. Matrigel™ can support the growth of a wide range of tumor cell types [30,73,80,92,145–147] and bears some tumor-specificity; for example, Matrigel™ can support the 3D migration of fibrosarcomas but not non-transformed fibroblasts [58].

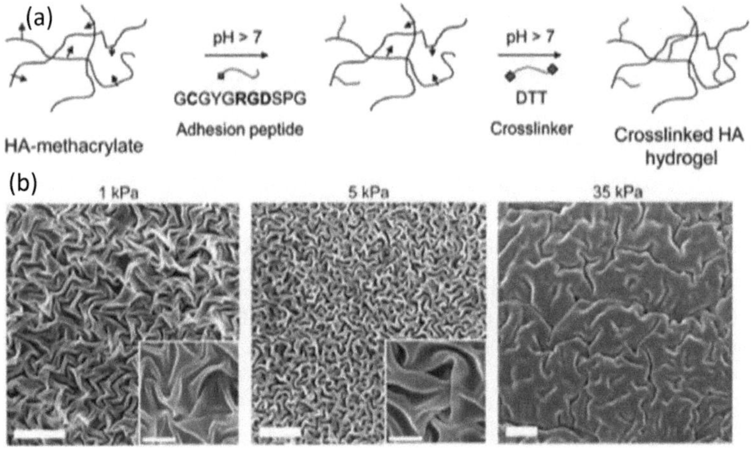


**Figure 11.1** Brain-mimetic in vitro environments created from modified hyaluronic acid. (a) Hyaluronic acid is first treated with methacrylic anhydride to create HA-methacrylate. HA-methacrylate is then reacted with an integrin-adhesive peptide containing a cysteine, causing the free thiol and methacrylate to undergo a Michael-type addition resulting in covalent attachment of RGD-containing peptide to the HA backbone. The matrix is then cross-linked with a bi-functional thiol-containing molecule, such as dithiothreitol (DTT). (b) The resulting matrix, as seen with SEM, is a dense, folded sheet with a range of mesh sizes similar to native brain. Adapted from Ananthanarayanan et al., *Biomaterials*, 2011 [9]. Scale bar = 20 μm; insert scale bar = 5 μm.

While collagen and Matrigel™ may be suitable platforms for investigating tumors that arise in connective tissues may thus have close association with epithelial basement membranes, these materials are less appropriate for modeling other tissue types. For example, brain ECM is composed largely of the glycosaminoglycan hyaluronic acid (HA), which is nonfibrillar and plays important structural and biochemical roles in normal brain function and tumor progression. Importantly, the highly malignant brain tumor glioblastoma multiforme (GBM) shows enhanced deposition of HA and expression of its cellular receptor CD44. Moreover, CD44 expression correlates with both tumor-initiating capacity and poor patient prognosis, suggesting that the CD44-HA interaction plays an integral role in the progression of the disease [94,96,129,138]. Reconstituted HA hydrogels are used extensively in the fields of tissue engineering, regenerative medicine, and stem cell biology; however, it is only recently that HA has been used to study the migratory properties of tumors originating in the brain. Often HA is modified to yield cross-linkable moieties so that it may produce a robust gel with tunable mechanical properties [3,59,68]. While a wide variety of chemical strategies could conceivably be used to introduce cross-links, in practice many are based on modification of the carboxyl or hydroxyl groups on the HA backbone to yield a reactive moiety (e.g., acrylate or thiol groups), which may then be conjugated to one another with a bifunctional cross-linker (Fig. 11.1) [3,59,68].

### 11.3.2 Chemotaxis

Migration away from a host tumor by a cell or group of cells often is influenced by chemotaxis, which is defined as the directed migration of a cell or a collection of cells down a gradient of a soluble molecule known as a chemokine. In tumors, these chemokines often take the form of growth or motility factors secreted by endothelial cells, stromal cells, or tumor cells at the invasive front [103].

Chemotaxis has been extensively recreated in vitro [54], with the Boyden chamber assay serving as perhaps the most widely used paradigm for studying this phenomenon [9]. In this system, cells are cultured atop a porous, compliant matrix platform that allows transmigration of cells. This device is then placed in a larger container such as a well of a tissue culture plate that contains culture medium with a defined concentration of serum or one or more specific chemokines. The presence of the chemokine causes the cells to migrate through the matrix and then through the pores of the flexible membrane. The number of cells that cross the membrane and time required to do so serve as quantitative measures of the chemotactic strength of the chemokine. While this method is simple and easy to use, it does have some drawbacks: First, because the axis of migration is in the vertical dimension, this method does not readily permit high-resolution optical imaging of the transmigration process. Second, the chemotactic gradients established in this method are not, in general, quantitatively well-defined, and the population-based nature of the readout averages out cell-to-cell variations in the population and time-dependent changes in cell morphology during the migration process.

There are other simple methods to establish chemokine gradients, such as the release of a chemoattractant from a microcapillary tube [112,124]. This method allows for direct visualization of the cell during chemotaxis; however, the gradient established is neither well-defined nor at steady state. Other simple methods to study chemotaxis include the Zigmond and Dunn chambers [149,150], in which two wells of differing chemokine concentration are connected by a bridge containing cells to establish a gradient, and the under-agarose method, where two solutions are separated by a block of agarose in which cells migrate under the agarose toward the solution of higher chemokine concentration. While more amenable to live cell microscopy than the Boyden chamber assay, these methods suffer from lack of reproducibility and the use of an ill-defined gradient [78]. Furthermore, all of the methods described thus far only established gradients at scales much larger than the cell.

Microfluidic devices, which enable control of flow at the microscale, offer impressively high parallelization and yield extremely precise chemical gradients (Fig. 11.2). One of the first microfluidic systems for the investigation of chemotaxis was developed by the laboratory of George Whitesides, and was subsequently adapted to cancer cells [51,72,133]. In this innovative study, cells were cultured on a microfabricated polydimethylsiloxane (PDMS) device in which cells were exposed to two distinct flow streams, one with and the other without the chemotokine CXCL12. Under laminar flow, two distinct fluid streams have a predictable, steady-state mixing pattern, permitting the investigation of chemotaxis in a well-defined chemokine gradient. This study revealed that cells were only responsive to the presence of CXCL12 gradient when EGF was present in the media. However, an important potential drawback of this system is that it exposes cells to shear stress, which can exert potentially confounding effects, such as altering cell proliferation and stiffness, complicating the interpretation of such experiments [78]. As this technology has progressed, microfluidic devices have also been developed that establish well-defined gradients based on diffusion rather than convection, which reduces the amount of shear stress experienced by cells [1,14].

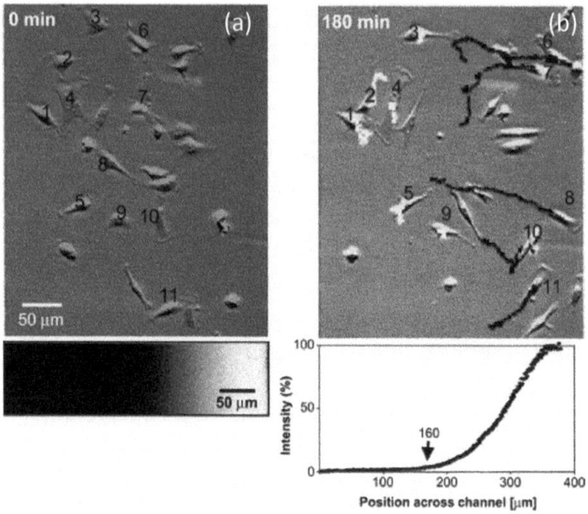


**Figure 11.2** Controlling MDA-MB-231 cell migration with a non-linear gradient of epidermal growth factor (EGF). MDA-MB-231 breast cancer cells were followed with phase-contrast microscopy over a period of three hours in a field with a non-linear gradient of EGF concentration. (a) Initially, cells were evenly distributed across the field of view. (b) After three hours, a percentage of the cells migrated persistently towards higher concentrations of EGF (b; black lines). Figure adapted from Wang et al., *Exp. Cell Res.*, 2004 [133].

Combining chemotactic cues with other matrix-associated cues promises to yield insight into the complex factors guiding tumor cell migration. For example, recognizing the importance of both the dimensionality of the matrix and the role of specific adhesion ligands in cell migration, it was shown that cells cultured in a modified microfluidic device containing 3D matrices of either collagen or Matrigel™ were able to respond to chemotactic gradients even in the presence of insoluble, ECM-based signals [2,18,39,40,105,113]. Engineering further complexity into these devices and improving the accessibility of them should greatly enhance the understanding of chemotaxis in tumor environments.

### 11.3.3 Interstitial Pressure and Flow

The effect of interstitial pressure and fluid flow on tumor cell migration has been convincingly established [44], which has fueled efforts to engineer in vitro model systems in which interstitial pressures and flows can be established and controlled. These approaches have exploited both gravity-driven and pump-driven pressure heads. In an especially elegant example of the former approach, Shields and colleagues cultured various breast cancer cell lines in the within a thin (~1 mm) layer of Matrigel™, placed this cell/gel composite in a Boyden Chamber, and then imposed a gravitational flow by placing culture medium atop the Matrigel™ [117].

Many systems have also been developed that use pumps to establish pressure gradients through matrices. By using a peristaltic pump to move fluid vertically through a circular collagen slice, Swartz and colleagues created a radially symmetric flow profile through the gel and discovered that dermal fibroblasts aligned perpendicularly to the flow direction [79]. Microfluidic devices can also be designed to establish interstitial flows, for example, by creating a device with a collagen matrix sandwiched between two reservoirs of culture media and applying a high pressure to one of the reservoirs. Breast carcinoma cells were found to migrate along the streamlines imposed by the flow field [90].

### 11.3.4 Mechanical Properties of the Tumor

In addition to active processes such as fluid flow and chemical stimulation, tumor cells respond to static mechanical properties of their microenvironment, including the stiffness and dimensionality of the surrounding matrix. In the past decades, many systems have been developed that can recapitulate important features of a tumor's mechanical environment in vitro.

In one of the first attempts to systematically study effects of ECM stiffness on tumor proliferation, it was discovered that *h-ras*-transformed fibroblasts could rescue the suppression of proliferation normally observed on soft ECMs [132]. This study took advantage of polyacrylamide-based matrices, in which stiffness is varied by adjusting the amount cross-linker, bis-acrylamide, while maintaining a constant acrylamide content. The surface is then chemically conjugated with a specific ECM protein to promote cell adhesion. Subsequent studies have used the same principle to vary the stiffness of PDMS gels [38]. PDMS gels are hydrophobic, but can be activated for protein adsorption by treatment with plasma.

Using PDMS as a cell culture system significantly simplifies the ECM attachment step but suffers from some loss of control of coating density. Polyacrylamide hydrogels and PDMS are still widely used for many experiments today. While the polyacrylamide system has long been assumed to afford orthogonal control over ECM stiffness and biochemistry, this has recently been called into question by the suggestion that cross-linker concentration alters surface porosity, and that this in turn influences cell behavior by affecting the density of conjugation points of the ECM protein to the polyacrylamide surface [130].

An important drawback of these approaches is that they only study the response of cells in two dimensions. The presence of a 3D ECM has not only been shown to influence the growth rate of tumor cells, [27] but it has also been shown capable of reversing a malignant and invasive phenotype in a breast cancer model [131,136,137]. Proper 3D matrix cues have also been shown to play an important role in maintaining the proper differentiation state of many cells, including hepatocytes [6]. Recent advances in 3D culture technology have allowed for the creation of defined 3D tissues in collagen gels, which has led to the discovery that the geometry of these tissues can affect the epithelial-to-mesenchymal transition [37,77]. Furthermore, the biophysical properties of the 3D matrices, especially stiffness and pore size, can be adjusted by a variety of techniques, including double-emulsion polymerization [88,122] and micropatterning [87]. High-throughput methods to systematically manipulate these and other properties offer promise for disentangling the complex inputs that cells receive from 3D matrices [95].

### 11.3.5 Tumor-Associated Fibroblasts

As noted previously, a significant portion of a solid tumor consists of tumor-associated fibroblasts, suggesting that these cells play a role in disease progression. Co-culturing tumor cells with normal fibroblasts has revealed that fibroblasts produce many soluble factors that promote tumor cell invasion [7,82,141]. Fibroblasts may also independently promote tumor cell migration away from the tumor through ECM remodeling. Using an in vitro mimic of tumor invasion in which fibroblasts and cancer cells were cultured simultaneously upon a layer of collagen gel, it has been observed that fibroblasts serve as leader cells—that is, they degrade and remodel the ECM to leave tracks that tumor cells migrate along as they leave the host tumor [32,116].

### 11.3.6 Angiogenesis

As described earlier, diffusion of material to and from nearby blood vessels is a critical bottleneck to the growth of solid tumors. Thus, the recruitment of new blood vessels from the host vasculature (angiogenesis) is key to continued tumor growth and has been extensively studied and modeled in vitro.

Recently, assays that simulate the formation of blood vessels in vitro have become increasingly popular. In these assays, endothelial cells are cultured on an ECM substrate—typically gels formed from fibrin, collagen I, Matrigel™, or a synthetic polymer such as polyacrylamide—and over time the cells are observed as they migrate, proliferate, and self-organize into structures that resemble vessels, including the formation of tight junctions between cells [4,52,60,98]. Similarly, cells encapsulated within 3D matrices also are able to self-organize and form 3D branched structures [33,34]. Modeling of angiogenic sprouting is often accomplished by culturing cells on latex or dextran "carrier beads," embedding them into a hydrogel (e.g., fibrin) and observing the rate at which the cells leave the beads and penetrate into the gels (Fig. 11.3) [76,148]. Recent work has focused on establishing highly defined gradients of soluble or ECM-bound endothelial chemoattractants to study angiogenesis. Angiogenesis can be spatially patterned by confining pro-angiogenic VEGF and anti-angiogenic VEGF antibodies in agarose hydrogels, which can create highly defined, and temporally stable regions of an ECM hydrogel amenable to angiogenesis [148]. Similar control of pro-angiogenic and anti-angiogenic growth factors can be achieved using microfluidics [22]. In addition to spatially and temporally controlling angiogenesis with soluble signals, it has been shown that patterning PEG hydrogels with light to contain spatially localized zones of the integrin-adhesive peptide RGD or bound VEGF can restrict the zone where tubule formation occurs [61].

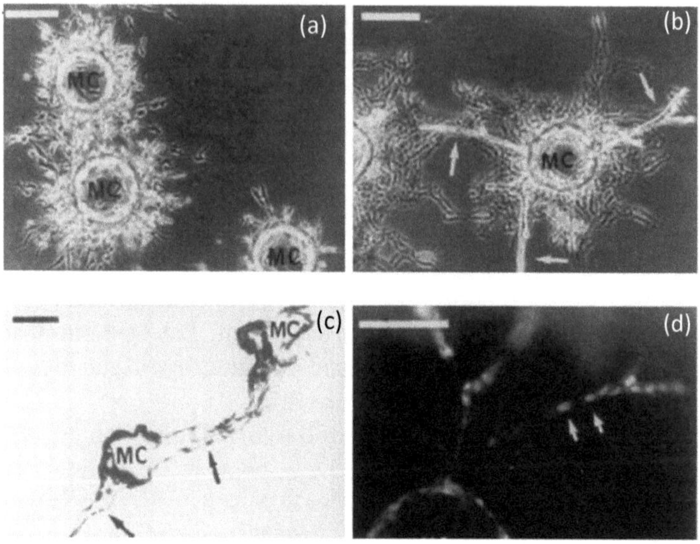


**Figure 11.3** Modeling angiogenesis with cell-laden carrier beads embedded in fibrin gels. Human endothelial cells were cultured on carrier beads (MC), embedded in fibrin hydrogels, and observed after three days. (a) Absent addition of growth factors, cells migrated away from the bead but did not form capillaries. The addition of basic fibroblast growth factor (bFGF) resulted in migration away from the bead and formation of vessel-like constructs (white arrows) as observed by phase contrast microscopy (b), methylene blue staining (c), and fluoresecent labeling of nuclei with bis-benzimide (d). Adapted from Nehls and Drenckhahn, *Microvasc. Res.*, 1995 [76]. Scale bars = 100 μm.

### 11.3.7 Extravasation

Model systems have also emerged to recapitulate an important and clinically undetectable early step in cancer progression: the extravasation of circulating tumor cells from the bloodstream through the endothelium. The simplest and most widely used system is a parallel plate flow chamber system in which endothelial cells are cultured on the glass coverslip that serves as a model endothelium [16,17,35]. A second glass coverslip is suspended above the epithelial layer and sealed at the edge to provide a closed flow chamber. Finally, inlet and outlet valves attached to a peristaltic pump are added to the chamber. Perfusion of cell culture medium containing suspended cancer cells through the chamber allows for observation with light microscopy of the cells as they adhere to, roll along, and finally extravasate through the endothelium.

## 11.4 Future Directions and Conclusions

Clearly, efforts to engineer the tumor microenvironment have produced tremendous insight into the role of individual microenvironmental cues in regulating cancer progression. However, the weaknesses of these approaches are derived from their strengths, in that cells in vivo do not experience environmental cues in isolation but rather as a cacophony of simultaneous inputs. By engineering additional complexity into these model systems, it will be possible to maintain the reproducibility of experiments while presenting cells with increasingly engaging environments.

### 11.4.1 Next-Generation Matrices

The most obvious room for improvement in the engineering of the tumor microenvironment is in the design of matrices that more closely mimic the ECM in which the tumor grows in vivo. Tissue engineers have developed an impressive collection of both synthetic and natural matrices that allow for increasingly precise and multiplexed manipulation of the extracellular environment. Perhaps one of the most powerful set of ECM systems in tissue engineering, given their ability to capture en masse the complex biochemical and architectural features of the ECM that guide cells to produce functional organs, are decellularized matrices. In these systems, ECMs are generated by removing cells with detergent treatment from a whole organ [5,83]. In the context of tumor biology, one could envision using this approach to remove cells from a primary or secondary tumor while retaining salient physical and chemical cues encoded or embedded in the matrix. A key limitation to the use of decellularized matrices is batch-to-batch variability, as every tumor environment is likely to differ in subtle but functionally important ways. Additionally, these matrices also may contain poorly defined quantities of soluble and matrix-bound factors, which may confound the interpretation of results from these studies. Despite these limitations, decellularized

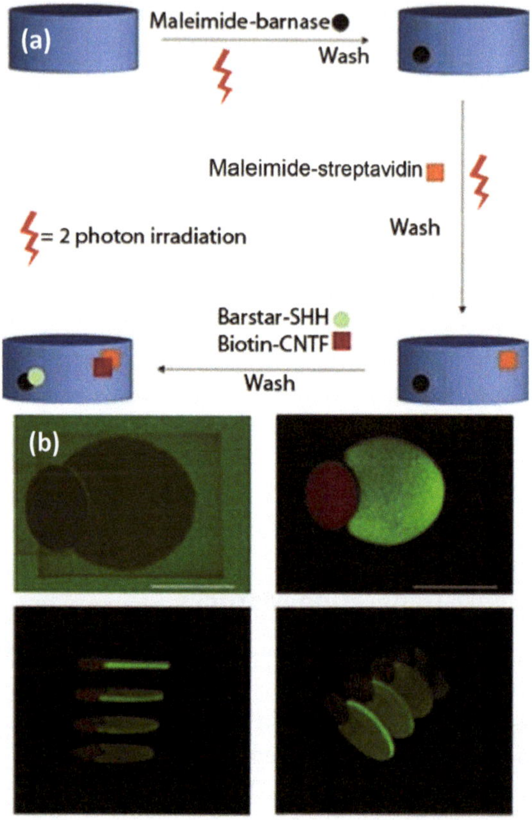


**Figure 11.4** Simultaneous patterning of multiple growth factors using two-photon deprotection of thiol groups. A coumarin-based modality protects thiol groups that were grafted onto the backbone of an agarose hydrogel (a). The hydrogel was incubated in a solution containing a maleimide-modified barnase group. Two-photon absorption was used to remove the protecting group, exposing free thiols to react with maleimide functional groups resulting in the covalent attachment of barnase to the agarose backbone. The unreacted maleimide–barnase complex was removed by thorough washing, and the process was repeated using a maleimide-streptavidin complex. By introducing the growth factors sonic hedgehog (SHH) and ciliary neurotrophic factor (CNTF) attached to barstar or biotin, the highly specific binding partners of barnase and streptavidin, respectively, there was high-precision localization of the two growth factors in three dimensions (b). Adapted from Wylie et al., *Nature Materials*, 2011 [144]. Scale bar = 100 μm.

matrices represent an important and relatively untapped resource in cancer biology and indeed are already beginning to be used in the study of cancer angiogenesis [10].

Alternatively, the tissue engineering field has produced an impressive cadre of modified natural matrices and fully synthetic matrices with many orthogonally tunable properties. Recent work has shown the ability to pattern stiffness in hyaluronic acid and polyacrylamide hydrogels [68,142]. Similarly, three-dimensional patterning of ligands has been shown in many different types of matrices using two-photon patterning and photocaging of specific reaction sites tethered to the polymer backbone (Fig. 11.4) [41,74,143,144].

A wide array of synthetic matrices has also been developed that afford an impressive amount of functionality and customizability [126]. Generally, these matrices involve the use of a bio-inactive polymer backbone that can be functionalized in a variety of ways and conjugated with ligands and cross-linked to control matrix stiffness. These polymer scaffolds may be functionalized by including cell adhesive domains such as the peptide sequence RGD [81,85], or by including cell-degradable peptide sequences [62,101,106]. Additionally, by controlling the starting polymer subunit and the polymerization conditions, one can actively control the material properties of these matrices, including scaffold pore size, structure, and stiffness [19,69,108]. Further spatiotemporal complexity can be engineered into the matrix by adding moieties that can undergo photopolymerization or photodegradation [29,126,127]. This rich degree of design flexibility should enable researchers to custom tailor the synthetic matrix to accurately model the complex features of individual tumor environments.

Finally, to fully capture the complexities of the in vivo microenvironment, it will be important to develop strategies for integrating these complex matrices with other reverse-engineered systems representative of the cancer environment. Once this is achieved, it is conceivable that these model tumors will provide the basis for high-throughput drug discovery platforms, along the lines of motivations for current organ-on-a-chip efforts. In an ambitious view, successful application of this technology will not only immensely accelerate the drug discovery process but should also significantly reduce discovery costs and increase the efficiency of the translational pipeline.

## 11.5 Conclusions

The tumor environment is a complex milieu that contains many extracellular signals that contribute to tumor growth, progression, angiogenesis, and metastasis. Over the past few decades, enormous effort has been devoted to developing reductionist platforms that have allowed investigators to dissect the contributions of individual physical and biochemical factors to tumor progression. Much work remains to integrate these technologies to produce truly tumor microenvironment-mimetic platforms, which should continue to facilitate both basic scientific investigation and high-throughput drug discovery.

### Acknowledgments

The authors gratefully acknowledge the support of the National Institutes of Health (1DP2OD004213, 1U54CA143836, and 1F32CA174361) and the National Science Foundation (CMMI 1105539).

### References

1. Abhyankar, V. V., Lokuta, M. A., Huttenlocher, A., and Beebe, D. J. (2006). Characterization of a membrane-based gradient generator for use in cell-signaling studies, *Lab Chip*, **6**, 389–393.
2. Abhyankar, V. V., Toepke, M. W., Cortesio, C. L., Lokuta, M. A., Huttenlocher, A., and Beebe, D. J. (2008). A platform for assessing chemotactic migration within a spatiotemporally defined 3D microenvironment, *Lab Chip*, **8**, 1507–1515.
3. Ananthanarayanan, B., Kim, Y., and Kumar, S. (2011). Elucidating the mechanobiology of malignant brain tumors using a brain matrix-mimetic hyaluronic acid hydrogel platform, *Biomaterials*, **32**, 7913–7923.
4. Auerbach, R., Lewis, R., Shinners, B., Kubai, L., and Akhtar, N. (2003). Angiogenesis assays: A critical overview, *Clin. Chem.*, **49**, 32–40.
5. Badylak, S. F., Taylor, D., and Uygun, K. (2011). Whole-Organ Tissue Engineering: Decellularization and Recellularization of Three-Dimensional Matrix Scaffolds, *Annu. Rev. Biomed. Eng.*, **13**, 27–53.

6. Berthiaume, F., Moghe, P. V., Toner, M., and Yarmush, M. L. (1996). Effect of extracellular matrix topology on cell structure, function, and physiological responsiveness: Hepatocytes cultured in a sandwich configuration, *FASEB J.*, **10**, 1471–1484.
7. Bhowmick, N. A., Neilson, E. G., and Moses, H. L. (2004). Stromal fibroblasts in cancer initiation and progression, *Nature*, **432**, 332–337.
8. Boyd, N. F., Dite, G. S., Stone, J., Gunasekara, A., English, D. R., McCredie, M. R. E., Giles, G. G., Tritchler, D., Chiarelli, A., Yaffe, M. J., and Hopper, J. L. (2002). Heritability of mammographic density, a risk factor for breast cancer, *N. Engl. J. Med.*, **347**, 886–894.
9. BOYDEN, S. (1962). The chemotactic effect of mixtures of antibody and antigen on polymorphonuclear leucocytes, *J. Exp. Med.*, **115**, 453–466.
10. Burns, J. S., Kristiansen, M., Kristensen, L. P., Larsen, K. H., Nielsen, M. O., Christiansen, H., Nehlin, J., Andersen, J. S., and Kassem, M. (2011). Decellularized matrix from tumorigenic human mesenchymal stem cells promotes neovascularization with galectin-1 dependent endothelial interaction, *PLoS ONE*, **6**, e21888.
11. Bussolati, B., Grange, C., Sapino, A., and Camussi, G. (2009). Endothelial cell differentiation of human breast tumour stem/progenitor cells, *J. Cell. Mol. Med.*, **13**, 309–319.
12. Butcher, D. T., Alliston, T., and Weaver, V. M. (2009). A tense situation: Forcing tumour progression, *Nat. Rev. Cancer*, **9**, 108–122.
13. Carney, D. N., Gazdar, A. F., and Minna, J. D. (1980). Positive correlation between histological tumor involvement and generation of tumor cell colonies in agarose in specimens taken directly from patients with small-cell carcinoma of the lung, *Cancer Res.*, **40**, 1820–1823.
14. Cheng, S.-Y., Heilman, S., Wasserman, M., Archer, S., Shuler, M. L., and Wu, M. (2007). A hydrogel-based microfluidic device for the studies of directed cell migration, *Lab Chip*, **7**, 763–769.
15. Chodak, G. W., Hospelhorn, V., Judge, S. M., Mayforth, R., Koeppen, H., and Sasse, J. (1988). Increased levels of fibroblast growth factor-like activity in urine from patients with bladder or kidney cancer, *Cancer Res.*, **48**, 2083–2088.
16. Chotard-Ghodsnia, R., Drochon, A., and Grebe, R. (2002). A new flow chamber for the study of shear stress and transmural pressure upon cells adhering to a porous biomaterial, *J. Biomech. Eng.*, **124**, 258–261.

17. Chotard-Ghodsnia, R., Haddad, O., Leyrat, A., Drochon, A., Verdier, C., and Duperray, A. (2007). Morphological analysis of tumor cell/endothelial cell interactions under shear flow, *J. Biomech.*, **40**, 335–344.

18. Cross, V. L., Zheng, Y., Won Choi, N., Verbridge, S. S., Sutermaster, B. A., Bonassar, L. J., Fischbach, C., and Stroock, A. D. (2010). Dense type I collagen matrices that support cellular remodeling and microfabrication for studies of tumor angiogenesis and vasculogenesis in vitro, *Biomaterials*, **31**, 8596–8607.

19. DeForest, C. A., Sims, E. A., and Anseth, K. S. (2010). Peptide-Functionalized Click Hydrogels with Independently Tunable Mechanics and Chemical Functionality for 3D Cell Culture, *Chem. Mater.*, **22**, 4783–4790.

20. Dingemans, K. P., Zeeman-Boeschoten, I. M., Keep, R. F., and Das, P. K. (1993). Transplantation of colon carcinoma into granulation tissue induces an invasive morphotype, *Int. J. Cancer*, **54**, 1010–1016.

21. Discher, D. E., Janmey, P., and Wang, Y. (2005). Tissue cells feel and respond to the stiffness of their substrate, *Science*, **310**, 1139–1143.

22. Farahat, W. A., Wood, L. B., Zervantonakis, I. K., Schor, A., Ong, S., Neal, D., Kamm, R. D., and Asada, H. H. (2012). Ensemble analysis of angiogenic growth in three-dimensional microfluidic cell cultures, *PLoS ONE*, **7**, e37333.

23. Ferrara, N. (2002). VEGF and the quest for tumour angiogenesis factors, *Nat. Rev. Cancer*, **2**, 795–803.

24. Fidler, I. J. (2003). The pathogenesis of cancer metastasis: The "seed and soil" hypothesis revisited, *Nat. Rev. Cancer*, **3**, 453–458.

25. Fokas, E., Engenhart-Cabillic, R., Daniilidis, K., Rose, F., and An, H.-X. (2007). Metastasis: The seed and soil theory gains identity, *Cancer Metastasis Rev.*, **26**, 705–715.

26. Folkman, J. (1971). Tumor Angiogenesis: Therapeutic Implications, *New Engl. J. Med.*, **285**, 1182–1186.

27. Folkman, J., and Greenspan, H. P. (1975). Influence of geometry on control of cell growth, *Biochim. Biophys. Acta (BBA)—Rev. Cancer*, **417**, 211–236.

28. Folkman, J., and Klagsbrun, M. (1987). Angiogenic factors, *Science*, **235**, 442–447.

29. Frey, M. T., and Wang, Y.-L. (2009). A photo-modulatable material for probing cellular responses to substrate rigidity, *Soft Matter*, **5**, 1918–1924.

30. Fridman, R., Giaccone, G., Kanemoto, T., Martin, G. R., Gazdar, A. F., and Mulshine, J. L. (1990). Reconstituted basement membrane (matrigel) and laminin can enhance the tumorigenicity and the drug resistance of small cell lung cancer cell lines, *Proc. Natl. Acad. Sci. U.S.A.*, **87**, 6698–6702.

31. Fujimoto, K., Ichimori, Y., Kakizoe, T., Okajima, E., Sakamoto, H., Sugimura, T., and Terada, M. (1991). Increased serum levels of basic fibroblast growth factor in patients with renal cell carcinoma, *Biochem. Biophys. Res. Commun.*, **180**, 386–392.

32. Gaggioli, C., Hooper, S., Hidalgo-Carcedo, C., Grosse, R., Marshall, J. F., Harrington, K., and Sahai, E. (2007). Fibroblast-led collective invasion of carcinoma cells with differing roles for RhoGTPases in leading and following cells, *Nat. Cell Biol.*, **9**, 1392–1400.

33. Gagnon, E., Cattaruzzi, P., Griffith, M., Muzakare, L., LeFlao, K., Faure, R., Béliveau, R., Hussain, S. N., Koutsilieris, M., and Doillon, C. J. (2002). Human vascular endothelial cells with extended life spans: In vitro cell response, protein expression, and angiogenesis, *Angiogenesis*, **5**, 21–33.

34. Ghosh, K., Thodeti, C. K., Dudley, A. C., Mammoto, A., Klagsbrun, M., and Ingber, D. E. (2008). Tumor-derived endothelial cells exhibit aberrant Rho-mediated mechanosensing and abnormal angiogenesis in vitro, *Proc. Natl. Acad. Sci. U. S. A.*, **105**, 11305–11310.

35. Giavazzi, R., Foppolo, M., Dossi, R., and Remuzzi, A. (1993). Rolling and adhesion of human tumor cells on vascular endothelium under physiological flow conditions, *J. Clin. Invest.*, **92**, 3038–3044.

36. Gilbert, M. R., Kuhn, J., Lamborn, K. R., Lieberman, F., Wen, P. Y., Mehta, M., Cloughesy, T., Lassman, A. B., Deangelis, L. M., Chang, S., and Prados, M. (2012). Cilengitide in patients with recurrent glioblastoma: The results of NABTC 03-02, a phase II trial with measures of treatment delivery, *J. Neurooncol.*, **106**, 147–153.

37. Gomez, E. W., Chen, Q. K., Gjorevski, N., and Nelson, C. M. (2010). Tissue geometry patterns epithelial-mesenchymal transition via intercellular mechanotransduction, *J. Cell. Biochem.*, **110**, 44–51.

38. Gray, D. S., Tien, J., and Chen, C. S. (2003). Repositioning of cells by mechanotaxis on surfaces with micropatterned Young's modulus, *J. Biomed. Mater. Res. Part A*, **66A**, 605–614.

39. Haessler, U., Kalinin, Y., Swartz, M. A., and Wu, M. (2009). An agarose-based microfluidic platform with a gradient buffer for 3D chemotaxis studies, *Biomed. Microdevices*, **11**, 827–835.

40. Haessler, U., Pisano, M., Wu, M., and Swartz, M. A. (2011). Dendritic cell chemotaxis in 3D under defined chemokine gradients reveals differential response to ligands CCL21 and CCL19, *PNAS*, **108**(14), pp. 5614–5619.

41. Hahn, M. S., Miller, J. S., and West, J. L. (2006). Three-Dimensional Biochemical and Biomechanical Patterning of Hydrogels for Guiding Cell Behavior, *Adv. Mater.*, **18**, 2679–2684.

42. Harris, A. L. (2002). Hypoxia—a key regulatory factor in tumour growth, *Nat. Rev. Cancer*, **2**, 38–47.

43. Heldin, C.-H., Rubin, K., Pietras, K., and Ostman, A. (2004). High interstitial fluid pressure—an obstacle in cancer therapy, *Nat. Rev. Cancer*, **4**, 806–813.

44. Hofmann, M., Guschel, M., Bernd, A., Bereiter-Hahn, J., Kaufmann, R., Tandi, C., Wiig, H., and Kippenberger, S. (2006). Lowering of tumor interstitial fluid pressure reduces tumor cell proliferation in a xenograft tumor model, *Neoplasia*, **8**, 89–95.

45. Hori, A., Sasada, R., Matsutani, E., Naito, K., Sakura, Y., Fujita, T., and Kozai, Y. (1991). Suppression of solid tumor growth by immunoneutralizing monoclonal antibody against human basic fibroblast growth factor, *Cancer Res.*, **51**, 6180–6184.

46. Huang, S., and Ingber, D. E. (2005). Cell tension, matrix mechanics, and cancer development, *Cancer Cell*, **8**, 175–176.

47. Huh, D., Hamilton, G. A., and Ingber, D. E. (2011). From 3D cell culture to organs-on-chips, *Trends Cell Biol.*, **21**, 745–754.

48. Humphries, J. D., Byron, A., and Humphries, M. J. (2006). Integrin ligands at a glance, *J. Cell Sci.*, **119**, 3901–3903.

49. Ingber, D. E., Madri, J. A., and Jamieson, J. D. (1981). Role of basal lamina in neoplastic disorganization of tissue architecture., *Proc. Natl. Acad. Sci. U. S. A.*, **78**, 3901–3905.

50. Jain, R. K. (1987). Transport of molecules in the tumor interstitium: A review, *Cancer Res.*, **47**, 3039–3051.

51. Jeon, N. L., Dertinger, S. K. W., Chiu, D. T., Choi, I. S., Stroock, A. D., and Whitesides, G. M. (2000). Generation of solution and surface gradients using microfluidic systems, *Langmuir*, **16**, 8311–8316.

52. Kanzawa, S., Endo, H., and Shioya, N. (1993). Improved in vitro angiogenesis model by collagen density reduction and the use of type III collagen, *Ann. Plast. Surg.*, **30**, 244–251.

53. Khaled, W., Reichling, S., Bruhns, O. T., Boese, H., Baumann, M., Monkman, G., Egersdoerfer, S., Klein, D., Tunayar, A., Freimuth, H.,

Lorenz, A., Pessavento, A., and Ermert, H. (2004). Palpation imaging using a haptic system for virtual reality applications in medicine, *Stud. Health Technol. Inform.*, **98**, 147–153.

54. Kim, B. J., and Wu, M. (2012). Microfluidics for mammalian cell chemotaxis, *Ann. Biomed. Eng.*, **40**, 1316–1327.
55. Klein, E. A., Yin, L., Kothapalli, D., Castagnino, P., Byfield, F. J., Xu, T., Levental, I., Hawthorne, E., Janmey, P. A., and Assoian, R. K. (2009). Cell-cycle control by physiological matrix elasticity and in vivo tissue stiffening, *Curr. Biol.*, **19**, 1511–1518.
56. Kleinman, H. K., and Martin, G. R. (2005). Matrigel: Basement membrane matrix with biological activity, *Semin. Cancer Biol.*, **15**, 378–386.
57. Kostic, A., Lynch, C. D., and Sheetz, M. P. (2009). Differential matrix rigidity response in breast cancer cell lines correlates with the tissue tropism, *PLoS ONE*, **4**, e6361.
58. Kramer, R. H., Bensch, K. G., and Wong, J. (1986). Invasion of reconstituted basement membrane matrix by metastatic human tumor cells, *Cancer Res.*, **46**, 1980–1989.
59. Kuo, J. W., Swann, D. A., and Prestwich, G. D. (1991). Chemical modification of hyaluronic acid by carbodiimides, *Bioconjug. Chem.*, **2**, 232–241.
60. Lawley, T. J., and Kubota, Y. (1989). Induction of morphologic differentiation of endothelial cells in culture, *J. Invest. Dermatol.*, **93**, 59S–61S.
61. Leslie-Barbick, J. E., Shen, C., Chen, C., and West, J. L. (2011). Micron-scale spatially patterned, covalently immobilized vascular endothelial growth factor on hydrogels accelerates endothelial tubulogenesis and increases cellular angiogenic responses, *Tissue Eng. Part A*, **17**, 221–229.
62. Lévesque, S. G., and Shoichet, M. S. (2007). Synthesis of enzyme-degradable, peptide-cross-linked dextran hydrogels, *Bioconjug. Chem.*, **18**, 874–885.
63. Liu, J., Tan, Y., Zhang, H., Zhang, Y., Xu, P., Chen, J., Poh, Y.-C., Tang, K., Wang, N., and Huang, B. (2012). Soft fibrin gels promote selection and growth of tumorigenic cells, *Nat. Mater.*, **11**, 734–741.
64. Lukashev, M. E., and Werb, Z. (1998). ECM signalling: Orchestrating cell behaviour and misbehaviour, *Trends Cell Biol.*, **8**, 437–441.
65. Lullo, G. A. D., Sweeney, S. M., Körkkö, J., Ala-Kokko, L., and Antonio, J. D. S. (2002). Mapping the ligand-binding sites and disease-

associated mutations on the most abundant protein in the human, type I collagen, *J. Biol. Chem.*, **277**, 4223–4231.

66. Lunt, S. J., Fyles, A., Hill, R. P., and Milosevic, M. (2008). Interstitial fluid pressure in tumors: Therapeutic barrier and biomarker of angiogenesis, *Future Oncol.*, **4**, 793–802.
67. Mammoto, A., Huang, S., and Ingber, D. E. (2007). Filamin links cell shape and cytoskeletal structure to Rho regulation by controlling accumulation of p190RhoGAP in lipid rafts, *J. Cell. Sci.*, **120**, 456–467.
68. Marklein, R. A., and Burdick, J. A. (2010). Spatially controlled hydrogel mechanics to modulate stem cell interactions, *Soft Matter*, **6**, 136–143.
69. Martens, P., and Anseth, K. (2000). Characterization of hydrogels formed from acrylate modified poly(vinyl alcohol) macromers, *Polymer*, **41**, 7715–7722.
70. Mbeunkui, F., and Johann, D. J., Jr. (2009). Cancer and the tumor microenvironment: A review of an essential relationship, *Cancer Chemother. Pharmacol.*, **63**, 571–582.
71. Mehlen, P., and Puisieux, A. (2006). Metastasis: A question of life or death, *Nat. Rev. Cancer*, **6**, 449–458.
72. Mosadegh, B., Saadi, W., Wang, S. J., and Jeon, N. L. (2008). Epidermal growth factor promotes breast cancer cell chemotaxis in CXCL12 gradients, *Biotechnol. Bioeng.*, **100**, 1205–1213.
73. Mullen, P., Ritchie, A., Langdon, S. P., and Miller, W. R. (1996). Effect of Matrigel on the tumorigenicity of human breast and ovarian carcinoma cell lines, *Int. J. Cancer*, **67**, 816–820.
74. Musoke-Zawedde, P., and Shoichet, M. S. (2006). Anisotropic three-dimensional peptide channels guide neurite outgrowth within a biodegradable hydrogel matrix, *Biomed. Mater.*, **1**, 162–169.
75. Nathan, S. S., Huvos, A. G., Casas-Ganem, J. E., Yang, R., Linkov, I., Sowers, R., DiResta, G. R., Gorlick, R., and Healey, J. H. (2008). Tumor interstitial fluid pressure may regulate angiogenic factors in osteosarcoma, *J. Orthop. Res.*, **26**, 1520–1525.
76. Nehls, V., and Drenckhahn, D. (1995). A novel, microcarrier-based in vitro assay for rapid and reliable quantification of three-dimensional cell migration and angiogenesis, *Microvasc. Res.*, **50**, 311–322.
77. Nelson, C. M., Vanduijn, M. M., Inman, J. L., Fletcher, D. A., and Bissell, M. J. (2006). Tissue geometry determines sites of mammary branching morphogenesis in organotypic cultures, *Science*, **314**, 298–300.

78. Nerem, R. M. (1991). Shear force and its effect on cell structure and function, *ASGSB Bull.*, **4**, 87–94.

79. Ng, C. P., and Swartz, M. A. (2003). Fibroblast alignment under interstitial fluid flow using a novel 3-D tissue culture model, *Am. J. Physiol. Heart Circ. Physiol.*, **284**, H1771–1777.

80. Noel, A., Simon, N., Raus, J., and Foidart, J. M. (1992). Basement membrane components (matrigel) promote the tumorigenicity of human breast adenocarcinoma MCF7 cells and provide an in vivo model to assess the responsiveness of cells to estrogen, *Biochem. Pharmacol.*, **43**, 1263–1267.

81. Nuttelman, C. R., Tripodi, M. C., and Anseth, K. S. (2005). Synthetic hydrogel niches that promote hMSC viability, *Matrix Biol.*, **24**, 208–218.

82. Orimo, A., and Weinberg, R. A. (2006). Stromal fibroblasts in cancer: A novel tumor-promoting cell type, *Cell Cycle*, **5**, 1597–1601.

83. Ott, H. C., Matthiesen, T. S., Goh, S.-K., Black, L. D., Kren, S. M., Netoff, T. I., and Taylor, D. A. (2008). Perfusion-decellularized matrix: Using nature's platform to engineer a bioartificial heart, *Nat. Med.*, **14**, 213–221.

84. Paget, S. (1989). The distribution of secondary growths in cancer of the breast. 1889, *Cancer Metastasis Rev.*, **8**, 98–101.

85. Park, K.-H., Na, K., and Chung, H.-M. (2005). Enhancement of the adhesion of fibroblasts by peptide containing an Arg-Gly-Asp sequence with poly(ethylene glycol) into a thermo-reversible hydrogel as a synthetic extracellular matrix, *Biotechnol. Lett.*, **27**, 227–231.

86. Paszek, M. J., Zahir, N., Johnson, K. R., Lakins, J. N., Rozenberg, G. I., Gefen, A., Reinhart-King, C. A., Margulies, S. S., Dembo, M., Boettiger, D., Hammer, D. A., and Weaver, V. M. (2005). Tensional homeostasis and the malignant phenotype, *Cancer Cell*, **8**, 241–254.

87. Pathak, A., and Kumar, S. (2012). Independent regulation of tumor cell migration by matrix stiffness and confinement, *PNAS*, **109**(26), pp. 10334–10039.

88. Peyton, S. R., Kalcioglu, Z. I., Cohen, J. C., Runkle, A. P., Van Vliet, K. J., Lauffenburger, D. A., and Griffith, L. G. (2011). Marrow-Derived stem cell motility in 3D synthetic scaffold is governed by geometry along with adhesivity and stiffness, *Biotechnol. Bioeng.*, **108**, 1181–1193.

89. Pezzolo, A., Parodi, F., Corrias, M. V., Cinti, R., Gambini, C., and Pistoia, V. (2007). Tumor origin of endothelial cells in human neuroblastoma, *J. Clin. Oncol.*, **25**, 376–383.

90. Polacheck, W. J., Charest, J. L., and Kamm, R. D. (2011). Interstitial flow influences direction of tumor cell migration through competing mechanisms, *PNAS*, **108**(27), pp. 11115–11120.
91. Polyak, K. (2007). Breast cancer: Origins and evolution, *J. Clin. Invest.*, **117**, 3155–3163.
92. Pretlow, T. G., Delmoro, C. M., Dilley, G. G., Spadafora, C. G., and Pretlow, T. P. (1991). Transplantation of human prostatic carcinoma into nude mice in Matrigel, *Cancer Res.*, **51**, 3814–3817.
93. Provenzano, P. P., Eliceiri, K. W., Campbell, J. M., Inman, D. R., White, J. G., and Keely, P. J. (2006). Collagen reorganization at the tumor-stromal interface facilitates local invasion, *BMC Med.*, **4**, 38.
94. Radotra, B., and McCormick, D. (1997). Glioma invasion in vitro is mediated by CD44-hyaluronan interactions, *J. Pathol.*, **181**, 434–438.
95. Ranga, A., and Lutolf, M. P. (2012). High-throughput approaches for the analysis of extrinsic regulators of stem cell fate, *Curr. Opin. Cell Biol.*, **24**, 236–244.
96. Ranuncolo, S. M., Ladeda, V., Specterman, S., Varela, M., Lastiri, J., Morandi, A., Matos, E., Bal de Kier Joffé, E., Puricelli, L., and Pallotta, M. G. (2002). CD44 expression in human gliomas, *J. Surg. Oncol.*, **79**, 30–35; discussion 35–36.
97. Reardon, D. A., Zalutsky, M. R., Akabani, G., Coleman, R. E., Friedman, A. H., Herndon, J. E., 2nd, McLendon, R. E., Pegram, C. N., Quinn, J. A., Rich, J. N., Vredenburgh, J. J., Desjardins, A., Guruangan, S., Boulton, S., Raynor, R. H., Dowell, J. M., Wong, T. Z., Zhao, X.-G., Friedman, H. S., and Bigner, D. D. (2008). A pilot study: 131I-antitenascin monoclonal antibody 81c6 to deliver a 44-Gy resection cavity boost, *Neuro-Oncol.*, **10**, 182–189.
98. Reinhart-King, C. A. (2011). How matrix properties control the self-assembly and maintenance of tissues, *Ann. Biomed. Eng.*, **39**, 1849–1856.
99. Ribatti, D., Mangialardi, G., and Vacca, A. (2006). Stephen paget and the "seed and soil" theory of metastatic dissemination, *Clin. Exp. Med.*, **6**, 145–149.
100. Rigolin, G. M., Fraulini, C., Ciccone, M., Mauro, E., Bugli, A. M., De Angeli, C., Negrini, M., Cuneo, A., and Castoldi, G. (2006). Neoplastic circulating endothelial cells in multiple myeloma with 13q14 deletion, *Blood*, **107**, 2531–2535.
101. Rizzi, S. C., and Hubbell, J. A. (2005). Recombinant protein-co-PEG networks as cell-adhesive and proteolytically degradable hydrogel

matrixes. Part I: Development and physicochemical characteristics, *Biomacromolecules*, **6**, 1226–1238.

102. Roose, T., Netti, P. A., Munn, L. L., Boucher, Y., and Jain, R. K. (2003). Solid stress generated by spheroid growth estimated using a linear poroelasticity model small star, filled, *Microvasc. Res.*, **66**, 204–212.

103. Roussos, E. T., Condeelis, J. S., and Patsialou, A. (2011). Chemotaxis in cancer, *Nat. Rev. Cancer*, **11**, 573–587.

104. Ruoslahti, E. (1996). Integrin Signaling and Matrix Assembly, *Tumor Biol.*, **17**, 117–124.

105. Saadi, W., Rhee, S. W., Lin, F., Vahidi, B., Chung, B. G., and Jeon, N. L. (2007). Generation of stable concentration gradients in 2D and 3D environments using a microfluidic ladder chamber, *Biomed. Microdevices*, **9**, 627–635.

106. Salinas, C. N., and Anseth, K. S. (2008). The enhancement of chondrogenic differentiation of human mesenchymal stem cells by enzymatically regulated RGD functionalities, *Biomaterials*, **29**, 2370–2377.

107. Sappino, A. P., Skalli, O., Jackson, B., Schürch, W., and Gabbiani, G. (1988). Smooth-muscle differentiation in stromal cells of malignant and non-malignant breast tissues, *Int. J. Cancer*, **41**, 707–712.

108. Sawhney, A. S., Pathak, C. P., and Hubbell, J. A. (1993). Bioerodible hydrogels based on photopolymerized poly(ethylene glycol)-co-poly(alpha-hydroxy acid) diacrylate macromers, *Macromolecules*, **26**, 581–587.

109. Schor, S. L. (1980). Cell proliferation and migration on collagen substrata in vitro, *J. Cell. Sci.*, **41**, 159–175.

110. Schor, S. L., Allen, T. D., and Harrison, C. J. (1980). Cell migration through three-dimensional gels of native collagen fibres: Collagenolytic activity is not required for the migration of two permanent cell lines, *J. Cell. Sci.*, **46**, 171–186.

111. Schor, S. L., Schor, A. M., Winn, B., and Rushton, G. (1982). The use of three-dimensional collagen gels for the study of tumour cell invasion in vitro: Experimental parameters influencing cell migration into the gel matrix, *Intl. J. Cancer*, **29**, 57–62.

112. Servant, G., Weiner, O. D., Herzmark, P., Balla, T., Sedat, J. W., and Bourne, H. R. (2000). Polarization of chemoattractant receptor signaling during neutrophil chemotaxis, *Science*, **287**, 1037–1040.

113. Shamloo, A., Ma, N., Poo, M., Sohn, L. L., and Heilshorn, S. C. (2008). Endothelial cell polarization and chemotaxis in a microfluidic device, *Lab Chip*, **8**, 1292.

114. Sherman-Baust, C. A., Weeraratna, A. T., Rangel, L. B. A., Pizer, E. S., Cho, K. R., Schwartz, D. R., Shock, T., and Morin, P. J. (2003). Remodeling of the extracellular matrix through overexpression of collagen VI contributes to cisplatin resistance in ovarian cancer cells, *Cancer Cell*, **3**, 377–386.

115. Shieh, A. C. (2011). Biomechanical forces shape the tumor microenvironment, *Ann. Biomed. Eng.*, **39**, 1379–1389.

116. Shieh, A. C., Rozansky, H. A., Hinz, B., and Swartz, M. A. (2011). Tumor cell invasion is promoted by interstitial flow-induced matrix priming by stromal fibroblasts, *Cancer Res.*, **71**, 790–800.

117. Shields, J. D., Fleury, M. E., Yong, C., Tomei, A. A., Randolph, G. J., and Swartz, M. A. (2007). Autologous chemotaxis as a mechanism of tumor cell homing to lymphatics via interstitial flow and autocrine CCR7 signaling, *Cancer Cell*, **11**, 526–538.

118. Siegel, R., Naishadham, D., and Jemal, A. (2012). Cancer statistics, 2012, *CA Cancer J. Clin.*, **62**, 10–29.

119. Simonsen, T. G., Gaustad, J.-V., Leinaas, M. N., and Rofstad, E. K. (2012). High interstitial fluid pressure is associated with tumor-line specific vascular abnormalities in human melanoma xenografts, *PLoS ONE*, **7**, e40006.

120. Slattery, M. J., Liang, S., and Dong, C. (2005). Distinct role of hydrodynamic shear in leukocyte-facilitated tumor cell extravasation, *Am. J. Physiol., Cell Physiol.*, **288**, C831–839.

121. Soda, Y., Marumoto, T., Friedmann-Morvinski, D., Soda, M., Liu, F., Michiue, H., Pastorino, S., Yang, M., Hoffman, R. M., Kesari, S., and Verma, I. M. (2011). Transdifferentiation of glioblastoma cells into vascular endothelial cells, *PNAS*, **108**(11) , pp. 4274–4280.

122. Stachowiak, A. N., Bershteyn, A., Tzatzalos, E., and Irvine, D. J. (2005). Bioactive hydrogels with an ordered cellular structure combine interconnected macroporosity and robust mechanical properties, *Adv. Mater.*, **17**, 399–403.

123. Streubel, B., Chott, A., Huber, D., Exner, M., Jäger, U., Wagner, O., and Schwarzinger, I. (2004). Lymphoma-specific genetic aberrations in microvascular endothelial cells in B-cell lymphomas, *N. Engl. J. Med.*, **351**, 250–259.

124. Tayalia, P., Mazur, E., and Mooney, D. J. (2011). Controlled architectural and chemotactic studies of 3D cell migration, *Biomaterials*, **32**, 2634–2641.

125. Terranova, V. P., Aumailley, M., Sultan, L. H., Martin, G. R., and Kleinman, H. K. (1986). Regulation of cell attachment and cell number by fibronectin and laminin, *J. Cell. Physiol.*, **127**, 473–479.

126. Tibbitt, M. W., and Anseth, K. S. (2009). Hydrogels as extracellular matrix mimics for 3D cell culture, *Biotechnol. Bioeng.*, **103**, 655–663.

127. Tibbitt, M. W., Kloxin, A. M., Dyamenahalli, K. U., and Anseth, K. S. (2010). Controlled two-photon photodegradation of PEG hydrogels to study and manipulate subcellular interactions on soft materials, *Soft Matter*, **6**, 5100.

128. Tien, J., Truslow, J. G., and Nelson, C. M. (2012). Modulation of invasive phenotype by interstitial pressure-driven convection in aggregates of human breast cancer cells, *PLoS ONE*, **7**, e45191.

129. Toole, B. P. (2004). Hyaluronan: From extracellular glue to pericellular cue, *Nat. Rev. Cancer*, **4**, 528–539.

130. Trappmann, B., Gautrot, J. E., Connelly, J. T., Strange, D. G. T., Li, Y., Oyen, M. L., Cohen Stuart, M. A., Boehm, H., Li, B., Vogel, V., Spatz, J. P., Watt, F. M., and Huck, W. T. S. (2012). Extracellular-matrix tethering regulates stem-cell fate, *Nat. Mater.*, **11**, 642–649.

131. Wang, F., Weaver, V. M., Petersen, O. W., Larabell, C. A., Dedhar, S., Briand, P., Lupu, R., and Bissell, M. J. (1998). Reciprocal interactions between beta1-integrin and epidermal growth factor receptor in three-dimensional basement membrane breast cultures: A different perspective in epithelial biology, *Proc. Natl. Acad. Sci. U. S. A.*, **95**, 14821–14826.

132. Wang, H.-B., Dembo, M., and Wang, Y.-L. (2000). Substrate flexibility regulates growth and apoptosis of normal but not transformed cells, *Am. J. Physiol. Cell Physiol.*, **279**, C1345–C1350.

133. Wang, S.-J., Saadi, W., Lin, F., Minh-Canh Nguyen, C., and Li Jeon, N. (2004). Differential effects of EGF gradient profiles on MDA-MB-231 breast cancer cell chemotaxis, *Exp. Cell Res.*, **300**, 180–189.

134. Wang, X., Ouyang, H., Yamamoto, Y., Kumar, P. A., Wei, T. S., Dagher, R., Vincent, M., Lu, X., Bellizzi, A. M., Ho, K. Y., Crum, C. P., Xian, W., and McKeon, F. (2011). Residual embryonic cells as precursors of a barrett's-like metaplasia, *Cell*, **145**, 1023–1035.

135. Watanabe, H., Hori, A., Seno, M., Kozai, Y., Igarashi, K., Ichimori, Y., and Kondo, K. (1991). A sensitive enzyme immunoassay for human basic fibroblast growth factor, *Biochem. Biophys. Res. Commun.*, **175**, 229–235.

136. Weaver, V. M., Fischer, A. H., Peterson, O. W., and Bissell, M. J. (1996). The importance of the microenvironment in breast cancer progression: Recapitulation of mammary tumorigenesis using a unique human mammary epithelial cell model and a three-dimensional culture assay, *Biochem. Cell Biol.*, **74**, 833–851.

137. Weaver, V. M., Lelièvre, S., Lakins, J. N., Chrenek, M. A., Jones, J. C. R., Giancotti, F., Werb, Z., and Bissell, M. J. (2002). beta4 integrin-dependent formation of polarized three-dimensional architecture confers resistance to apoptosis in normal and malignant mammary epithelium, *Cancer Cell*, **2**, 205–216.

138. Wei, K.-C., Huang, C.-Y., Chen, P.-Y., Feng, L.-Y., Wu, T.-W. E., Chen, S.-M., Tsai, H.-C., Lu, Y.-J., Tsang, N.-M., Tseng, C.-K., Pai, P.-C., and Shin, J.-W. (2010). Evaluation of the prognostic value of CD44 in glioblastoma multiforme, *Anticancer Res.*, **30**, 253–259.

139. Weidner, N. (1993). Tumor angiogenesis: Review of current applications in tumor prognostication, *Semin. Diag. Pathol.*, **10**, 302–313.

140. Weigelt, B., Peterse, J. L., and Veer, L. J. van't. (2005). Breast cancer metastasis: Markers and models, *Nat. Rev. Cancer*, **5**, 591–602.

141. De Wever, O., and Mareel, M. (2003). Role of tissue stroma in cancer cell invasion, *J. Pathol.*, **200**, 429–447.

142. Wong, J. Y., Velasco, A., Rajagopalan, P., and Pham, Q. (2003). Directed movement of vascular smooth muscle cells on gradient-compliant hydrogels, *Langmuir*, **19**, 1908–1913.

143. Wosnick, J. H., and Shoichet, M. S. (2007). Three-dimensional chemical patterning of transparent hydrogels, *Chem. Mater.*, **20**, 55–60.

144. Wylie, R. G., Ahsan, S., Aizawa, Y., Maxwell, K. L., Morshead, C. M., and Shoichet, M. S. (2011). Spatially controlled simultaneous patterning of multiple growth factors in three-dimensional hydrogels, *Nat. Mater.*, **10**, 799–806.

145. Wyrick, B. J., Ozawa, T., Lamborn, K. R., Bollen, A. W., and Deen, D. F. (1997). Effects of Matrigel on the SF-767 malignant glioma athymic mouse tumor model, *Anticancer Res.*, **17**, 2419–2425.

146. Yamaura, T., Murakami, K., Doki, Y., Sugiyama, S., Misaki, T., Yamada, Y., and Saiki, I. (2000). Solitary lung tumors and their spontaneous metastasis in athymic nude mice orthotopically implanted with human non-small cell lung cancer, *Neoplasia*, **2**, 315–324.

147. Yoshida, Y., Kamitani, N., Sasaki, H., and Kotsuji, F. (1998). Establishment of a cell line from a neuroendocrine carcinoma of the endometrium and an invasion model, *Anticancer Res.*, **18**, 1569–1577.

148. Yuen, W. W., Du, N. R., Chan, C. H., Silva, E. A., and Mooney, D. J. (2010). Mimicking nature by codelivery of stimulant and inhibitor to create temporally stable and spatially restricted angiogenic zones, *PNAS*, **107**(42), pp. 17933–17938.

149. Zicha, D., Dunn, G. A., and Brown, A. F. (1991). A new direct-viewing chemotaxis chamber, *J. Cell. Sci.*, **99**(Pt 4), 769–775.

150. Zigmond, S. H. (1977). Ability of polymorphonuclear leukocytes to orient in gradients of chemotactic factors, *J. Cell Biol.*, **75**, 606–616.

151. Alberts B., Johnson A., Lewis J., et al (2002). The Extracellular Matrix of Animals—Molecular Biology of the Cell, *Garland Science*. New York.

# Chapter 12

# Micropost Methods for Cell Biomechanics of the Cardiovascular System

**Kevin S. Bielawski and Nathan J. Sniadecki**

*Department of Mechanical Engineering, University of Washington, Seattle, WA 98195, USA*

nsniadec@uw.edu

The cardiovascular system is more than a biological pump and a system of pipes. Unlike a typical plumbing system, the human body actively changes the size of its pump and pipes through a process of mechanotransduction. Understanding this process at a cellular level enables a greater understanding of the entire system at a fundamental level. This chapter focuses on studies using flexible polymer microposts, which can measure the contractile forces of cardiovascular cells with nanoscale precision. Specifically, microposts have been used to understand how cardiomyocytes, endothelial cells, and platelets change their function in response to react to changes in their microenvironment and how these changes relate to diseases and disorders of the cardiac, vascular, and coagulation systems.

*Cells, Forces, and the Microenvironment*
Edited by Charles M. Cuerrier and Andrew E. Pelling

ISBN 978-981-4613-36-1 (Hardcover), 978-981-4613-37-8 (eBook)
www.panstanford.com

## 12.1 Background on Cardiovascular System

The cardiovascular system pumps blood throughout the body to regulate the supply and removal of cells, proteins, or molecules that regulate metabolism, growth, healing, and the immune response. The function of the cardiovascular system is classically understood on a macroscale level, but there are opportunities to learn more about its physiology at a microscale level. For example, the function of the heart is to pump blood, but its individual cells must contract with ample power in order to achieve this effect.

Three major cell types in the cardiovascular system are highlighted in this chapter: cardiomyocytes, endothelial cells, and platelets. These cells serve a unique set of functions in the cardiovascular system: pumping, containment, and wound healing. Cardiomyocytes contract together, allowing the heart to beat and force blood through the vascular system. Endothelial cells line the inside of vessel walls, serving as a barrier between blood and the underlying vascular tissue. Platelets act as first-responders in wound healing by attaching to a damaged vessel wall and preventing further blood loss.

Both muscle and nonmuscle cells are able to produce contractile forces via their actin and myosin proteins. In particular, cellular forces are important for adhesion, proliferation, migration, contraction, wound healing, and mechanotransduction. For cardiomyocytes, endothelial cells, and platelets, their cellular forces are important for pumping blood, maintaining their barrier function, or forming stable clots. The study of cell forces is a relatively new field and the techniques to fabricate devices have improved over the years, producing refined tools on the micro- and nanoscale. There is a growing understanding on the relationship between cellular forces and cellular function, which has enabled new technologies and treatments for cardiovascular disease.

## 12.2 Background on Microposts

The study of cellular traction forces was pioneered by Harris et al. in 1980 using thin films of silicone rubber [6]. Cells seeded on the thin films pulled on the surface of the silicone and produced wrinkles that were visible under an optical microscope. The number and size of these wrinkles indicated the magnitude of the tract force generated by the cells. Although it was a novel and exciting development, the initial tool was qualitative measurement of cellular traction forces and not a quantitative one.

In response to the shortcomings with wrinkling silicone membranes, traction force microscopy was developed [7]. In order to quantitatively determine cellular traction forces, beads were placed onto a sheet of silicone rubber and when cells exerted forces, the beads moved with the deformation of the silicone sheet. The motion of the beads coupled with the elasticity of the silicone enabled a quantitative measurement of the cellular traction forces. This technique was improved upon with the use of polyacrylamide gels instead of silicone rubber, which was softer and enabled larger deformations that were easier to track [8].

Measuring cellular traction forces took another step forward when cantilevers were used for cell migration studies [9]. In the first incarnation, a cantilever was oriented horizontally and situated underneath a flat surface in which there was a small opening where the cell could attach to the tip of the cantilever. As a cell crawled across the surface, its traction force on the tip of the cantilever caused entire cantilever to bend. The cantilever acted like a spring where the movement of its tip was proportional to the applied force. This tool was novel for it gave insight into the dynamics of cellular traction forces during migration, but it suffered from being able to only report force at one location, rather than at each point of adhesion that a cell has with a substrate. Recently, vertical cantilevers made from silicone rubber were developed to determine cellular traction forces [10]. The mechanics and techniques behind using vertical cantilevers will be discussed in detail in this chapter.

### 12.2.1 Overview of Micropost Technology

Micropost technology consists of an array of vertical posts that bend under traction forces of cells attached onto the tips of the posts. The posts are made out of polydimethylsiloxane (PDMS) and are functionalized with extracellular matrix protein to enhance cell adhesion (Fig. 12.1). Once a cell has attached to the tips of the posts, they are able to contract, thereby deflecting the tip of each post while the bottom of each post remains stationary. Consequently, the total deflection of each post can be determined by comparing the location of the tip of each post with the location of the bottom of each post. Typically, microposts are stained with fluorescent dye and bending of the posts is measured using custom image analysis software. Each post deforms independently of other posts in the array and the force at the tip acts as a single point load such that each post can be treated as a fixed beam cantilever with a point force load on the end. Microposts are commonly made with a circular cross section, which gives the relation between the force and the deflection of the top of the beam in Eq. 12.1, where $F$ is the force, $E$ is Young's modulus of the material, $d$ is the diameter of the post, $L$ is the height of the post, and $\delta$ is the deflection of the top of the post relative to the bottom (Fig. 12.1c).

$$F = \frac{64E\pi d^4}{3L^3}\delta \tag{12.1}$$

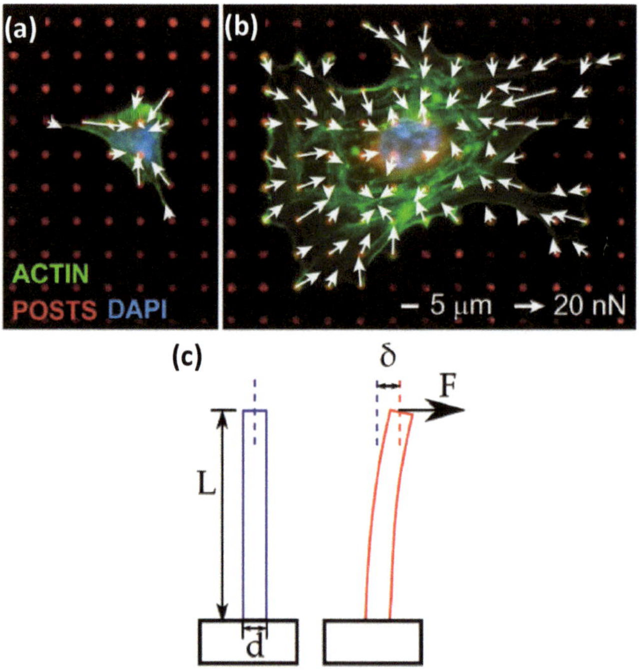


**Figure 12.1** (a) and (b) Cells seeded onto micropost arrays with force vectors. (c) Diagram of the deflection of a post with relevant variables.

### 12.2.2 Manufacturing Methods of Microposts

Microposts are created using a soft lithography process. First, a master structure can be fabricated using SU-8 photoresist in a clean room (Fig. 12.2). First, SU-8 is spun onto a clean silicon wafer to the desired height of the posts (typically 5–10 μm). Next, the SU-8 is exposed to light in a pattern of dots and then developed to produce a master array of posts. In order to prevent PDMS from adhering to the master when fabricating negative molds, a vapor fluorosilane treatment is performed on the posts. The fluorosilane creates a self-assembled monolayer of silane on the SU-8 master, preventing adhesion of PDMS to the master. After silane treatment, a PDMS negative mold is fabricated from the master and treated subsequently with silane. The negative mold is then used with a ratio of 10:1 base to cross-linking agent of PDMS to create the final version of posts, typically casting against a glass slide activated with plasma and baking in an oven at 110°C to fully cross-link the PDMS microposts. After the array has been peeled from the negative mold, they are ready for functionalization with protein to enable adhesion of the cells to the surface.

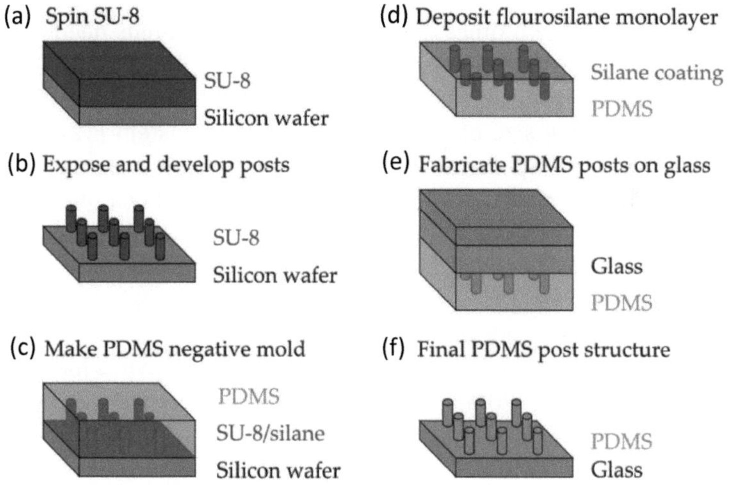


**Figure 12.2** Fabrication of PDMS microposts starting with an SU-8 master.

Functionalizing the PDMS microposts for cell adhesion involves cleaning the posts, adhering protein to the tips, and blocking the remaining areas of the posts. First, the posts are cleaned with UV-

Ozone to ready them for protein adhesion. Rather than directly seeding protein on the tips of the posts, protein is absorbed onto a stamp that is also made out of PDMS by using a 30:1 base to curing agent ratio to make it more flexible and offset any small difference in height of the post. After the protein has absorbed onto the stamp in a monolayer, the stamp is dried and then placed in direct contact with the tops of the posts. After removing the stamp, the posts are dipped in ethanol before they are placed in water in order to make them more hydrophilic, preventing collapse. Finally, the posts are placed in media, and then cells are seeded on top of them in a ratio appropriate to the specific experiment being conducted.

Although fabricating PDMS posts from an SU-8 master is fairly straightforward, there has been some research regarding the repeatability of each set of posts in terms of dimensions and properties. A negative mold made out of PDMS is highly desired as it enables the user to create a thin layer of PDMS under the posts and to more easily peel the mold from the glass without collapsing the posts. Unfortunately, thermal fracture of the necessary silane layer on the PDMS has been discovered to create problems when fabricating more than one set of posts from a single negative mold [11]. Because of the thermal fracture, molds can only be used for a maximum of two times and then a new set of molds needs to be made which involves another treatment of silane. Additionally, the silane treatment that is required on the negative mold can affect the actual size of the post if it is not consistent [12], which is particularly important when fabricating posts in the nanoscale range. Furthermore, the properties of PDMS have been questioned as factors such as mixing ratio, mixing time and curing time have been found to affect Young's modulus [13–15]. Ensuring the same properties of PDMS on each array of posts is vital to ensuring accurate experiments and comparisons.

Aside from a simply passive environment, nanowires have been embedded in the PDMS microposts to apply forces to a discrete region of focal adhesions in a cell and determine how the cell responds to the motion of the post [16]. By embedding cobalt wires into the microposts and applying a uniform magnetic field across the substrate, a single post under a cell was moved up to 0.8 μm. The novel approach of actuating single focal adhesions showed a connection between the cell and the other regions of the cell. The traction forces were either reduced or increased throughout the cell depending on the proximity to the magnetic posts and whether the point of measurement was on the periphery of the cell or the interior.

## 12.3 Cardiomyocytes Contract on Microposts

Cardiomyocytes are vital to the cardiovascular system for they work to pump blood throughout the body. Cardiomyocytes beat constantly and do not typically regenerate once the body has reached maturity. Significant research is focused around how to develop cardiomyocytes to be strong enough in order to graft them into damaged regions of a heart. Additionally, the stiffness of the heart changes significantly from the time of birth until maturity as well as after infarctions [17,18]. The change in stiffness is thought to have an effect on many properties, and microposts were used to study the effects of stiffness on the force and power generation of cardiomyocytes.

Several researchers have looked at the forces during beating of cardiomyocytes, and have successfully seeded them onto PDMS microposts and had them beat. Zhao and Zhang put cardiomyocytes onto posts and measured the displacement of cardiomyocytes during a beating cycle [19]. After converting the displacement of the posts to forces per Eq. 12.1, they measured the maximum contraction forces relative to the static force of the cardiomyocytes. Kim et al. measured the forces throughout the spontaneous beating cycle [20] of rat neonatal cardiomyocytes cultured on PDMS posts and showed the time-varying nature of the forces. These studies showed the ability to measure time-varying forces on microposts and looked at the forces while the cell was beating in an attempt to measure the peak contractile force.

### 12.3.1 Power of Neonatal Cardiomyocytes Measured with Microposts

Rodriguez, et al. observed rat neonatal cardiomyocytes on microposts of varying stiffnesses to study the effects of stiffness on intracellular calcium, force, and power of cardiomyocytes [4]. The microposts the researchers used had different heights and diameters to change the stiffness of the posts as given by Eq. 12.1.

They discovered that increasing the stiffness of the posts increased the force generation of the cardiomyocytes (Fig. 12.3d). Furthermore, the researchers developed a line-scanning program to determine the velocity of each post while the cell was contracting. By analyzing the velocity, the researchers determined the power of the cardiomyocyte and discovered that power also increases with stiffness (Fig. 12.3d). Finally, the researchers looked at the intracellular calcium levels and found that calcium was also increased as stiffness was increased (Fig. 12.3b). The novel findings of this study show a correlation between the stiffness of the substrate and three main factors of the health of neonatal cardiomyocytes.

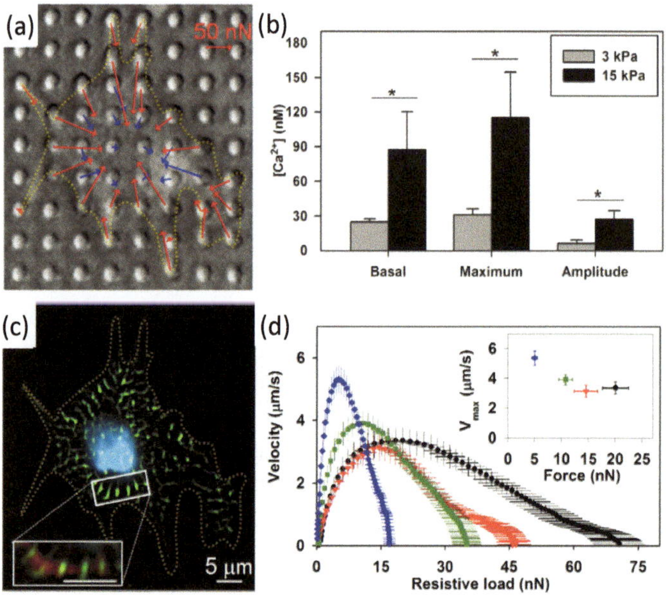


**Figure 12.3** Stiffness affects neonatal cardiomyocytes. (a) The force is determined by micropost deflection with stronger forces on the periphery (red) than the interior (blue). (b) The basal, maximum, and amplitude of the intracellular calcium concentration are higher on stiffer substrates. (c) The z-bands are characterized by staining for α-actinin (green) and actin filaments (red). (d) Microposts were used to determine the power generation, which also increased for cells on increasingly stiff substrates of 3 kPA (blue), 8kPa (green), 10kPa (red), and 15 kPA (black). Adapted from Rodriguez, et al. [4].

## 12.4 Endothelial Cells Produce Force on Microposts

Endothelial cells in the cardiovascular system are vital to withstanding the pressure of the blood flow and regulating the transport of cells and molecules to the underlying tissue. The endothelium is a monolayer of cells and is characterized by tight junctions between neighboring cells. Without the barrier function of endothelial cells, there can be uncontrolled regulation of transport into the walls of the arteries, which is typically thought to be the onset to atherosclerosis [21,22]. In addition to complications with atherosclerosis, the endothelium plays a role in the growth and spread of cancer [23,24]. Because of the unique nature of the role of cell force, particularly intercellular forces, microposts have been used with endothelial cells in novel patterns and experiments to study how the endothelial forces are affected through external factors.

### 12.4.1 Endothelial Cell Patterning on Microposts

Previous studies have shown that both cell area and cell stiffness affect the ability of cells to produce force [25,26]. Recently, a hypothesis has suggested that the spread area of a cell and the traction force the cell generates are closely related. The spread area affects the average focal adhesion size, which in turn affects the maximum force generation of the cell (Fig. 12.4). Additionally, the substrate stiffness also affects the spread area and the force generation [1]. In order to determine the relation between force generation and spread area, the cells needed to be patterned to a specific size rather than allowing the cell to freely spread on a surface. During the stamping step, patterns of squares were stamped off of a 30:1 base to curing agent PDMS stamp and were then stamped onto the tops of the micropost arrays. By stamping the posts in a specific pattern, more information about the cells was obtained. Microcontact printing is a common method to pattern cells, and it has also been successfully applied to microposts to conduct studies on specific patterns and sizes of cells and groups of cells.

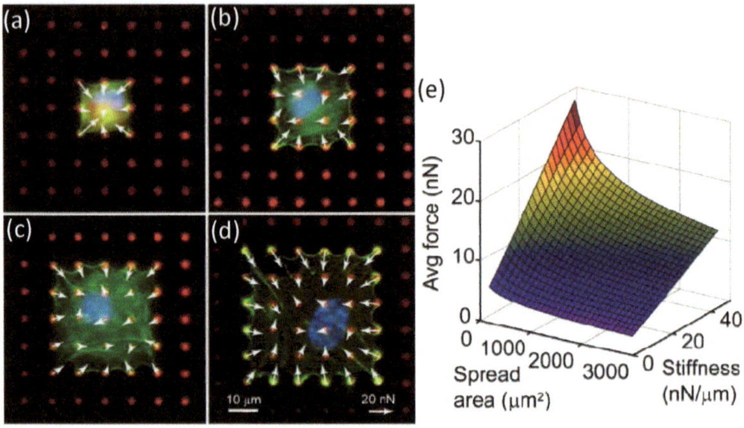


**Figure 12.4** Endothelial cells patterned on microposts. (a–d) Cell spread area on microposts can be controlled by microcontact printing. (e) Traction forces change in response cell spread area and micropost stiffness. Adapted from Han, et al. [1].

### 12.4.2 Endothelial Cells in Pairs Studied Using a Bowtie Pattern

If one were to measure force generation in a gel with pairs of cells, it would be challenging to determine which forces were generated at the intercellular connections, and which forces were generated at the cell-substrate interface. All of the forces are coupled and the motion of the underlying layer can be affected at a distance from where the force was generated, leaving the force generation in each region inconclusive. After utilizing the point-force measurement of microposts, one can determine the total force associated with each cell and the imbalance is the force generated by the intercellular adhesions. Because the sum of forces of each cell must equal zero due to the actin-myosin nature of force generation, the intercellular force is equal to the imbalance in each cell's force vector. All of the remaining forces are transmitted through the cytoskeleton to the micropost substrate, so the remaining forces must be to the other cell.

Liu et al. described how to calculate and determine the intercellular forces of a pair of cells, then patterned endothelial cells as bowties to study how they were affected when in pairs or alone (Fig. 12.5) [3]. Each cell took up one half of the bowtie and then joined to the other cell at the point of the bowtie in a location that was not directly above a micropost adhesion site. The study showed that the average force of each cell remained the same regardless of whether a cell was in contact with another cell or by itself. In fact, the average traction force of a cell was not regulated by the size of the intercellular connection, but the intercellular connection did have a linear relationship with the intercellular force. As the junction size increased, so did the tugging force between cells.

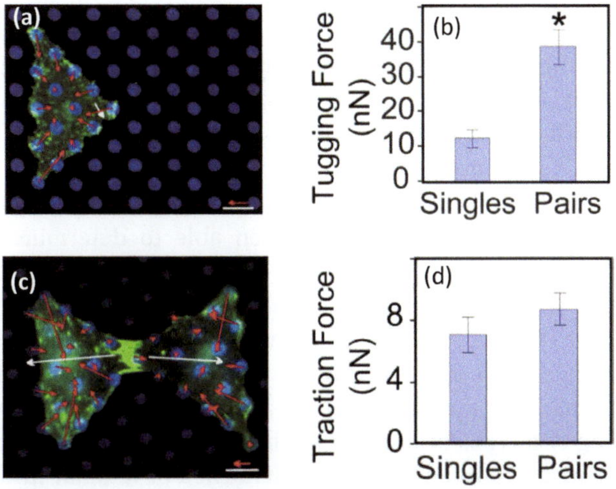


**Figure 12.5** Endothelial cells produce more tugging force and more overall traction force in pairs than as single cells. (a) Single cells were compared with (b) pairs of cells and assessed for adherens junctions through β-catenin staining (green), while the posts are stained blue. (b) The total force of pairs of cells was much greater than single cells, while (d) the average force per post remained consistent for either single cells or pairs of cells. Distance scale bar is 10 μm, and force reference arrow is 10 nN. Adapted from Liu et al. [3].

## 12.4.3 Transmigration of Leukocytes

In another study by Liu et al., endothelial cells were patterned on microposts to form a small region of a monolayer [27]. After treating the endothelial cells to promote adhesion, leukocytes were distributed in the media and allowed to adhere and transmigrate through the monolayer. Because the microposts were used to measure the traction forces, there was a region for the leukocytes to migrate below the layer, and the posts were able to report the underlying forces of the layer. The transmigration process is important for immune system response, and the authors found that endothelial monolayers increased in total force when leukocytes were adhered and transmigrating through the layer compared to when the endothelial cells were simply in a monolayer.

### 12.4.4 Endothelial Cells Are Affected by the Type of Flow

An endothelial cell in vivo generally has blood flowing across its apical surface. Not surprisingly, the speed and type of flow across the endothelium has been shown to have an effect on the ability of the cells to create tight junctions and remain attached. Several studies have looked at the role of flow across a monolayer of endothelial cells, but have only been able to determine protein expression, not cell tugging forces [28,29]. Although this method is useful to show the effect of laminar or disturbed flow, it does not give more than a suggestion about how cell and inter-cellular forces are affected. Ting, et al. has shown the effects of monolayer forces and intercellular forces by seeding a monolayer on microposts and applying two types of flow above the cells [2]. In a single chamber, they used a step to apply either a disturbed or laminar flow across a patterned monolayer seeded on microposts (Fig. 12.6). The disturbed flow was defined by a recirculation region to mimic a region of disturbed flow in the body, typically at bifurcations or downstream of stents. They found that in the region of disturbed flow, the cells had lower total force, and lower directionality in the force as compared with monolayers exposed to laminar flow.

In order to determine the direction of intercellular force, Ting, et al. showed that by analyzing the resultant force of a single cell, the vector sum of its tugging forces can be determined (Fig. 12.7) [2]. Individual interactions between any two cells are mathematically indeterminate, but the total imbalance of each cell can be used to determine its force on its neighboring cells. In this manner, information about the intercellular forces can be determined even if each specific interaction cannot. Information about the tugging force vector is a valuable tool to determine force alignment and intercellular tugging based on different types of flows.

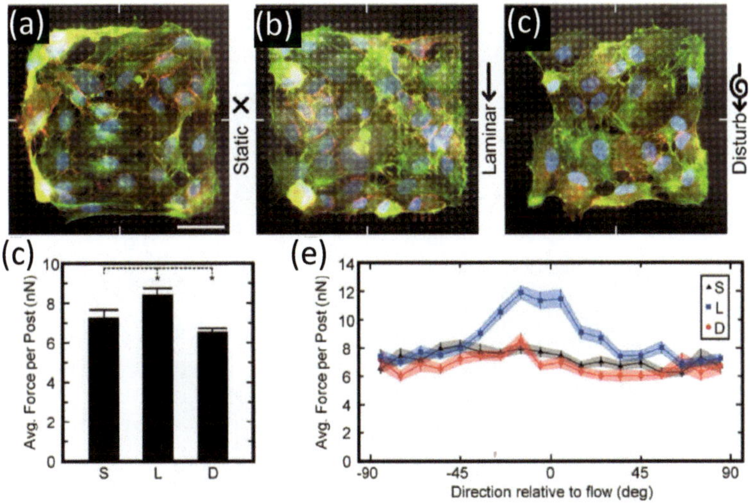


**Figure 12.6** Endothelial cell monolayers were seeded on microposts and exposed to (a) static, (b) laminar, and (c) disturbed flow. Cells are stained for actin (green), β-catinen (red), nucleus (blue), and posts are white. (d) Average force per post, and (e) direction of force per post of cell monolayers exposed to laminar (L) flow was greater than for either static (S) or disturbed (D) flow. Scale bar is 40 μm. Adapted from Ting, et al. [2].

When exposed to laminar flow, the actin in an endothelial cell monolayer is aligned in the direction of flow when compared with actin in monolayers under either static conditions or disturbed flow. Furthermore, the traction forces of the monolayer on the substrate are also aligned in the direction of flow under laminar flow (Fig. 12.6). Intercellular forces and the junction protein β-catenin are also affected by the type of flow and are stronger when exposed to a laminar flow than when exposed to disturbed flow or static conditions (Fig. 12.7). Because of the novel use of measuring the vector sum of intercellular forces, Ting, et al. were able to determine that the intercellular forces were indeed affected by flow conditions as to be expected from previous studies showing an increase in junction protein levels [2]. The micropost method enhanced the previous results to show the actual forces rather than solely the protein expression or the degree to which cells were connected together.

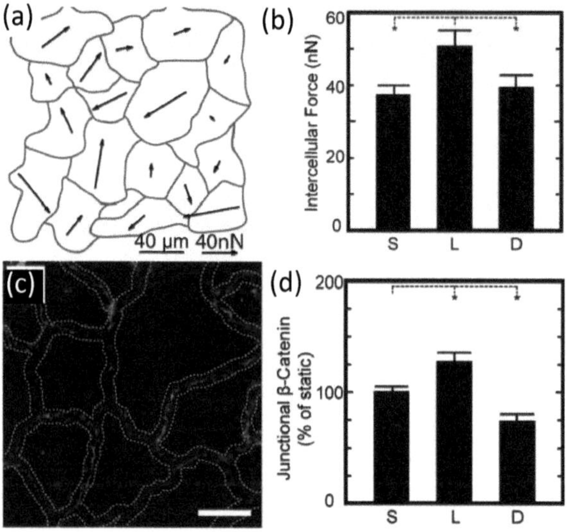


**Figure 12.7** (a) Each cell tugging force can be mapped as the resultant force for each cell outlined in a monolayer. After cell monolayers are exposed to either laminar (L), static (S), or disturbed (D) flow, the intercellular cell tugging force is greater for cell monolayers exposed to laminar flow, scale bar is 40 μm, force reference arrow is 40 nN. (c) Junctional protein β-catenin was measured as the intensity per unit length and divided by the number of cells in each monolayer, with (D) the finding of an increase in monolayers exposed to laminar flow, scale bar in (c) is 20 μm. Values in (d) are intensity normalized by static condition intensity. Adapted from Ting, et al. [2].

## 12.5 Platelet Contraction with Microposts

Platelets serve a vital role in the cardiovascular system, and much of the function is mechanical. Platelets are found throughout the body and play a critical role in wound healing by adhering to damaged endothelial cells and signaling other platelets to adhere to cover the damage to the endothelium [30,31]. Although an agglomeration of platelets is the first and necessary step to heal a damaged area, the thrombus needs to contract in order to continue the process of tissue repair. A few methods to measure clot strength have been used including thromboelastography (TEG), clot retraction assay, and platelet clot strips, but none of them provide a direct measurement of platelet strength in clots [32]. Micropost arrays are able to use a platelet rich mixture and measure force without further straining the microclot that has formed on the tips.

### 12.5.1 Formation of Microclots on Microposts

In order for platelets to adhere to the tops of the microposts, a specific matrix coating for the platelets was required. In the study performed by Liang, et al., both human fibrinogen and human fibronectin were used as a matrix coating on the surface of the microposts as described earlier in this chapter with no significant difference in thrombus force between coatings [33]. After centrifuging human blood to separate platelets, the platelets were suspended in Tyrode buffer and seeded onto microposts. After 25 min had passed, the platelets were stimulated with thrombin to form the microclots and allow them to generate force. The microclots were fixed and stained for actin and imaged.

### 12.5.2 Platelet Force Increases with Time and Thrombin Concentration

Platelet forces can be affected by a variety of factors, and the amount of thrombin can affect the contractility of a microclot on microposts. Thrombin enhances platelet contractility and signals them to start contracting. The study by Liang, et al. showed that an increase in thrombin concentration will enhance platelet force generation in a non-linear fashion (Fig. 12.8) [33]. As higher concentrations of thrombin are used, both the total force of the clots increased and the average force based on clot size increased.

Unlike many aspects of cell signaling and dynamics, platelet force generation takes place over a period of time. A series of microclots were analyzed after specific time points up to an hour after seeding adding thrombin to the platelet solution and seeding on microposts. The researchers found that microclots continued to have increased forces all the way up to sixty minutes after adding thrombin.

The role of thrombin and time in how platelet microclots generate force is important to wound healing and treatment. Platelet force generation is also significant in hemostasis and thrombosis. A clot without significant force may break away from the damaged region and catch onto another point in the body.

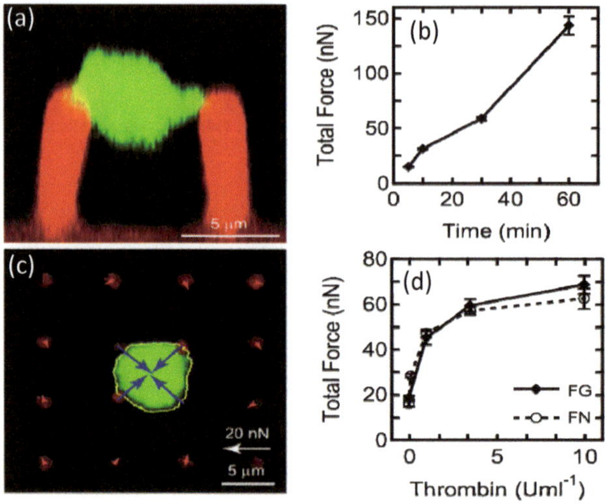


**Figure 12.8** Platelets form microclots on microposts. (a) Confocal image of a microclot stained for actin (green) suspended between and bending two microposts (red). (b) Thrombin is added to platelets on microposts at $t$ = 0, and platelet forces increase with time. (d) Thrombin concentration affects platelet forces, measured 30 min after thrombin addition. Adapted from Liang, et al. [5].

## 12.6 Future Directions in Micropost Methods for Cardiovascular Cell Studies

Microposts have been used for a variety of studies of cells, in particular cardiovascular cells and will further the understanding of mechanical forces in the cardiovascular system. New techniques are being developed to both measure and manipulate forces on cardiovascular cells. Future developments in this field may see a trend toward nanoscale dimensions to understand single platelets as well as a higher density of force measurements. Furthermore, tools may be developed out of the current methods to be utilized in the hospital to diagnose disorders that primarily rely on mechanical force deficiencies. Ultimately, the cardiovascular system will continue to be studied as a mechanical system due to the analogous system of a pump and piping system. Micropost methods enable a look at the individual components of the system in the same way that studying the grain structure of a pipe identifies causes of a crack that led to a leak and future cardiovascular diagnoses will depend on understanding all of the factors that went into instigating a particular disorder.

## References


1. Han, S. J., et al., Decoupling substrate stiffness, spread area, and micropost density: A close spatial relationship between traction forces and focal adhesions. *Biophys. J.*, 2012. **103**(4), 640–648.
2. Ting, L. H., et al., Flow mechanotransduction regulates traction forces, intercellular forces, and adherens junctions. *Am. J. Physiol.-Heart Circ. Phys.*, 2012. **302**(11), H2220–H2229.
3. Liu, Z. J., et al., Mechanical tugging force regulates the size of cell-cell junctions. *Proc. Natl. Acad. Sci. U. S. A.*, 2010. **107**(22), 9944–9949.
4. Rodriguez, A. G., et al., Substrate stiffness increases twitch power of neonatal cardiomyocytes in correlation with changes in myofibril structure and intracellular calcium. *Biophys. J.*, 2011. **101**(10), 2455–2464.
5. Liang, X. M., et al., Platelet retraction force measurements using flexible post force sensors. *Lab Chip*, 2010. **10**(8), 991–998.
6. Harris, A. K., Wild, P., and Stopak, D. Silicone-rubber substrata - new wrinkle in the study of cell locomotion. *Science*, 1980. **208**(4440), 177–179.
7. Lee, J., et al., Traction forces generated by locomoting keratocytes. *J. Cell Biol.*, 1994. **127**(6), 1957–1964.
8. Dembo, M., and Wang, Y. L. Stresses at the cell-to-substrate interface during locomotion of fibroblasts. *Biophys. J.*, 1999. **76**(4), 2307–2316.
9. Galbraith, C. G., and Sheetz, M. P. A micromachined device provides a new bend on fibroblast traction forces. *Proc. Natl. Acad. Sci.*, 1997. **94**(17), 9114–9118.
10. Tan, J. L., et al., Cells lying on a bed of microneedles: An approach to isolate mechanical force. *Proc. National Acad. Sci. U. S. A.*, 2003. **100**(4), 1484–1489.
11. Tooley, W. W., et al., Thermal fracture of oxidized polydimethylsiloxane during soft lithography of nanopost arrays. *J. Micromechan. Microeng.*, 2011. **21**(5), 054013 doi:10.1088/0960-1317/21/5/054013.
12. Ting, L. H., et al., Effect of silanization film thickness in soft lithography of nanoscale features. *J. Nanotechnol. Eng. Med.*, 2011. **2**(4), 041006.

13. Fuard, D., et al., Optimization of poly-di-methyl-siloxane (PDMS) substrates for studying cellular adhesion and motility. *Microelectron. Eng.*, 2008. **85**(5–6), 1289–1293.

14. Khanafer, K., et al., Effects of strain rate, mixing ratio, and stress-strain definition on the mechanical behavior of the polydimethylsiloxane (PDMS) material as related to its biological applications. *Biomed. Microdevices*, 2009. **11**(2), 503–508.

15. Mata, A., Fleischman, A. J., and Roy, S. Characterization of polydimethylsiloxane (PDMS) properties for biomedical micro/nanosystems. *Biomed. Microdevices*, 2005. **7**(4), 281–293.

16. Sniadecki, N. J., et al., Magnetic microposts as an approach to apply forces to living cells. *Proc. Natl. Acad. Sci. U. S. A.*, 2007. **104**(37), 14553–14558.

17. Jacot, J. G., Martin, J. C., and Hunt, D. L. Mechanobiology of cardiomyocyte development. *J. Biomechan.*, 2010. **43**(1), 93–98.

18. Litwin, S. E., et al., Contractility and stiffness of noninfarcted myocardium after coronary ligation in rats—Effects of chronic angiotensin converting enzyme-inhibition. *Circulation*, 1991. **83**(3), 1028–1037.

19. Zhao, Y., and Zhang, X. Cellular mechanics study in cardiac myocytes using PDMS pillars array. *Sens. Actuators Phys.*, 2006. **125**(2), 398–404.

20. Kim, K., et al., Calibrated micropost arrays for biomechanical characterisation of cardiomyocytes. *Micro Nano Lett.*, 2011. **6**(5), 317–322.

21. Ross, R., Mechanisms of disease—Atherosclerosis—An inflammatory disease. *N. Engl. J. Med.*, 1999. **340**(2), 115–126.

22. Balda, M. S., and Matter, K. Tight junctions at a glance. *J. Cell Sci.*, 2008. **121**(22), 3677–3682.

23. Hill, J. M., et al., Circulating endothelial progenitor cells, vascular function, and cardiovascular risk. *N. Engl. J. Med.*, 2003. **348**(7), 593–600.

24. Carmeliet, P., and Jain, R. K. Angiogenesis in cancer and other diseases. *Nature*, 2000. **407**(6801), 249–257.

25. Ghibaudo, M., et al., Traction forces and rigidity sensing regulate cell functions. *Soft Matter*, 2008. **4**(9), 1836–1843.

26. Califano, J. P., and Reinhart-King, C. A. Substrate stiffness and cell area predict cellular traction stresses in single cells and cells in contact. *Cell. Mol. Bioeng.*, 2010. **3**(1), 68–75.

27. Liu, Z. J., Sniadecki, N. J., and Chen, C. S. Mechanical forces in endothelial cells during firm adhesion and early transmigration of human monocytes. *Cell. Mol. Bioeng.*, 2010. **3**(1), 50–59.

28. Chien, S., Effects of disturbed flow on endothelial cells. *Ann. Biomed. Eng.*, 2008. **36**(4), 554–562.

29. Chiu, J. J., and Chien, S. effects of disturbed flow on vascular endothelium: Pathophysiological basis and clinical perspectives. *Physiol. Rev.*, 2011. **91**(1), 327–387.

30. Jackson, S. P., The growing complexity of platelet aggregation. *Blood*, 2007. **109**(12), 5087–5095.

31. George, J. N., Platelets. *Lancet*, 2000. **355**(9214), 1531–1539.

32. Feghhi, S., and Sniadecki, N. J. Mechanobiology of platelets: Techniques to study the role of fluid flow and platelet retraction forces at the micro- and nano-scale. *Intl. J. Mol. Sci.*, 2011. **12**(12), 9009–9030.

33. Liang, X. M., et al., Platelet retraction force measurements using flexible post force sensors. *Lab Chip*, 2010. **10**(8), 991–998.

# Chapter 13

# Dynamic Mechanical Environments to Quantify and Control Cellular Dynamics

**Luke MacQueen,[a] Christopher Moraes,[b] Yu Sun,[a,c] and Craig A. Simmons[a,c]**

[a]*Department of Mechanical and Industrial Engineering, University of Toronto, Toronto, Ontario, Canada*
[b]*Department of Biomedical Engineering and Biointerfaces Institute, University of Michigan, Ann Arbor, Michigan, USA*
[c]*Institute of Biomaterials and Biomedical Engineering, University of Toronto, Toronto, Ontario, Canada*

c.simmons@utoronto.ca

Experimental models of physiological processes enable investigations of complex cellular dynamics in controlled conditions. Dynamic mechanical forces that result from tissue deformation or fluid flow are key factors influencing cellular fate in living tissues. In this chapter, we review recent efforts in our research group to apply dynamic mechanical forces to cultured cells and tissue constructs. We describe microfabricated platforms for cell stretching, biomaterial compression, and cell co-culture under fluid flow conditions. We discuss the benefits of arrayed and microfabricated platforms and demonstrate their utility for experimental mechanobiology. These physiologically relevant cell culture strategies enable investigations of processes for which in vivo access is limited. Their arrayed and scalable formats increase

*Cells, Forces, and the Microenvironment*
Edited by Charles M. Cuerrier and Andrew E. Pelling

ISBN 978-981-4613-36-1 (Hardcover), 978-981-4613-37-8 (eBook)
www.panstanford.com

experimental throughput and show promise for drug screening that is predictive of in vivo results.

## 13.1 Introduction

Dynamic mechanical forces are ubiquitous physiological and pathological regulators of cell behavior but they are not recapitulated in traditional static cell culture platforms. Although traditional culture platforms lack mechanically dynamic aspects of in vivo environments, they often include arrayed formats in which many independent culture conditions are varied systematically (Fig. 13.1a). These platforms enable high-throughput screening (HTS) of cell responses to soluble factors that are added to the culture media [1]. HTS has become a central part of the drug discovery process, but the fraction of seemingly promising HTS leads that result in clinically effective therapeutics is notoriously low [2]. For this reason, emphasis is shifting from developing higher screening throughput capabilities toward improving the content and quality of such biological test systems [2,3]. Improved in vitro models are sought to ensure that HTS assays are relevant and predictive of in vivo results [2,4]. These will include application-specific co-culture of multiple cell types, fluid flow, and a variety of dynamic mechanical forces that are found in the body.

In contrast to array-based static culture platforms, customized "mechano-culture" systems that apply fluid flow and dynamic mechanical forces to cells and tissues have relatively low experimental throughput [5]. Hence, these technologies typically enable researchers to replicate dynamic aspects of physiological or pathological processes while sacrificing the advantages associated with high-throughput experimental platforms. This limitation is significant because combinatorial experiments are required to examine the effects of mechanical loading parameters and their modulation of other (e.g., soluble) factors. Although conventional macroscale techniques require a large number of cells per set of environmental conditions, emerging array-based mechano-culture platforms (Fig. 13.1b) are well suited to study the response of rare cell types to combinatorially manipulated mechanobiological cues. As outlined in Fig. 13.1c, cells and tissues are subjected to a variety of mechanical forces that include hydrostatic pressure, fluid shear, and mechanical strain in shear, tension, and compression. Emerging mechanically active culture platforms apply these forces to cells and tissues to improve the physiological relevance of such culture systems, with the potential for high experimental throughput [6].

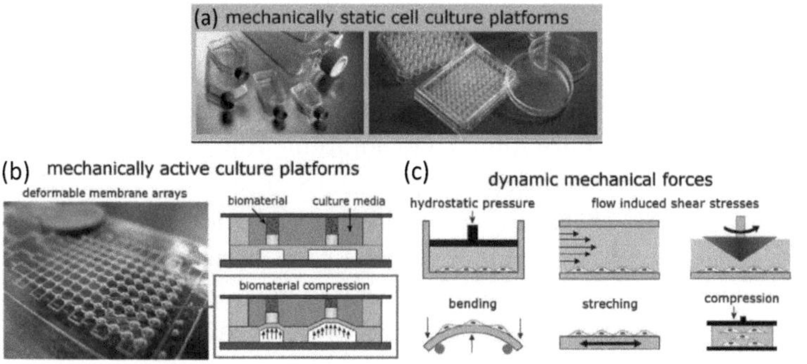


**Figure 13.1** Dynamic mechanical forces and arrayed culture platforms. (a) Traditional static cell culture platforms include flasks (left), well-plates and petri dishes (right). (b) Mechanically active culture platforms include arrayed deformable membranes for cell stretching [7] and biomaterial compression [8] Reprinted from [7] with permission from the Royal Society of Chemistry and [8] with permission from Elsevier. (c) Various types of mechanical forces, adapted from [6] with permission from the Royal Society of Chemistry.

In this chapter, we describe recent efforts in our research group to develop mechanically active culture platforms for applications in experimental mechanobiology. We focus on the use of deformable elastomeric membranes to apply tensile or compressive forces to cells and tissues, and multilayer microfluidic devices to apply fluid shear forces. To motivate our focus on deformable membrane technologies, we describe their capabilities within the context of spatial dimensions and force magnitudes that are encountered in experimental mechanobiology and biomechanics. We then describe their use for cell stretching and biomaterial compression. Microfluidic devices for fluid shear experiments are then examined with a focus on endothelial cell mechanobiology. We conclude by discussing challenges and future directions for these technologies.

## 13.2 Deformable Elastomeric Membranes

Cells cultured on deformable substrates can be mechanically stimulated by applying bending moments or in-plane deformations to the culture substrates. These methods can be extended to cells cultured in three-dimensional (3D) scaffolds and can be applied in combination to mechanically manipulate multiple microenvironmental parameters [6]. Deformable membranes enable mechanical loading in arrayed formats and generate enough force to actuate a wide variety of samples. They are used in commercial cell stretching and tissue compression platforms (e.g., FlexCell International Corporation, Hillsborough, NC, USA) and in emerging microfabricated HTS formats [7,8,12,13]. The broad range of forces applicable using deformable substrates enable experiments using individual cells or relatively large tissue samples. They are therefore well suited for integrative approaches that link cellular mechanobiology with macroscopic tissue-level biomechanical observables.

Mechanical aspects of human physiology are studied throughout a range of spatial scales from molecules to whole organisms. Macroscopic properties are described in biomechanical terms that relate applied stress to measured strain. At the microscopic level, mechanobiology describes biochemical changes within cells that are elicited by mechanical forces. Merryman and Engler [14] distinguished biomechanics from mechanobiology using the following statement: "*biomechanics is the application of the principles of mechanics to study living organisms and their components*" whereas "*mechanobiology is the application or analysis of the role of mechanical forces in eliciting a molecular response, leading to a quantifiable change in form and/or function.*" These two fields necessarily overlap and an integrative approach is required to describe how molecular responses give rise to macroscopic tissue properties.

### 13.2.1 Biomechanics and Mechanobiology: There Is Plenty of Room in the Middle

The combined potential for arrayed fabrication (discussed below) and a broad range of applicable forces motivates our focus on deformable membranes for cell and tissue mechanobiology. Spatial scales and force magnitudes associated with human physiology and achieved experimentally using various mechanical testing platforms are illustrated in Fig. 13.2. Although Fig. 13.2 includes only a small sample of available methods, it illustrates the middle-range of sample sizes and applied forces that are studied using deformable elastomeric membranes. This middle range is essential to small (mm-sized) biomaterial and tissue studies that bridge cell- and tissue-level phenomena. For a description of other mechanical testing methods, the reader is referred to Brown [5] and, for a focus on microtechnology platforms, to Kim et al. [15], Moraes et al. [6], and Polacheck et al. [16].

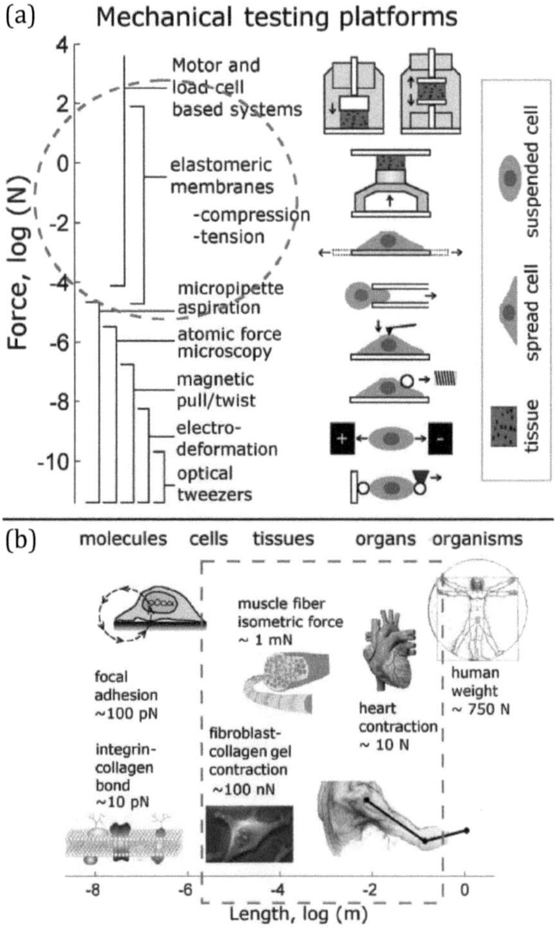


**Figure 13.2** Force and spatial scales accessible using deformable elastomeric membranes (highlighted areas). (a) Ranges of force applied to samples using various mechanical testing platforms. (b) Spatial scales and force magnitudes associated with human physiology.

### 13.2.2 Deformable Elastomeric Membranes: Operating Principles

The use of deformable membranes for cell stretching or biomaterial compression is illustrated in Fig. 13.3. Pressure is supplied through underlying channels and membrane deformation results in vertical displacement of attached loading posts. Miniaturized versions of the membrane + loading post configuration [12] are referred to as "micro Braille" [17]. For cell stretching applications, cells are cultured on an additional thin membrane placed above the posts and vertical displacement of the posts generates tensile strain in the overlying membranes (Fig. 13.3a, left) [7]. For sample compression, the loading posts apply compression directly to the sample (Fig. 13.3a, right) [8,18]. Multiple actuation heights and strain magnitudes can be achieved

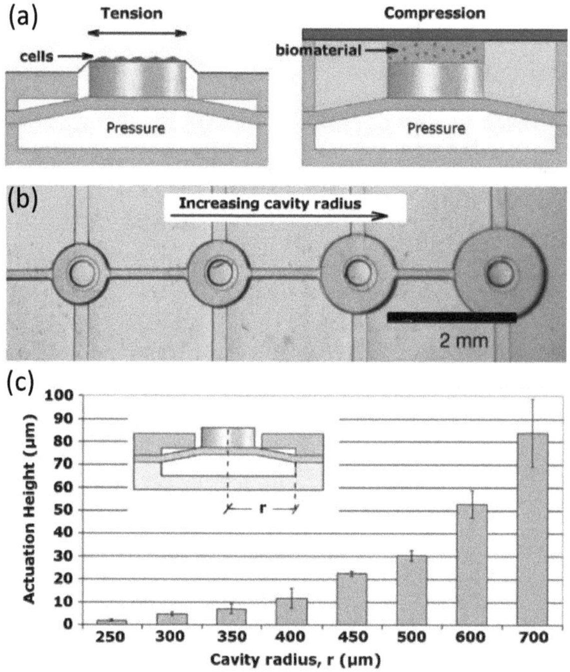


**Figure 13.3** (a) Micro Braille actuators used for cell stretching and biomaterial compression. (b) Pressure supplied through an underlying cavity deforms the membranes and (c) the cavity radius determines membrane actuation height. Reprinted from [17] with permission from IOP Publishing.

using a single driving pressure by varying the pressure cavity radius (Figs. 13.3b,c). Thus, the effects of multiple strain values are screened while maintaining the operational simplicity of a single driving pressure supply.

### 13.2.3 Deformable Elastomeric Membranes: Microfabrication

Canonical advantages of microfabricated platforms include the use of smaller quantities of samples and reagents, reduced device footprint, and the ability to vary a large number of experimental conditions without manual intervention. Microfabricated platforms are well suited for experiments in which reduced sample size does not alter relevant physiological functions.

Deformable membrane arrays are fabricated using standard soft lithography techniques. Unless specified otherwise, all membranes described here were fabricated using Sylgard 184 PDMS kits (Dow Corning) and SU-8 (MicroChem) for mold masters. Fabrication of micro Braille arrays requires three SU-8 mold masters. The process is illustrated in Fig. 13.4a as described in [17] and demonstrated step-by-step online in [19]. Three mold films are fabricated using the SU-8 masters. The first film is bonded to a glass substrate, and the backing film is peeled away. The second film acts as a mold for an additional layer of PDMS. This mold is silanized and an 80 μm thick layer of uncured PDMS is spin coated onto it and partially cured in an oven for 15 min at 80°C. This mold is then inverted, aligned and plasma bonded with the first film. A second partial curing step of 15 min at 80°C strengthens the bond and the silanized PDMS mold is peeled away from the structure. Alternative loading post shapes can be produced by modifying a single mask in this stage of the fabrication process. For example, rectangular and elliptical posts with changing aspect ratios can create a range of anisotropic biaxial strains. The third layer is then aligned and plasma bonded to complete the structure. Access ports are cored into the PDMS device, fitted with tubing, and connected to a pressure supply.

For cell stretching applications, an additional membrane is bonded to the third layer and suspended over the post (Fig. 13.4b). Actuating the post generates tensile strain in the suspended membrane, as shown previously in Fig. 13.3a. In this configuration, a lubricant is required to prevent adhesion of the post and membrane. For biomaterial compression, hydrogel precursors are photo-polymerized in situ by masked UV exposure (Fig. 13.4c).

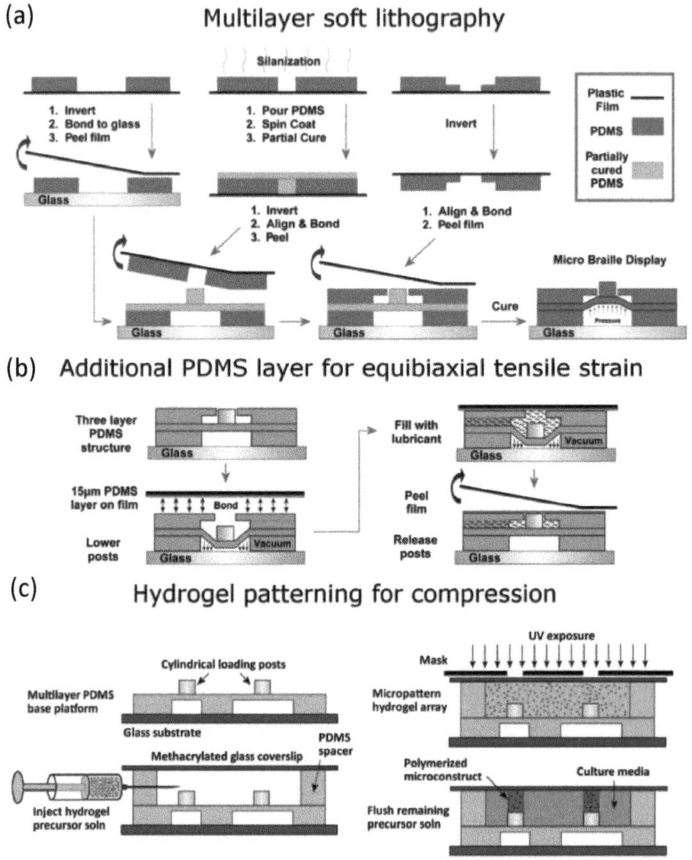


**Figure 13.4** Schematic outline of the micro Braille fabrication process. (a) Micro Braille fabrication steps, reprinted from [17] with permission from IOP. Modifications required for applications requiring (b) in-plane substrate tension, reprinted from [7] with permission from the Royal Society of Chemistry, or (c) three-dimensional sample compression, reprinted from [8] with permission from Elsevier.

### 13.2.4 Cell Stretching

In 1985, Banes et al. [20] introduced the first flexible-bottomed circular cell culture plates that were stretched by negative pressure (vacuum) applied to a manifold underlying the cell culture substrate. This system's simplicity of operation and versatility led to commercialization and widespread adoption within the mechanobiological research community (FlexCell). The simplicity of this platform derives from its minimal number of moving parts and limited calibration requirements. Programmable vacuum magnitude, waveform, and duty cycle enable a broad range of mechanical loading regimes. However, permanent changes to the tensile properties of macroscale membranes during extended use have raised concerns related to potential measurement inaccuracies [21].

In contrast with commercial systems, Moraes et al. [7] showed that applied strain levels in a microscale stretching platform did not change during extended use ($10^5$ continuous cycles). These microfabricated arrays were capable of simultaneously applying cyclic equibiaxial substrate strains ranging in magnitude from 2% to 15% to small populations of adherent cells. The platform's utility for mechanobiology experiments was demonstrated by studying mechanical activation of the canonical Wnt/β-catenin signaling pathway in cardiac valve mesenchymal progenitor cells. Nuclear accumulation of β-catenin was found to depend both on equibiaxial strain magnitude and stimulation time. More recently, Moraes et al. [34] used their cell stretching platform to determine the combined effects of dynamic substrate stretch, soluble cues and matrix proteins on small populations of primary aortic valvular interstitial cells (VICs). VICs are found in spatially distinct and mechanically dynamic layers of the valve leaflet and disease progression is associated with the pathological differentiation of VICs to myofibroblasts. Using the mechanosensitive material screening array, they demonstrated that spatially distinct VIC subpopulations respond differentially to microenvironmental cues and that mechanically induced myofibroblast differentiation of side-specific VICs is differentially modulated by matrix and soluble cues. Myofibroblast differentiation was most significantly influenced by cell origin, despite the presence of potent mechanobiological cues such as applied strain and TGF-β1 (Fig. 13.5). This platform overcomes a critical limitation of conventional macroscale culture technologies by reducing the number of primary cells required to obtain mechanobiological response profiles.

This increases the range of biological phenomena that can be studied in vitro and the results obtained help to generate rational hypotheses for more specific mechanistic research.

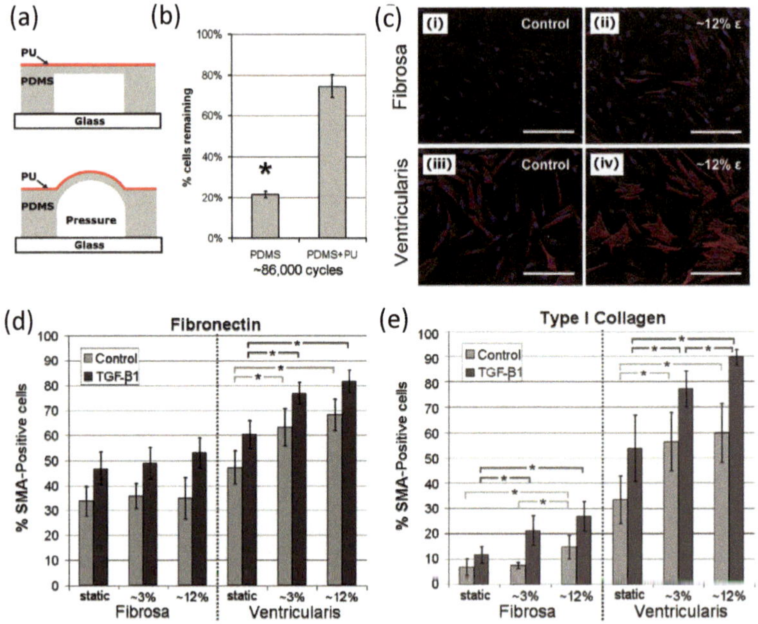


**Figure 13.5** Myofibroblast differentiation of side-specific VICs in response to cyclic strain, matrix proteins, and TGF-β1. (a) Cells were stretched on composite films consisting of a thick layer of PDMS and a thin layer of polyurethane (PU) suspended over arrays of actuation cavities (top panel) and distended by the application of a pressure differential (bottom panel); (b) VICs adhered significantly better to PU compared to PDMS after 24 h of mechanical stimulation (1 Hz) at ~12% strain. (c) Myofibroblast differentiation of side-specific VICs cultured on type I collagen-coated surfaces. Representative images of cells from (i, ii) the fibrosa and (iii, iv) the ventricularis, cultured with 5 ng/mL TGF-β1 supplemented media under (i, iii) no mechanical stimulation, and (ii, iv) ~12% cyclic strain. Scale bar = 100 μm; red = α-smooth muscle actin stress fibers; blue = nuclei. (d, e) Myofibroblast differentiation of side-specific VICs is sensitive to *combination* of matrix protein, strain level, and TGF-β1 treatment. Selected statistical comparisons are highlighted (*$p < 0.01$). Reprinted from [22] with permission from the Royal Society of Chemistry.

### 13.2.5 Biomaterial Compression

In tissues, cells are embedded in 3D mechanically dynamic environments where cell behavior differs substantially compared to 2D monolayer cultures [23–25]. Dynamic mechanical strains that are present throughout all stages of development in living tissues modulate a variety of cellular functions that include matrix synthesis by chondrocytes [26–28] and multipotent stem cells (MSCs) [29], matrix mineralization by osteogenic MSCs [30] and altered solute transport kinetics [31]. Dynamic mechanical forces are likely needed for full differentiation and committal of MSCs within mature tissues [32] and mechanical stimulation (via oscillatory fluid flow) induces epigenetic changes that can be passed to daughter cells, suggesting that "downstream" effects of mechanical loading persist long after forces are applied [33]. Physiologically relevant 3D tissue models are therefore a key component of emerging culture platforms [34,35].

Extending deformable membrane arrays to achieve mechanical stimulation of arrayed 3D tissue constructs was recently demonstrated by Moraes et al. [8]. They used a microfabricated micro Braille device to apply compressive strains ranging from 6% to 26% simultaneously to cells encapsulated in an array of micropatterned biomaterials (Fig. 13.6a,b). Nuclear and cellular deformation in response to applied compression was assessed in C3H10T1/2 mouse MSCs encapsulated within poly(ethylene glycol) hydrogels. Biomaterial, cellular, and nuclear deformations were non-linearly related. Cellular deformation only changed significantly at the highest strain levels (Fig. 13.6c) and nuclear deformation was insignificant regardless of applied mechanical strain (Fig. 13.6d). Parametric finite element simulations suggested that this phenomenon was due to the relative stiffness differences between the hydrogel matrix and that of the encapsulated cell and nucleus and to strain stiffening of the matrix with increasing compression [8].

Visual observation of strain in 3D biomaterials is difficult to achieve using arrayed formats and this difficulty is compounded by the need to incubate tissue samples during culture. Optical measurement of tissue deformation during culture in opaque chambers is especially prohibitive. To address the need for in situ strain measurements, MacQueen et al. [9] embedded elastomeric

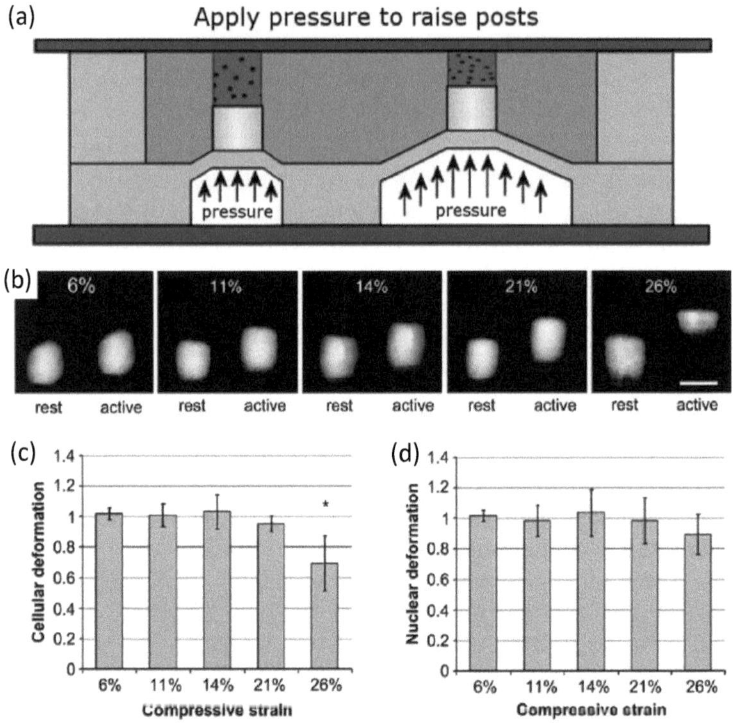


**Figure 13.6** Compression of cells encapsulated in 3D hydrogels. (a) Schematized cross-section of a micro Braille device. Pressure supplied to chamber cavities deforms overlying membranes, which raises the loading posts and compresses 3D biomaterials against a fixed surface. (b) Orthogonally re-sliced confocal images of 3D poly(ethylene glycol) hydrogels undergoing increasing levels of compression. Scale bar = 20 μm. (c) Cellular deformation observed across the array. Whole cell deformation increases non-linearly with increasing compressive strains (*$p < 0.001$ as compared to all other compression levels). (d) Nuclear deformation observed across the array. No significant changes in nuclear shape were observed at these strain levels. Reprinted from Moraes et al. [8] with permission from Elsevier.

strain sensors within deformable membranes to monitor membrane deflection amplitude (Fig. 13.7). The sensors were fabricated using blends of carbon nanotubes (CNT) in a silicone elastomer base to form a composite material that exhibits strain-dependent electrical conductivity. During sample compression, membrane deflection is reduced in proportion to the sample's stiffness and the magnitude of the sensor's resistive strain, $|\Delta R/R_0|$, versus input pressure permits determination of the sample stiffness. Ongoing efforts aim to correlate stiffness changes with functional maturation of MSC-based 3D tissue constructs.

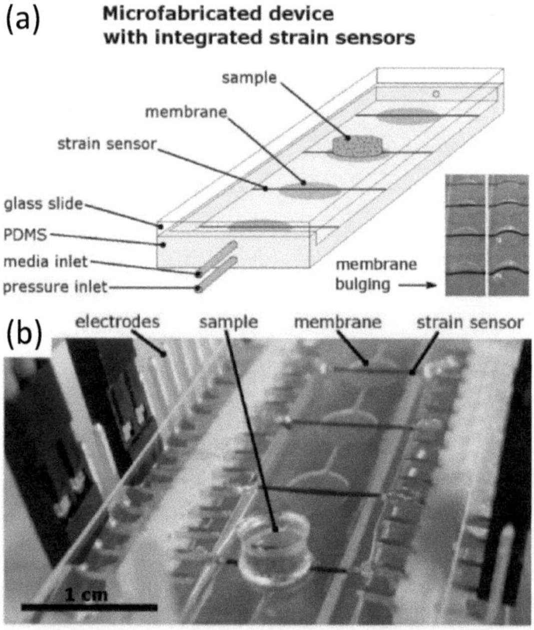


**Figure 13.7** Micro device array for material mechanical testing. (a) Schematic showing strain sensors integrated in a deformable membrane based mechanical testing platform. (b) A four-element device containing a single sample: electrodes are used to acquire the sensor signal; scale bar = 1 cm. Reprinted from [9] with permission from the Royal Society of Chemistry.

## 13.3 Fluid Shear and Co-Cultures

Shear forces that result from fluid flow occur within a variety of physiological processes but are not recapitulated within traditional static cell culture platforms. The study of fluid shear forces and their effects on cell biology in vitro represent a key application area for microfluidic platforms [36]. Multilayer microfluidics is particularly well suited to study cell–cell interactions under flow: For example, we use bilayer membrane microfluidic platforms to study endothelial adhesion [37] and permeability [10], and cell–cell interactions that include endothelial-leukocyte [11] and valvular endothelial–interstitial [38] co-cultures. Bifurcating or multilayer microfluidic platforms are shown in Figs. 13.8 and 13.9, respectively.

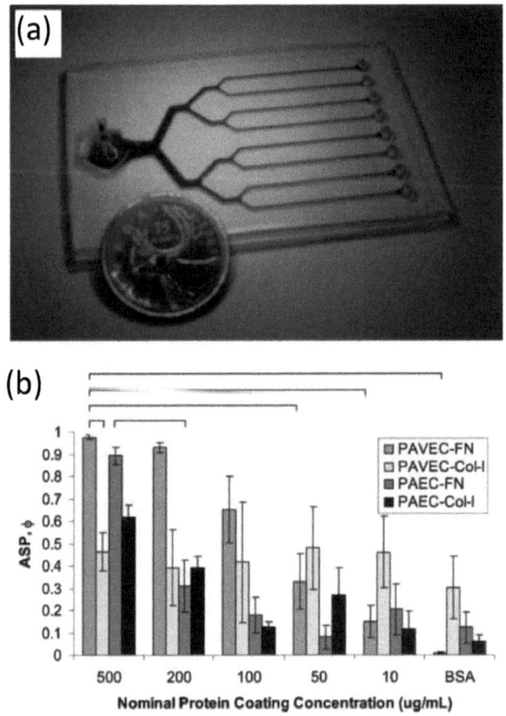


**Figure 13.8** Microfluidic shear device for endothelial adhesion studies. (a) Device with branching fluidic channels. Total device footprint is 2″ × 3″. (b) Adhesion strength parameter (ASP) for different cell-protein combinations at various levels of protein coating concentration. Data presented as mean ± SE ($n = 3$) and brackets indicate statistical significance, ($P < 0.01$). For clarity, some significant comparisons have not been shown. Reprinted from [37] with permission from the Royal Society of Chemistry.

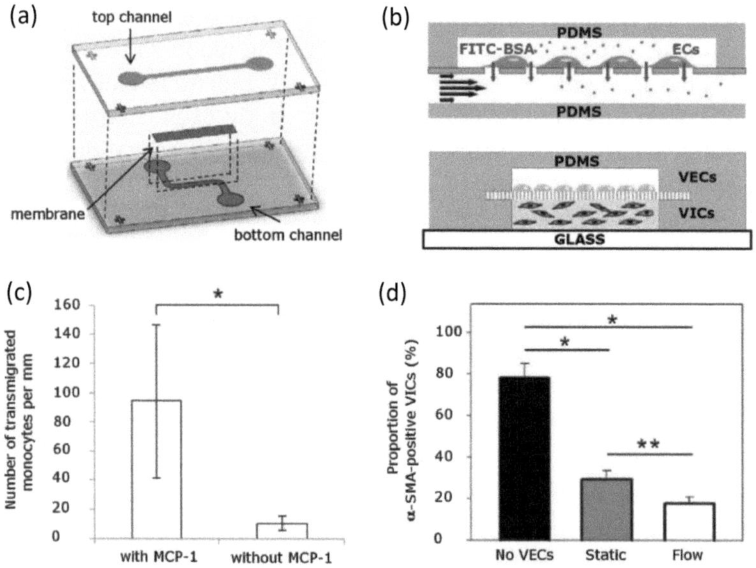


**Figure 13.9** Membrane microfluidic devices to study endothelial cell permeability, monocyte-valvular endothelial-interstitial cell co-culture, and endothelial co-culture. (a) Illustration of the membrane microfluidic device. Two PDMS channels were used to sandwich a porous membrane. Reprinted from [11] with permission from AIP. (b, top panel): cross section of a two-layer membrane microfluidic device used to study endothelial cell permeability. Reprinted from [10] with permission from ACS. (b, bottom panel) Cross-sectional schematic illustrating the spatial arrangement of the cells in the device. Reprinted from [27] with permission from the Royal Society of Chemistry. (c) Monocyte adhesion and transmigration. Addition of MCP-1 in the lower channel increased monocyte transmigration. Data presented as mean ± SD (*$p < 0.05$). Reprinted from [11] with permission from AIP. (d) Quantification of αSMA-positive VICs (myofibroblasts) in the hydrogels under various co-culture and shear stress conditions (*$p < 0.001$; **$p = 0.066$). Reprinted from [27] with permission from the Royal Society of Chemistry.

### 13.3.1 Fluid Shear Forces and Endothelial Adhesion

Endothelial cells (ECs) cover all blood-contacting surfaces of the cardiovascular system, including the internal lining of blood vessels [39] and the outer coating of heart valves [40]. Because of their anatomic location, ECs are constantly exposed to hemodynamic shear stress generated by flowing blood. Shear stress is known to regulate many endothelial cell functions and applying shear on ECs in vitro is critical to the study of EC function [41]. Microfluidic technologies enable systematic investigation of EC mechanobiology under controlled flow conditions and support emerging in vivo-like drug screening platforms for pharmacological applications [42].

Interactions between endothelial cells and the underlying extracellular matrix regulate adhesion and cellular responses to microenvironmental stimuli, including flow-induced shear stress [43]. Taking advantage of the parallelization possibilities in fabricating microfluidic devices, Young et al. [37] compared adhesion of endothelial cells to extracellular matrix proteins (fibronectin and type I collagen) over a broad range of concentrations. As shown in Fig. 13.8, porcine aortic valve endothelial cells (PAVECs) showed increased adhesion on fibronectin compared to type I collagen, particularly for coating concentrations of 100, 200, and 500 μg/mL. PAVECs also withstood shear significantly better on fibronectin than on collagen for 500 μg/mL. Porcine aortic endothelial cells (PAECs) showed increased spreading on collagen compared to PAVECs but did not have significantly better adhesion strength. These results reaffirmed that cell adhesion is both cell-type and matrix dependent and revealed important phenotypic differences between vascular and valvular endothelium.

### 13.3.2 Fluid Shear Forces and Endothelial Permeability

A major function of the endothelium is selective trafficking of solutes between blood and the tissue interstitium [39]. Transport processes through the endothelial monolayer involve both transcellular and paracellular mechanisms that are regulated by chemical and mechanical factors in the surrounding microenvironment. Characterizing endothelial permeability is required to evaluate uptake of drugs and other biomolecules but current techniques rarely account for the effects of fluid flow-induced shear stress on endothelial permeability or have low experimental throughput [42].

To recapitulate dynamic in vivo conditions where ECs are constantly exposed to shear stress, Young et al. [10] developed a multilayer microfluidic platform to study EC permeability under flow. As shown in Fig. 13.9 (a and top panel of b), the device consists of two fluidic channels that are separated by a porous membrane. Fluorescently tagged solutes introduced in the top channel, permeated through the membrane by pressure driven flow, and their concentration in the bottom channel was measured by laser induced fluorescence. Analyte concentration at the point of detection varied with channel flow rates in agreement with theoretical analyses using a pore flow model. On the basis of the model, a sequential flow rate stepping scheme was developed and applied to obtain the permeability of cell-free and cell-bound membrane layers. The membrane-based microfluidic device enabled endothelial permeability measurements under shear stress conditions.

### 13.3.3 Cell–Cell Interactions and Fluid Flow

The adhesion and subsequent diapedesis of monocytes into the vascular endothelial lining is one of the earliest changes detectable in inflammation, immune responses, and atherosclerosis [44]. This process can be studied efficiently in membrane microfluidic platforms, as demonstrated by Srigunapalan et al. [11]. In this platform, the membrane in the top channel was coated with 100 μg/mL bovine plasma fibronectin and seeded with primary PAECs. Addition of an inflammatory cytokine, tumor necrosis factor-α (TNF-α; 25 ng/mL), significantly increased monocyte-EC adhesion and simultaneous application of TNF-α and 20 dynes/cm$^2$ shear stress caused increased monocyte adhesion compared with the static case. For monocyte transmigration assays, nuclear-stained monocytes were injected into the top channel with the PAECs and their migration to the bottom layer was tracked using fluorescence microscopy. The lower channel contained medium with or without 500 ng/mL monocyte chemoattractant protein-1 (MCP-1). As shown in Fig. 13.9c, MCP-1 induced a five-fold increase in monocyte transmigration.

Building on the bilayer membrane microfluidic platform, Chen et al. [38] incorporated a 3D cell-laden hydrogel in the bottom channel to mimic the vascular/valvular microenvironment. Cardiac valves contain VICs that populate valve tissue and a valvular endothelial cell (VEC) monolayer that covers the valve. To mimic valve physiology in vitro, VECs were cultured in the top channel and VICs were encapsulated in 3D hydrogels in the bottom channel (Fig. 13.9b, bottom panel). As shown in Fig. 13.9d, the presence of VECs suppressed VIC pathological differentiation into α-smooth-muscle-actin (α-SMA)-expressing myofibroblasts. This effect was enhanced when the endothelium was exposed to flow-induced shear stress.

Taken together, the results described above demonstrate that a broad range of co-culture and fluid shear force studies can be achieved using these microfluidic platforms. Their small device footprints and straightforward fabrication offer great potential for scalability and ease-of-use.

## 13.4 Conclusion

In this chapter, we presented recent efforts in our laboratory to microfabricate in vitro models of mechanically active physiologies. These included deformable membrane arrays for tensile and compressive loading of cells and tissues, and multilayer microfluidic devices for co-culture and fluid shear experiments.

Challenges associated with these platforms typically involve a trade-off between device complexity and functionality. For example, micro braille arrays that incorporate a thin overlying membrane to achieve reasonable strain uniformity during cell stretching require lubrication to reduce sticking between membranes and loading posts (Fig. 13.5b). Deformation of simple membranes without the addition of loading posts and secondary membranes simplifies device fabrication and operation, but introduces some variation in strain across the device surface [22]. In future designs, stiction between the loading post and culture film might be reduced by replacing PDMS with alternative elastomeric materials such as photocurable perfluoropolyethers [45].

During long-term cell stretching experiments, cell detachment from the substrate must be minimized. A variety of surface treatments have been developed to improve upon PDMS's hydrophobicity, fast hydrophobic recovery after surface hydrophilization, and low cell adhesion strength [46]. Surface modification strategies such as silane/glutaraldehyde-based covalent protein cross-linking [47] and antibody-mediated adhesion of specific binding proteins [48] can improve cell adhesion. In our studies [22,49], we found that thin polyurethane films provided a well-studied and clinically relevant biomaterial that enabled longer-term cell culture than was possible with PDMS substrates.

Platforms that include 3D tissue models are associated with a variety of challenges particular to the relatively nascent field of 3D cell culture. For example, scaling up to mm-sized tissue constructs requires careful consideration of scaffold porosity and oxygen and solute diffusion to preserve cell viability [50]. In 3D, cell fate also depends on scaffold stiffness gradients [51], nonlinear strain-stiffening [52], and soluble growth factors that can be trapped in the scaffold [53]. In cases where dynamic mechanical loading is present, transfer of mechanical loads depends on the relative stiffness of cells and scaffold [8] and cell-based scaffold remodeling [54,55] changes these properties during culture. Cell-based scaffold remodeling alters properties of the cell niche that feedback to influence cell fate and scaffold mechanical properties that are measured macroscopically (e.g., stiffness) may therefore not accurately represent conditions seen locally by cells. Fluid flow patterns and solute diffusion kinetics differ in 3D hydrogels or other porous scaffolds compared with those seen in 2D (monolayer) geometries and they show a strong dependence on mechanical loading conditions [31]. In these environments, the effects of mechanical loading can appear contradictory due to the diversity of loading protocols and interacting factors such as fluid flow [32].

Despite these challenges, the platforms described here demonstrate progress toward in vitro model physiologies and pathologies. Broadly speaking, future efforts will increase experimental throughput, ease of use, and the complexity and quality of physiologically realistic model culture systems. Dynamic mechanical stretching and fluidics have already been combined in various platforms, for example to mimic lung physiology [13] and a wider effort is underway to combine various human organ models on-chip [56–59]. These platforms will increase the efficiency of HTS by providing physiologically relevant alternatives to costly animal and clinical studies. More generally, they will enable a wide variety of in vitro experiments that have thus far been hampered by a lack of suitable culture platforms.

## References


1. Pereira, D. A., Williams, J. A. (2007). Origin and evolution of high throughput screening. *Br. J. Pharmacol.*, **152**, 53–61.
2. McKim, J. M., Jr. (2010). Building a tiered approach to in vitro predictive toxicity screening: A focus on assays with in vivo relevance. *Comb. Chem. High Throughput Screen*, **13**, 188–206.
3. Mayr, L. M., Fuerst, P. (2008). The future of high-throughput screening. *J. Biomol. Screen*, **13**, 443–448.
4. Fox, S., Farr-Jones, S., Sopchak, L., Boggs, A., Nicely, H. W., Khoury, R., et al. (2006). High-throughput screening: Update on practices and success. *J. Biomol. Screen*, **11**, 864–869.
5. Brown, T. D. (2000). Techniques for mechanical stimulation of cells in vitro: A review. *J. Biomech.*, **33**, 3–14.
6. Moraes, C., Sun, Y., Simmons, C. A. (2011). (Micro)managing the mechanical microenvironment. *Integr. Biol. (Camb)*, **3**, 959–971.
7. Moraes, C., Chen, J. H., Sun, Y., Simmons, C. A. (2010). Microfabricated arrays for high-throughput screening of cellular response to cyclic substrate deformation. *Lab Chip*, **10**, 227–234.
8. Moraes, C., Wang, G., Sun, Y., Simmons, C. A. (2010). A microfabricated platform for high-throughput unconfined compression of micropatterned biomaterial arrays. *Biomaterials*, **31**, 577–584.
9. MacQueen, L., Chebotarev, O., Simmons, C. A., Sun, Y. (2012). Miniaturized platform with on-chip strain sensors for compression testing of arrayed materials. *Lab Chip*, **12**, 4178–4184.
10. Young, E. W., Watson, M. W., Srigunapalan, S., Wheeler, A. R., Simmons, C. A. (2010). Technique for real-time measurements of endothelial permeability in a microfluidic membrane chip using laser-induced fluorescence detection. *Anal. Chem.*, **82**, 808–816.
11. Srigunapalan, S., Lam, C., Wheeler, A. R., Simmons, C. A. (2011). A microfluidic membrane device to mimic critical components of the vascular microenvironment. *Biomicrofluidics*, **5**, 13409.
12. Kamotani, Y., Bersano-Begey, T., Kato, N., Tung, Y. C., Huh, D., Song, J. W., et al. (2008). Individually programmable cell stretching microwell arrays actuated by a Braille display. *Biomaterials*, **29**, 2646–2655.
13. Huh, D., Matthews, B. D., Mammoto, A., Montoya-Zavala, M., Hsin, H. Y., Ingber, D. E. (2010). Reconstituting organ-level lung functions on a chip. *Science*, **328**, 1662–1668.

14. David Merryman, W., Engler, A. J. (2010). Innovations in cell mechanobiology. *J. Biomech.*, **43**, 1.

15. Kim, D. H., Wong, P. K., Park, J., Levchenko, A., Sun, Y. (2009). Microengineered platforms for cell mechanobiology. *Annu. Rev. Biomed. Eng.*, **11**, 203–233.

16. Polacheck, W. J., Li, R., Uzel, S. G., Kamm, R. D. (2013). Microfluidic platforms for mechanobiology. *Lab Chip*, **13**, 2252–2267.

17. Moraes, C., Sun, Y., Simmons, C. A. (2009). Solving the shrinkage-induced PDMS alignment registration issue in multilayer soft lithography. *J. Micromech. Microeng.*, **19**, 065015.

18. Moraes, C., Zhao, R., Likhitpanichkul, M., Simmons, C. A., Sun, Y. (2011). Semi-confined compression of microfabricated polymerized biomaterial constructs. *J. Micromech. Microeng.*, **21**, 054014.

19. Moraes, C., Sun, Y., Simmons, C. A. (2010). Microfabricated platforms for mechanically dynamic cell culture. *J. Vis. Exp.* (doi:2224 [pii], 10.3791/2224).

20. Banes, A. J., Gilbert, J., Taylor, D., Monbureau, O. (1985). A new vacuum-operated stress-providing instrument that applies static or variable duration cyclic tension or compression to cells in vitro. *J. Cell Sci.*, **75**, 35–42.

21. Bieler, F. H., Ott, C. E., Thompson, M. S., Seidel, R., Ahrens, S., Epari, D. R., et al. (2009). Biaxial cell stimulation: A mechanical validation. *J. Biomech.*, **42**, 1692–1696.

22. Moraes, C., Likhitpanichkul, M., Lam, C. J., Beca, B. M., Sun, Y., Simmons, C. A. (2013). Microdevice array-based identification of distinct mechanobiological response profiles in layer-specific valve interstitial cells. *Integr. Biol. (Camb.)*, **5**, 673–680.

23. Cukierman, E., Pankov, R., Stevens, D. R., Yamada, K. M. (2001). Taking cell-matrix adhesions to the third dimension. *Science*, **294**, 1708–1712.

24. Larsen, M., Artym, V. V., Green, J. A., Yamada, K. M. (2006). The matrix reorganized: Extracellular matrix remodeling and integrin signaling. *Curr. Opin. Cell Biol.*, **18**, 463–471.

25. Pedersen, J. A., Swartz, M. A. (2005). Mechanobiology in the third dimension. *Ann. Biomed. Eng.*, **33**, 1469–1490.

26. Buschmann, M. D., Gluzband, Y. A., Grodzinsky, A. J., Hunziker, E. B. (1995). Mechanical compression modulates matrix biosynthesis in chondrocyte/agarose culture. *J. Cell Sci.*, **108 (Pt 4)**, 1497–1508.

27. Quinn, T. M., Grodzinsky, A. J., Buschmann, M. D., Kim, Y. J., Hunziker, E. B. (1998). Mechanical compression alters proteoglycan deposition and matrix deformation around individual cells in cartilage explants. *J. Cell Sci.*, **111 (Pt 5)**, 573–583.
28. Hung, C. T., Mauck, R. L., Wang, C. C., Lima, E. G., Ateshian, G. A. (2004). A paradigm for functional tissue engineering of articular cartilage via applied physiologic deformational loading. *Ann. Biomed. Eng.*, **32**, 35–49.
29. Ku, C. H., Johnson, P. H., Batten, P., Sarathchandra, P., Chambers, R. C., Taylor, P. M., et al. (2006). Collagen synthesis by mesenchymal stem cells and aortic valve interstitial cells in response to mechanical stretch. *Cardiovasc. Res.*, **71**, 548–556.
30. Simmons, C. A., Matlis, S., Thornton, A. J., Chen, S., Wang, C. Y., Mooney, D. J. (2003). Cyclic strain enhances matrix mineralization by adult human mesenchymal stem cells via the extracellular signal-regulated kinase (ERK1/2) signaling pathway. *J. Biomech.*, **36**, 1087–1096.
31. Albro, M. B., Chahine, N. O., Li, R., Yeager, K., Hung, C. T., Ateshian, G. A. (2008). Dynamic loading of deformable porous media can induce active solute transport. *J. Biomech.*, **41**, 3152–3157.
32. Potier, E., Noailly, J., Ito, K. (2010). Directing bone marrow-derived stromal cell function with mechanics. *J. Biomech.*, **43**, 807–817.
33. Arnsdorf, E. J., Tummala, P., Castillo, A. B., Zhang, F., Jacobs, C. R. (2010). The epigenetic mechanism of mechanically induced osteogenic differentiation. *J. Biomech.*, **43**, 2881–2886.
34. Reilly, G. C., Engler, A. J. (2010). Intrinsic extracellular matrix properties regulate stem cell differentiation. *J. Biomech.*, **43**, 55–62 (doi:S0021-9290(09)00500-4.
35. Griffith, L. G., Swartz, M. A. (2006). Capturing complex 3D tissue physiology in vitro. *Nat. Rev. Mol. Cell Biol.*, **7**, 211–224.
36. Whitesides, G. M. (2006). The origins and the future of microfluidics. *Nature*, **442**, 368–373.
37. Young, E. W., Wheeler, A. R., Simmons, C. A. (2007). Matrix-dependent adhesion of vascular and valvular endothelial cells in microfluidic channels. *Lab Chip*, **7**, 1759–1766.
38. Chen, M. B., Srigunapalan, S., Wheeler, A. R., Simmons, C. A. (2013). A 3D microfluidic platform incorporating methacrylated gelatin hydrogels to study physiological cardiovascular cell–cell interactions. *Lab Chip*, **13**, 2591–258.

39. Sumpio, B. E., Riley, J. T., Dardik, A. (2002). Cells in focus: Endothelial cell. *Int. J. Biochem. Cell Biol.*, **34**, 1508–1512.
40. Armstrong, E. J., Bischoff, J. (2004). Heart valve development: Endothelial cell signaling and differentiation. *Circ. Res.*, **95**, 459–470.
41. Davies, P. F. (1995). Flow-mediated endothelial mechanotransduction. *Physiol. Rev.*, **75**, 519–560.
42. Young, E. W., Simmons, C. A. (2010). Macro- and microscale fluid flow systems for endothelial cell biology. *Lab Chip*, **10**, 143–160.
43. Butcher, J. T., Penrod, A. M., Garcia, A. J., Nerem, R. M. (2004). Unique morphology and focal adhesion development of valvular endothelial cells in static and fluid flow environments. *Arterioscler. Thromb. Vasc. Biol.*, **24**, 1429–1434.
44. Wojciak-Stothard, B., Williams, L., Ridley, A. J. (1999). Monocyte adhesion and spreading on human endothelial cells is dependent on Rho-regulated receptor clustering. *J. Cell Biol.*, **145**, 1293–1307.
45. Rolland, J. P., Van Dam, R. M., Schorzman, D. A., Quake, S. R., DeSimone, J. M. (2004). Solvent-resistant photocurable liquid fluoropolymers for microfluidic device fabrication [corrected]. *J. Am. Chem. Soc.*, **126**, 2322–2323.
46. Zhou, J., Ellis, A. V., Voelcker, N. H. (2010). Recent developments in PDMS surface modification for microfluidic devices. *Electrophoresis*, **31**, 2–16.
47. Wipff, P. J., Majd, H., Acharya, C., Buscemi, L., Meister, J. J., Hinz, B. (2009). The covalent attachment of adhesion molecules to silicone membranes for cell stretching applications. *Biomaterials*, **30**, 1781–1789.
48. Bellin, R. M., Kubicek, J. D., Frigault, M. J., Kamien, A. J., Steward, R. L., Jr., Barnes, H. M., et al. (2009). Defining the role of syndecan-4 in mechanotransduction using surface-modification approaches. *Proc. Natl. Acad. Sci. U. S. A.*, **106**, 22102–22107.
49. Moraes, C., Kagoma, Y. K., Beca, B. M., Tonelli-Zasarsky, R. L., Sun, Y., Simmons, C. A. (2009). Integrating polyurethane culture substrates into poly(dimethylsiloxane) microdevices. *Biomaterials*, **30**, 5241–5250.
50. Randall, C. L., Kalinin, Y. V., Jamal, M., Manohar, T., Gracias, D. H. (2011). Three-dimensional microwell arrays for cell culture. *Lab Chip*, **11**, 127–131.
51. Tse, J. R., Engler, A. J. (2011). Stiffness gradients mimicking in vivo tissue variation regulate mesenchymal stem cell fate. *PLoS One*, **6**, e15978.

52. Winer, J. P., Oake, S., Janmey, P. A. (2009). Non-linear elasticity of extracellular matrices enables contractile cells to communicate local position and orientation. *PLoS One*, **4**, e6382.

53. Simmons, C. A., Alsberg, E., Hsiong, S., Kim, W. J., Mooney, D. J. (2004). Dual growth factor delivery and controlled scaffold degradation enhance in vivo bone formation by transplanted bone marrow stromal cells. *Bone*, **35**, 562–569.

54. Huebsch, N., Arany, P. R., Mao, A. S., Shvartsman, D., Ali, O. A., Bencherif, S. A., et al. (2010). Harnessing traction-mediated manipulation of the cell/matrix interface to control stem-cell fate. *Nat. Mater.*, **9**, 518–526.

55. Khetan, S., Guvendiren, M., Legant, W. R., Cohen, D. M., Chen, C. S., Burdick, J. A. (2013). Degradation-mediated cellular traction directs stem cell fate in covalently crosslinked three-dimensional hydrogels. *Nat. Mater.* (doi:nmat3586 [pii], 10.1038/nmat3586).

56. Ghaemmaghami, A. M., Hancock, M. J., Harrington, H., Kaji, H., Khademhosseini, A. (2012). Biomimetic tissues on a chip for drug discovery. *Drug Discov. Today*, **17**, 173–181.

57. Huh, D., Hamilton, G. A., Ingber, D. E. (2011). From 3D cell culture to organs-on-chips. *Trends Cell Biol.*, **21**, 745–754.

58. Zhang, C., Zhao, Z., Abdul Rahim, N. A., van Noort, D., Yu, H. (2009). Towards a human-on-chip: Culturing multiple cell types on a chip with compartmentalized microenvironments. *Lab Chip*, **9**, 3185–3192.

59. Moraes, C., Mehta, G., Lesher-Perez, S. C., Takayama, S. (2012). Organs-on-a-chip: A focus on compartmentalized microdevices. *Ann. Biomed. Eng.*, **40**, 1211–1227.

## Chapter 14

# Tip Growth in Walled Cells: Cellular Expansion and Invasion Mechanisms

**Amir Sanati Nezhad[a] and Anja Geitmann[b]**

[a]*Department of Biomedical Engineering,*
*McGill University and Génome Québec Innovation Centre,*
*740 Dr. Penfield Avenue, Montréal, Québec H3A 0G1, Canada*
[b]*Département de sciences biologiques,*
*Institut de recherche en biologie végétale, Université de Montréal,*
*4101 Rue Sherbrooke est, Montréal, Québec H1X 2B3, Canada*

anja.geitmann@umontreal.ca

Tip growth in walled cells is regulated by a spatially confined yielding of the cell wall under the effect of the turgor pressure. This mechanism is displayed by a variety of cell types such as fungal hyphae, pollen tubes and root hairs. The common feature of these cells is their ability to invade a substrate or matrix. Here we review the cellular and biomechanical underpinnings of tip growth compared to growth in other walled cells. The physical challenges associated with an invasive life style are assessed and explained.

*Cells, Forces, and the Microenvironment*
Edited by Charles M. Cuerrier and Andrew E. Pelling

ISBN 978-981-4613-36-1 (Hardcover), 978-981-4613-37-8 (eBook)
www.panstanford.com

## 14.1 Introduction

Growth and movement in walled cells such as those of plants, fungi, and bacteria are achieved through a mechanism that is very different from those in most animal cells. Whereas the forces necessary for cellular expansion in wall-less cells are generated by the cytoskeleton, the cell walls enveloping walled cells prevent any cytoskeleton mediated deformation of the plasma membrane. Morphogenetic changes to the cell shape must therefore be accomplished by global or local yielding of the wall to the internal hydrostatic pressure, the turgor. The mechanical principles governing this mechanism are covered by Ortega (see Chapter 15 of this book). In the present chapter, the particular case of polar growth or tip growth will be examined. The general mechanical principles of the morphogenetic process in polarly growing cells such as root hairs, pollen tubes, and fungal hyphae are distinguished from those occurring in most other walled cells. Furthermore, the mechanics of the particular life style of tip-growing cells will be illustrated—the ability to invade other tissues or substrates.

## 14.2 Tip Growth and Diffuse Growth: Two Examples on a Continuous Spectrum of Morphogenetic Patterns

### 14.2.1 The Geometry of Cellular Growth in Walled Cells

Cellular growth in walled cells can occur over large portions of the cellular surface or it can be a spatially confined process. Two typical examples of a morphogenetic pattern include "tip growth" and a process that is termed "diffuse" or "intercalary growth." Typically, these two types are presented to be completely different from each other, and, somewhat misleadingly, "diffuse growth" seems to encompass anything that is not "tip growth." In the following, we will show that these two types of cellular expansion are really only two examples on a continuous, two-dimensional spectrum, and that this spectrum deserves a more precise characterization.

Tip growth is generally defined as highly localized expansion of the cellular surface, typically giving rise to a cylindrical protuberance. Only the cell wall at the very tip of the cell, typically a hemisphere-shaped dome, expands, whereas once the wall has matured to form the cylindrical portion of the protuberance it ceases all expansion (Fig. 14.1). Tip growth is thus generated by an extremely spatially confined expansion activity of the cellular surface.

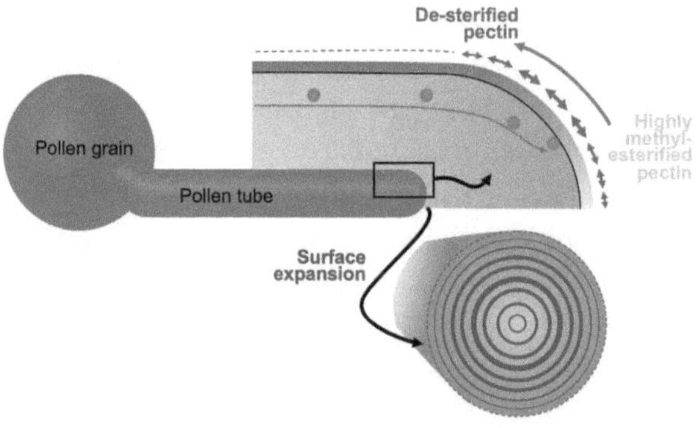


**Figure 14.1** Surface expansion in a tip growing cell, such as this pollen tube, is confined to the apical region of the cell. Thicker red arrows and rings indicate regions of highest expansion rate. Expansion occurs only in the apical region because new, softer cell wall material is delivered to this site by exocytosis (green arrow in the cytoplasm). In the periphery of the soft spot, the pectinaceous cell wall material stiffens through calcium gelation as it becomes enzymatically de-esterified (acidic).

Diffuse growth, on the other hand, implies that cellular expansion occurs over large portions or over the entire surface of the cell. The cylindrical cells forming the plant root and shoot tissues display this pattern (Fig. 14.2, right), as well as those of cylindrical algae. All of these cell types typically elongate over their entire length. However, it has to be pointed out that even if surface expansion in these cells occurs over the entire lateral surface, they usually do not expand at the two base areas. Hence, expansion occurs non-uniformly over a selected surface, albeit a large portion of the total surface. Moreover, expansion of the lateral surface occurs in highly anisotropic manner with a much higher expansion rate in the direction parallel to the long axis of the cell compared to the circumferential direction. A modification in this degree of anisotropy is typical for certain mutants displaying swollen cells and, consequently, swollen roots or shoots. To summarize, even in cells that are typically identified as undergoing "diffuse growth," surface expansion is neither truly uniformly distributed nor is it isotropic. The attribute "diffuse" as it is commonly used, therefore, refers to the fact the expansion occurs over a large portion of the surface. Truly uniform and isotropic growth would therefore be only a special case of diffuse growth. Uniformly distributed isotropic cellular expansion activity eventually and inevitably generates a sphere, independent of the initial shape of the cell, although the presence of neighboring cells can press this spherical shape into that of a polyhedron (similar to aggregates of soap bubbles), a type of cell shape typical for fruit parenchyma tissue.

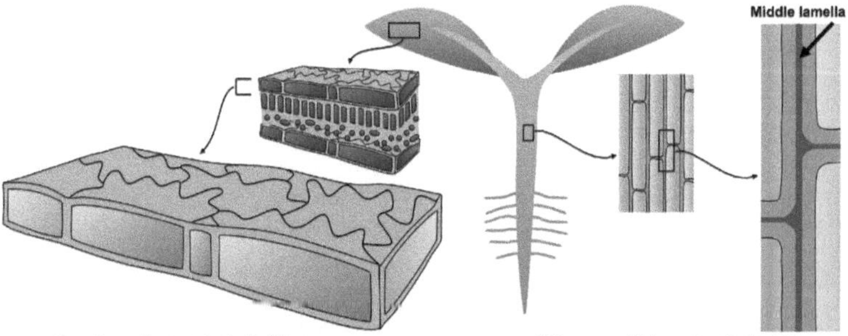


**Figure 14.2** The cells in the plant body grow according to a variety of patterns. Most cells in the root and shoot elongate in diffuse, but highly anisotropic manner (right). Particularly in the leaf, cell expansion can display complex geometry, such as in the epidermis (left). Plant cells are attached to each other by the middle lamella, a pectinaceous material (far right).

Although tip growth and diffuse growth are useful distinctions, there are many cell shapes in the plant body that do not seem to fall in these categories and that are generated by geometrically more complex expansion processes. These include for example the star-shaped cells of the aerenchyma or the jigsaw puzzle shaped cells of the leaf epidermis (Fig. 14.2, left). To generate complex shapes either the spatially uniform growth rate or isotropic expansion behavior of the wall must be abandoned temporarily and/or locally (Fig. 14.3). In other words, local variations must appear in the growth rate, or deformability of the cell wall material must become anisotropic. Both can be combined and occur simultaneously or subsequently during cellular development. Shape generation in cells with complex geometry is thus accomplished by temporally and spatially regulated expansion activities that are either non-uniform or anisotropic or both and the growth pattern can change during the different phases of cellular differentiation. While the distinction between tip growth and diffuse growth may therefore be useful, these terms do not exhaustively cover all types of shape generation and it might be helpful to use more specific vocabulary such as that proposed by Geitmann and Ortega [3].

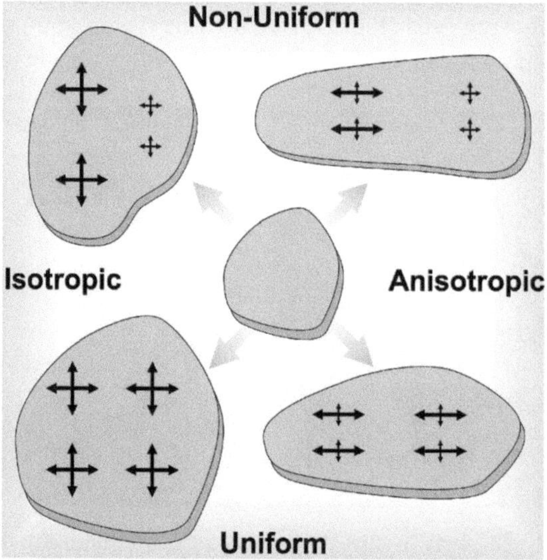


**Figure 14.3** A section of cell wall under tensile stress can deform in various ways that can be described using a two-dimensional spectrum. In anisotropic expansion the degree of deformation in one direction is larger than in the perpendicular direction. A non-uniform distribution causes different areas of the section to be stretched to a different degree.

As illustrated by the soap bubble shapes of the fruit parenchyma tissue, shape formation of individual cells within a plant tissue is influenced significantly by the immediate surrounding (micro-environment). Neighboring cells can limit the free unfolding of a given cell even when its wall relaxes. More important, not only do neighboring cells limit a cell's expansion, but also does their tight adhesion to the expanding cell prevent the latter from sliding to avoid the compression by neighboring cells [4]. As a result, the lobed anticlinal surfaces of two adjacent jigsaw puzzle shaped cells of the leaf epidermis are shaped simultaneously and the direct neighbors cannot shape independently from each other. The reason for this dependence is the tight adherence between plant cells by a structure called the middle lamella—a pectin-rich layer that operates as a glue (Fig. 14.2). In some tissues, this gluing layer is dissolved partially to give rise to intercellular spaces leaving the individual cell to shape independently (e.g., aerenchyma), whereas in others the middle lamella tightly connects all cellular surfaces and a morphogenetic process occurring in one cell inevitably involves a shape change in the neighboring cells (Fig. 14.4). How this tissue-wide coordination of cell shaping is regulated at single cell level is poorly understood.

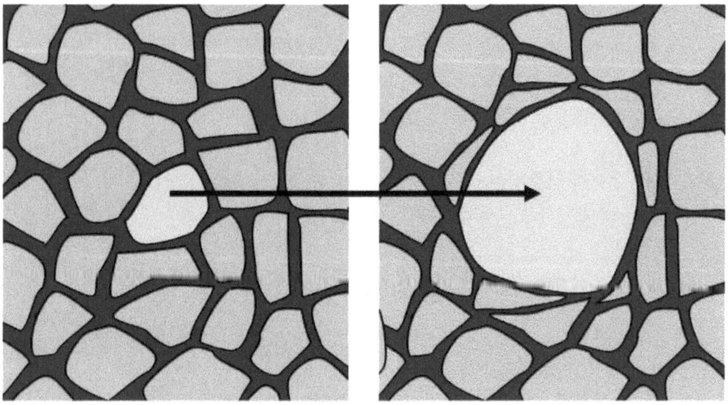

**Figure 14.4** In plant tissues the growth of individual cells influences the shapes of the surrounding cells. Since the cells generally do not slide against each other, the neighboring cells are flattened and stretched by the growing cell.

### 14.2.2 Spatial Regulation of Cell Wall Properties

How can cell wall expansion be regulated to be non-uniform or anisotropic? It is important to realize that the driving force of cellular growth in walled cells is the turgor, the hydrostatic pressure inside the cell. It is generated by water uptake that follows an osmotic gradient. In plant cells, it is typically between 0.1 and 0.8 MPa, whereas certain fungal cells are able to build up to 8 MPa. However, pressure is a scalar and is uniformly exerted onto all surfaces of a closed compartment. To accomplish that a selected area on the cellular surface yields more than others, or for a surface to yield more in one direction than the other, the mechanical properties of the cell wall must be regulated such that they promote preferential yielding. In the case of tip growth, this essentially necessitates the presence of a single, small, softer region in the cell wall. The generation of a tip-growing protuberance is therefore typically preceded by local changes in the cell wall properties of the cell body giving rise to the protuberance [5–7]. Even once the protuberance is established, the tip needs to be maintained softer, since the tensile stress at the hemisphere-shaped tip of a cylindrical protuberance is only half that of the stress in hoop direction in the cylindrical shank [8]. The gradient necessary to generate a perfectly tubular shape with the strain pattern observed experimentally has been modeled using finite element analysis [9]. The "soft spot" at the pollen tube tip consists of a small region of methyl-esterified pectins. At the periphery of this spot, the pectins are gradually de-esterified to become acidic pectins (Fig. 14.1). This chemical change brings about a change in mechanical properties since the negatively charged, acidic pectin is able to cross-link in the presence of calcium ions thus forming a gel that is significantly stiffer than the ungelated methyl-esterified version of the polysaccharide. The pectins at the periphery of the soft spot therefore do not yield anymore and instead form a sleeve-like tube. The center of the soft spot is maintained soft by a continuous supply of new methyl-esterified pectins delivered by exocytosis. In cells with more complex, non-uniform distribution of cell expansion such as those of the leaf epidermis, it is poorly understood whether localized exocytosis contributes to the maintenance of softer cellular regions although it seems likely [10].

Anisotropy in the mechanical properties of the cell wall is generated by the inclusion of structural features that confer a directionality to cell wall deformability. Similar to a composite material, the plant cell wall comprises both matrix and fibrillar components [11]. If fibrillar components are oriented randomly, isotropy is maintained, but when the net orientation of a population of fibrils is different from random, the wall displays higher stiffness in the direction parallel to this net orientation [3,12]. In plant cells, these fibrillar structures are typically cellulose microfibrils, whereas in the fungal cell wall chitin fibrils are employed to confer anisotropy.

### 14.2.3 Measuring Cell Wall Properties

The phenomenon of tip growth has been addressed by numerous theoretical and numerical modeling approaches [13–15]. In particular, those approaches that include mechanical aspects of tube growth have been faced with the challenge that absolute numbers for the mechanical properties of the structural cellular features such as the cell wall are difficult to obtain experimentally. However, since educated guesses based on material properties obtained from in vitro polysaccharide extracts have limited validity, the development of biomechanical approaches at single cell level has surged in recent years. In order to determine the mechanical cell wall properties of living pollen tubes, two different approaches based on MEMS (micro-electromechanical systems) techniques have been developed: Using a micro-indentation technique combined with finite element modeling, Vogler et al. [16] were able to determine Young's modulus of the lily pollen tube cell wall. In this assay the cell was compressed locally by a stylus (Fig. 14.5a) and the measured apparent stiffness was used to calculate the mechanical properties of the cell wall. This is not trivial since the cellular envelope is under pressure resulting from the turgor and

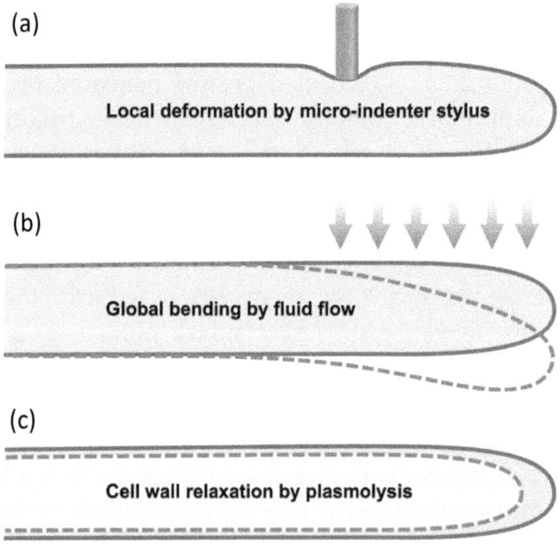


**Figure 14.5** The mechanical properties of the pollen tube have been measured using a micro-indentation technique (a), cell bending by fluid flow (b) and through shrinkage after plasmolysis (c).

much of the measured stiffness is due to the pressure exerted by the hydroskeleton. In a completely different approach, Sanati Nezhad et al. [17] bent *Camellia* pollen tubes using a microfluidic device in which fluid flow was employed to accomplish the different experimental steps: First, a pollen grain was positioned at the entrance of a microscopic channel into which the pollen tube elongated. Second, a side inlet was used to position the elongating tube against the side wall of the microfluidic network to provide it with a stable support. Third, fluid flow was used to bend the tube (Fig. 14.5b). Finite element modeling was then used to determine Young's modulus of the cell wall based on the bending stiffness of the tube. The values were consistent with a third, completely different approach in which the wall stress was relaxed by removing the turgor and both the longitudinal and horizontal Young's moduli could be determined from the shrinkage that occurred as a result (Fig. 14.5c) [17].

### 14.2.4 Targeted Intracellular Transport: A Necessity for Cellular Morphogenesis

Both the generation of soft spots and the strengthening of the cell wall by microfibrils require the spatially targeted delivery of cargo to the cellular surface. Soft spots are normally generated by enzymatic linkage breaks or by the insertion of new cell wall polymers that are weakly linked, such as methyl-esterified pectins. These are delivered to the cell wall by exocytosis (Fig. 14.1). To maintain the limited dimensions of the soft spot, the material at its periphery must be hardened, which in the case of pectins occurs through demethyl-esterification. Therefore, the enzymes responsible for this process (pectin methyl esterase; PME) as well as the proteic inhibitors of PME require targeted delivery as well [18]. Other cell wall modifying proteins as well as synthases responsible for the production of cell wall polymers rigidifying the cell wall (e.g., cellulose and callose synthases) are also delivered by exocytosis at or near the growing tip of the pollen tube. This illustrates that the targeted delivery of cargo carrying vesicles to selected cellular surfaces is crucial for correct morphogenesis. In tip-growing cells, the vesicle transport pathways are well described [19].

The cytoskeletal contribution to morphogenesis in walled cells resides, therefore, in orchestrating this logistics of intracellular cargo transport to the cell surface. A geometrical change in the morphogenetic process requires a modulation of cytoskeletal activities. This is illustrated by the processes that lead up to a change in the growth direction in pollen tubes. Such a change in growth direction can be brought about by exposing the cells to an electrical field [20]. This method enables the researcher to switch on the directional trigger at a defined time, unlike the exposure to a chemical gradient that establishes gradually. The deviation from the axially symmetric growth pattern is clearly preceded by an asymmetric polymerization activity of the actin cytoskeleton at the apical growth region of the tube (Fig. 14.6). As a result, the principal region of exocytosis, an annular zone around the pole of the cell, is shifted off axis and, eventually, the growing tip continues elongating at an angle. The ability to reorient cellular expansion upon perception of a directional trigger is a pivotal requirement for chemotropism and thigmotropism, the response mechanisms that allow tip-growing cells to grow toward an attractant, away from a repulsive agent, or to avoid obstacles. This behavior is important for cellular functioning of many tip-growing cells. Fungal hyphae for example must identify sources of nutrients, whereas pollen tubes need to find the site of sperm delivery—the ovule housing the female gametophyte, nestled deep within the tissues of the flower.

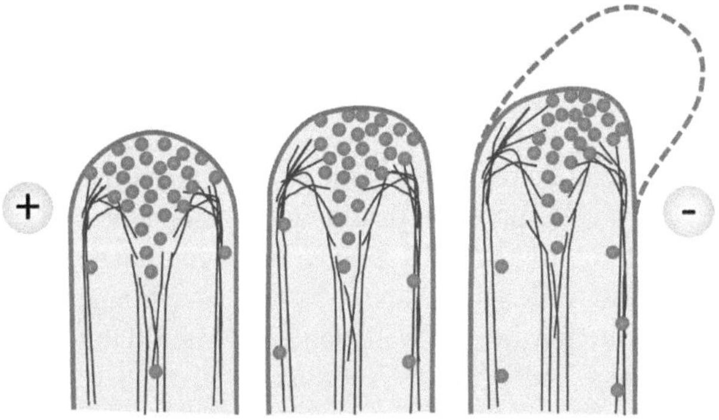


**Figure 14.6** To initiate a change in growth direction, the apical actin fringe polymerizes asymmetrically. Consequently, the delivery of cell wall material to the apical region is shifted to an off-axis location. This change in growth direction can be triggered by application of an electrical field. Based on [20].

## 14.3 Tip Growth as a Tool for an Invasive Life Style

### 14.3.1 Plant Cell Growth within a Microenvironment

Responsiveness to guidance cues, be they chemotropic or thigmotropic, is not the only skill that tip-growing cells require for their functioning. The second competence enables the cells to physically breach the impedance posed by the biotic or abiotic matrices in their path. The nature of these matrices depends on the type of the tip-growing cell. Fungal hyphae are able to penetrate anything from the soft parenchyma tissue of a mature fruit or the mucous membranes of an animal to soil and rocks. Root hairs are equally able to penetrate the soil, whereas pollen tubes have to cope with the different tissues composing the flower tissue. These include the cuticle covering the stigma of some flowers and the nucellus, the tissue surrounding the female gametophyte. Penetration of a tip-growing cell into a plant tissue generally entails separating the cell–cell connections of the invaded tissue at the pectinaceous middle lamella. Additionally, plasmodesmata, the linkages between the protoplasts of two neighboring plant cells, may pose a mechanical obstacle. The same challenges have to be met by a number of plant cells whose tip-growing nature is not firmly established, but whose invasive properties are a necessity for their differentiation: sclerenchyma fibers and laticifers. The necessity for invasion in these cell types results from the fact that the final length of these cells is much longer than that of the surrounding cells. Since the precursors of both cell types typically have simple polyhedric shapes, their elongation growth during differentiation must entail the invasive penetration of the surrounding tissues, similar to the activities pollen tubes and fungal hyphae have to deploy to invade plant tissues.

Coping with a mechanical impedance to cellular growth entails several mechanical challenges that describe the interaction of the invasive cell with its micro-environment [21]:

1. **Driving force:** The cell must somehow generate the force that is necessary to either displace or breach the mechanical obstacle. The driving force must be adequate to handle both the toughness and the hardness of the obstacle. The hardness describes the yield strain of the material and indicates how

easily an invaded substrate can be compressed or dilated to make room for the invading cell. The toughness is a measure for the fracture resistance of a material. It is determined by the adhesion or cohesion between the molecules forming the invaded substrate. The formation of fractures in the substrate can facilitate the further invasion by the tip-growing cell.

2. **Friction:** Friction is both a necessary requirement and inevitable impediment for invasive growth. The friction between the distal region of the tip-growing cell and its surroundings is necessary to prevent pushback (Fig. 14.7). If the distal end of the cell was not immobilized by friction, pushback would cause the growing tip to push its own distal region backwards. As a result, rather than exerting an invasive force, the cell would simply move backwards. This can be likened to the ineffectiveness of a pushing motion when one is on roller skates, a principle that was immortalized by Charlie Chaplin in his movie "The Rink" [22].

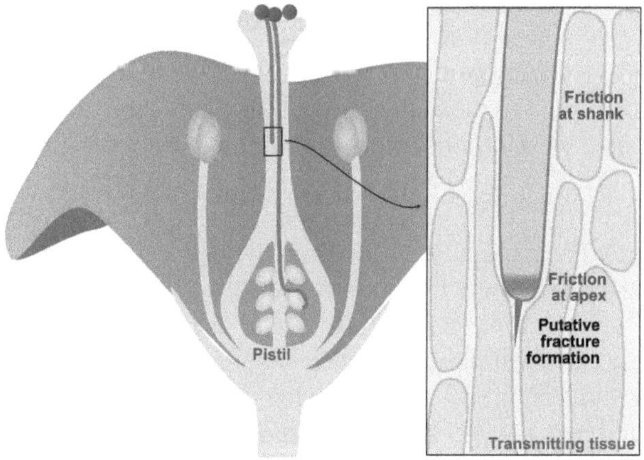


**Figure 14.7** During the growth through the pistil, the pollen tube needs to invade the apoplast of the transmitting tissue. Friction at the apical end of the tube influences growth negatively, whereas friction between the distal region of the tube and the surrounding tissue is important to prevent pushback.

While friction in the distal region of the tip-growing cell is crucial for force exertion, at the apical growth zone this effect is counterproductive to invasion. Inevitably, the apical surface of the invading cell experiences some friction against the substrate it is invading. The resulting shear forces need to be overcome by the advancing cell and affect the speed of invasion negatively. To limit the amount of friction at the apex, tip-growing cells have reduced the surface area on which it occurs by expanding only at the apex of the cell. One of the reasons for the spatial confinement of cell wall expansion in tip-growing cells is thus likely the reduction of friction forces between cell and substrate (Fig. 14.7).

## 14.3.2 Driving Force for Invasion

Invasive growth or migration in animal cells is based on forces generated by the cytoskeleton—either through polymerization of cytoskeletal arrays or through contraction of cytoskeletal elements against each other in combination with adhesion to the growth substrate [21]. In walled cells, the amounts of force potentially generated by the cytoskeleton are very small compared to those created by the hydrostatic pressure of the turgor. Any invasive force generated in walled cells is therefore thought to be based on the turgor although exceptions may exist [23]. If turgor is the driving force, then the maximum pressure a cell can exert is that of its turgor. Experimentally, the invasive pressure exerted by tip-growing cells has been found to be smaller than the turgor, amounting to about 10–50% in fungal hyphae [24] and 75% in pollen tubes [25]. This can be explained with the fact that the cell wall proper absorbs some of the pressure to expand. To measure the invasive force of tip-growing cells, different methods have been employed. Fungal hyphae have been assayed using a strain gauge that was placed perpendicularly in front of the elongating apex [26]. Alternatively, the invasion pegs of fungal appressoria of *Colletotrichum graminicola*, a plant pathogen, were placed to deform a waveguide [27]. For the latter this has yielded growth forces in the range of several μN for individual hyphae, corresponding to a pressure of several MPa.

In pollen tubes of *Camellia*, a different method was employed. Rather than exposing the apex to a perpendicular cantilever, the cell was guided into a wedge shaped obstacle made from elastic polymer material [25] using an experimental platform based on microfluidics [28]. This method had the advantage that the cell did not attempt to avoid the obstacle by changing its growth direction.

The calibrated material properties of the elastic polymer were used to calculate the dilating force exerted by the pollen tube which in turn was used to obtain the dilating pressure. With 150 kPa, the dilating pressure was found to be close to the typical turgor pressure of pollen tubes which in lily pollen tubes was measured to be between 0.1 and 0.4 MPa [29].

### 14.3.3 Supplementary Tools to Overcome Mechanical Impedance

Force generation is essential for an invasive life style, but invasive cells generally use additional, sophisticated means to facilitate intrusion, especially if the invaded substrate is a living tissue. Fungal hyphae that invade human or animal tissues produce a battery of proteinases that are meant to digest the molecular links between the cells composing the attacked tissue [26]. Plant pathogenic fungi or pollen tubes, on the other hand, produce enzymes with activities that digest polysaccharides-with the aim to break down the middle lamella connecting the cells of the plant tissue [24,30,31]. These digestive activities likely reduce the toughness of the tissue and facilitate fracture formation. The breakdown products of these enzymatic reactions may possibly have the additional benefit of lubricating the interface between invading cell and tissue thus reducing friction at the apical region of the tip-growing cell. Experimental evidence for the lubrication effect is elusive, but breakdown products of cell wall polysaccharide polymers are typically mucilagineous, thus supporting this notion.

An even more artful strategy to aid invasion is that of certain pollen tubes which induce the invaded pistillar tissue to essentially commit suicide, or programmed cell death [32,33]. This process likely reduces the mechanical impedance against invasion and the cellular breakdown products might even supply nutrients to the growing tube.

### 14.3.4 Target Finding

As mentioned earlier, tip-growing cells are able to actively redirect their growth direction following external directional cues. The investigation of this guidance mechanism is being addressed experimentally with increasing precision aided by microfluidic set-ups [28,34]. Many aspects of this tropic behavior have been covered elsewhere [35–39] and are beyond the scope of the present chapter. However, it is worth comparing how different tip-growing cell types deal with an obstacle course that is placed in their way and that imposes a redirection of their growth direction. Experimentally this can be achieved by letting the cells grow against a single obstacle [40] or within a more complex, microstructured environment that can be designed to answer specific questions [2,25,28,41].

Interestingly, some tip-growing cells possess an internal memory that dictates their preferred growth direction independently of other guidance signals, whereas others do not possess this memory and fully rely on external guidance cues. When root hairs or fungal hyphae are presented with an obstacle that forces them to change their growth direction, they will usually return into the original direction once the obstacle has been passed [41–44]. Pollen tubes, on the other hand, do not possess this internal memory and in the absence of a chemical signal their default growth direction will typically be the one that was last imposed on them by a mechanical constraint [28]. When pollen tubes elongate through a winding microstructured microfluidic network, they maintain the growth direction of the last portion of the passage (Fig. 14.8). These behavioral differences between the different cell types make sense when one compares the cellular behavior in situ. A pollen tube must follow the guidance cues of the pistil and the female gametophyte to find its target. For its success it is not important where, respective to the elongating apex, the base of the cell (the pollen grain) is located. The initial direction of tube elongation upon germination is therefore completely irrelevant for the elongating tube. Root hairs and fungal hyphae, on the other hand, typically grow away from their basal regions for optimal efficiency. Root hairs grow away radially from the central root axis to explore the surrounding soil in optimal manner, and the exploration of new territory by a fungal colony is most efficient when all hyphae radiate out away from the center of the colony. Transient deviations in the growth direction enforced by an obstacle may be a nuisance, but are dealt with efficiently if the general growth direction is re-established after passing the obstacle. This requires that the growing apex is somehow informed about the original direction. How this memory function operates in terms of the signaling between the basal and apical portion of the cell is poorly understood, but mutant studies should hopefully shed light on this phenomenon [41].

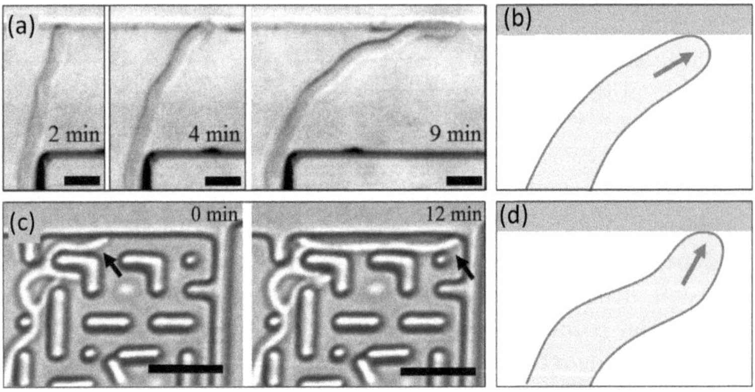


**Figure 14.8** Different types of tip growing cells navigate microstructured environments differently. The pollen tube (a, b) maintains the direction that was imposed by the last obstacle, whereas fungal hyphae (c, d) return to the original growth direction after passing an obstacle. The tip of a hypha deviated by an obstacle remains with its "nose down." a shows a *Camellia* pollen tube (from [1] with permissions). c shows the fungal hypha of *Neurospora crassa* (from [2] with permissions). Scale bars = 25 μm.

An entry point to understanding the presence or absence of internal memory might be provided by the different ways in which pollen tubes and fungal hyphae respond initially when they encounter a mechanical obstacle in their path. Both will likely stop growth temporarily if the obstacle is presented perfectly perpendicularly. They then resume growth at an angle or even split and form two growing apices [2,40,41]. However, if the obstacle surface is at least slightly tilted relative to the long axis of the cell, the behavior differs between these cell types. The pollen tube initially simply keeps elongating in axially symmetric manner and as a consequence the subapical region starts buckling under this compressive force (Fig. 14.8a). The apex then slides along the surface and once it is at a small angle with the wall it will reorient to realign its central axis with the wall [1]. This direction is maintained until the next obstacle is encountered. Hyphae instead grow along the wall but keep their apex oriented at the original angle pointing into the wall [2,41,43]. This "nose-down" effect might be the key to the internal memory that allows the fungal hypha to resume the original direction once the obstacle is bypassed (Fig. 14.8b).

## 14.4 Conclusion

Cellular morphogenesis is strongly influenced by neighboring cells or by the nature of the surrounding growth substrate. This microenvironment influences cellular differentiation and behavior. While this is true for all cells, the constraints exerted by the microenvironment are particularly important in the case of tip-growing cells. These cells need to grow rapidly and over long distances. They need to navigate obstacles and have to perceive and respond to guidance and repulsive cues. How tip-growing cells succeed in integrating different directional signals while focusing their energy on producing rapid and forceful cellular elongation against mechanical impedance represents an excellent challenge for an interdisciplinary approach to cell biology. In particular, the integration of mechanical concepts and mathematical modeling in combination with high-resolution imaging strategies and microfluidic engineering have aided this field in moving toward a more comprehensive understanding of the multiple facets characterizing the phenomenon "tip growth."

### Acknowledgments

Research in the Geitmann lab is supported by grants from the Natural Sciences and Engineering Research Council of Canada (NSERC) and the Fonds Québécois de la Recherche sur la Nature et les Technologies (FQRNT).

### References

1. Agudelo, C., Sanati Nezhad, A., Ghanbari, M., Packirisamy, M., and Geitmann, A. (2012). A microfluidic platform for the investigation of elongation growth in pollen tubes. *J. Micromechan. Microeng.*, **22**, 115009.

2. Hanson, K., Nicolau Jr., D., Filipponi, L., Wang, L., Lee, A., and Nicolau, D. (2006). Fungi use efficient algorithms for the exploration of microfluidic networks. *Small*, **2**, 1212–1220.
3. Geitmann, A., and Ortega, J. K. E. (2009). Mechanics and modeling of plant cell growth. *Trends Plant Sci.*, **14**, 467–478.
4. Knox, J. (1992). Cell adhesion, cell separation and plant morphogenesis. *Plant J.*, **2**, 137–141.
5. Geitmann, A. (2010). How to shape a cylinder: Pollen tube as a model system for the generation of complex cellular geometry. *Sex. Plant Reprod.*, **23**, 63–71.
6. Parre, E., and Geitmann, A. (2005). Pectin and the role of the physical properties of the cell wall in pollen tube growth of Solanum chacoense. *Planta*, **220**, 582–592.
7. Baluška, F., Salaj, J., Mathur, J., Braun, M., Jasper, F., Šamaj, J., Chua, N. H., Barlow, P. W., and Volkmann, D. (2000). Root hair formation: F-actin-dependent tip growth is initiated by local assembly of profilin-supported F-actin meshworks accumulated within expansin-enriched bulges. *Dev. Biol.*, **227**, 618–632.
8. Geitmann, A., and Steer, M. W. (2006). The architecture and properties of the pollen tube cell wall, in: *The Pollen Tube: A Cellular and Molecular Perspective, Plant Cell Monographs* (Malhó, R., ed.), Springer Verlag: Berlin Heidelberg. p. 177–200.
9. Fayant, P., Girlanda, O., Chebli, Y., Aubin, C. E., Villemure, I., and Geitmann, A. (2010). Finite element model of polar growth in pollen tubes. *Plant Cell*, **22**, 2579–2593.
10. Smith, L. G., and Oppenheimer, D. G. (2005). Spatial control of cell expansion by the plant cytoskeleton. *Ann. Rev. Cell Dev. Biol.*, **21**, 271–295.
11. Geitmann, A. (2010). Mechanical modeling and structural analysis of the primary plant cell wall. *Curr. Opin. Plant Biol.*, **13**, 693–699.
12. Baskin, T. (2005). Anisotropic expansion of the plant cell wall. *Ann. Rev. Cell Dev. Biol.*, **21**, 203–222.
13. Kroeger, J., and Geitmann, A. (2011). Modeling pollen tube growth: Feeling the pressure to deliver testifiable predictions. *Plant Signal. Behav.*, **6**, 1828–1830.
14. Kroeger, J., and Geitmann, A. (2012). Pollen tube growth: Getting a grip on cell biology through modeling. *Mechan. Res. Commun.*, **42**, 32–39.

15. Kroeger, J., and Geitmann, A. (2012). The pollen tube paradigm revisited. *Curr. Opin. Plant Biol.*, **15**, 618–624.

16. Vogler, H., Draeger, C., Weber, A., Felekis, D., Eichenberger, C., Routier-Kierzkowska, A.-L., Boisson-Dernier, A., Ringli, C., Nelson, B. J., Smith, R. S., and Grossniklaus, U. (2013). The pollen tube: A soft shell with a hard core. *Plant J.*, **73**, 617–627.

17. Sanati Nezhad, A., Naghavi, M., Packirisamy, M., Bhat, R., and Geitmann, A. (2013). Quantification of the Young's modulus of the primary plant cell wall using Bending-Lab-On-Chip (BLOC). *Lab Chip*, **13**, 2599–2608.

18. Röckel, N., Wolf, S., Kost, B., Rausch, T., and Greiner, S. (2008). Elaborate spatial patterning of cell-wall PME and PMEI at the pollen tube tip involves PMEI endocytosis, and reflects the distribution of esterified and de-esterified pectins. *Plant J.*, **53**, 133–143.

19. Chebli, Y., Kroeger, J., and Geitmann, A. (2013). Transport logistics in pollen tubes. *Mol. Plant*, **6**, 1037–1052.

20. Bou Daher, F., and Geitmann, A. (2011). Actin is involved in pollen tube tropism through redefining the spatial targeting of secretory vesicles. *Traffic*, **12**, 1537–1551.

21. Sanati Nezhad, A., and Geitmann, A. (2013). The cellular mechanics of an invasive life style. *J. Exp. Bot.*, **64**, 4709–4728.

22. Chaplin, C., *The Rink*, 1916. Film.

23. Suei, S., and Garrill, A. (2008). An F-actin-depleted zone is present at the hyphal tip of invasive hyphae of Neurospora crassa. *Protoplasma*, **232**, 165–172.

24. Bastmeyer, M., Deising, H. B., and Bechinger, C. (2002). Force exertion in fungal infection. *Ann. Rev. Biophys. Biomol. Struct.*, **31**, 321–342.

25. Sanati Nezhad, A., Naghavi, M., Packirisamy, M., Bhat, R., and Geitmann, A. (2013). Quantification of cellular penetrative forces using Lab-on-a-Chip technology and finite element modeling. *PNAS*, **110**, 8093–8098

26. Money, N. P. (2007). Biomechanics of invasive hyphal growth, in: *Biology of the Fungal Cell* (Howard, R. J., and Gow, N. A. R., ed.), Springer-Verlag: Berlin Heidelberg. p. 237–249.

27. Bechinger, C., Giebel, K.-F., Schnell, M., Leiderer, P., Deising, H. B., and Bastmeyer, M. (1999). Optical measurements of invasive forces exerted by appressoria of a plant pathogenic fungus. *Science*, **285**, 1896–1899.

28. Agudelo, C. G., Sanati Nezhad, A., Ghanbari, M., Naghavi, M., Packirisamy, M., and Geitmann, A. (2013). Tip Chip: A modular, MEMS-based platform for experimentation and phenotyping of tip-growing cells. *Plant J.*, **73**, 1057–1068.

29. Benkert, R., Obermeyer, G., and Bentrup, F. W. (1997). The turgor pressure of growing lily pollen tubes. *Protoplasma*, **198**, 1–8.

30. Mendgen, K., and Deising, H. (1993). Infection structures of fungal plant pathogens—a cytological and physiological evaluation. *N. Phytol.*, **124**, 193–213.

31. Dearnaley, J. D. W., and Daggard, G. A. (2001). Expression of a polygalacturonase enzyme in germinationg pollen of Brassica napus. *Sex. Plant Reprod.*, **13**, 265–271.

32. Cheung, A. (1995). Pollen-pistil interactions in compatible pollination. *PNAS*, **92**, 3077–3080.

33. Herrero, M., and Dickinson, H. G. (1979). Pollen-pistil incompatibility in *Petunia hybrida*: Changes in the pistil following compatible and incompatible intraspecific crosses. *J. Cell Sci.*, **36**, 1–18.

34. Yetisen, A., Jiang, L., Cooper, J., Qin, Y., Palanivelu, R., and Zohar, Y. (2011). A microsystem-based assay for studying pollen tube guidance in plant reproduction. *J. Micromechan. Microeng.*, **21**, e054018.

35. Márton, M., and Dresselhaus, T. (2010). Female gametophyte-controlled pollen tube guidance. *Biochem. Soci. Trans.*, **38**, 627–630.

36. Geitmann, A., and Palanivelu, R. (2007). Fertilization requires communication: Signal generation and perception during pollen tube guidance. *Floriculture Ornamental Biotechnol.*, **1**, 77–89.

37. Palanivelu, R., and Tsukamoto, T. (2011). Pathfinding in angiosperm reproduction: Pollen tube guidance by pistils ensures successful double fertilization. *WIREs Dev. Biol.*, **1**, 96–113.

38. Brand, A. Hyphal growth in human fungal pathogens and its role in virulence. *Intl. J. Microbiol.*, **2012**, 11.

39. Brand, A., and Gow, N. A. R. (2009). Mechanisms of hypha orientation of fungi. *Curr. Opin. Microbiol.*, **12**, 350–357.

40. Gossot, O., and Geitmann, A. (2007). Pollen tube growth—Coping with mechanical obstacles involves the cytoskeleton. *Planta*, **226**, 405–416.

41. Held, M., Edwards, C., and Nicolau, D. (2011). Probing the growth dynamics of *Neurospora crassa* with microfluidic structures. *Fungal Biol.*, **115**, 493–505.

42. Hanson, K., Filipponi, L., Lee, A., and Nicolau, D. (2005). Negotiation of obstacles by fungi in micro-fabricated structures: To turn or not to turn? in *3rd Annual International IEEE EMBS Special Topic Conference on Microtechnologies in Medicine and Biology*. Kahuku, Oahu, Hawaii.
43. Brand, A. (2013). Calcium-signalling in fungi: Key to the fungal lifestyle. in *SEB Meeting*. Valencia.
44. Bibikova, T. N., Zhigilei, A., and Gilroy, S. (1997). Root hair growth in *Arabidopsis thaliana* is directed by calcium and an endogenous polarity. *Planta*, **203**, 495–505.

# Chapter 15

# Expansive Growth of Cells with Walls: Force Generation and Growth Regulation

**Joseph K. E. Ortega, Jason T. Truong, Cindy M. Munoz, and Elena L. Ortega**

*Bioengineering Laboratory, Department of Mechanical Engineering, University of Colorado Denver, Denver, Colorado 80217-3364, USA*

joseph.ortega@ucdenver.edu

All algal, fungal, and plant cells have walls. These cells with walls undergo expansive growth that can enlarge their volume by as much as 30,000 times their initial volume. Expansive growth is fundamental to the development of these cells and essential to producing growth responses to environment signals that enhance their survival. The cell wall is an exoskeleton that provides shape, physical support, and protection for these cells. However, the cell wall also represents an obstacle to expansive growth. The cell wall is a composite of networks of complex polysaccharides and proteins. Generally, the expanding cell wall is strong and varies in thickness (0.1–1.0 μm). During expansive growth, the cells must produce a significant force to stretch, or deform, their walls to

*Cells, Forces, and the Microenvironment*
Edited by Charles M. Cuerrier and Andrew E. Pelling

ISBN 978-981-4613-36-1 (Hardcover), 978-981-4613-37-8 (eBook)
www.panstanford.com

accommodate larger and larger volumes. All of these cells with walls produce the same force to deform their walls, pressure force. The pressure force produces stresses within the wall causing it to deform. These cells regulate the mechanical properties of their walls to control wall deformation and expansive growth. Experimental and theoretical investigations are beginning to elucidate the many biological, chemical, and physical processes involved in expansive growth and its regulation. This chapter presents conceptual and mathematical frameworks, or models, that are consistent with experimental evidence. These models are used to study and explore the underlying processes and mechanisms involved in expansive growth and regulation of expansive growth.

## 15.1 Introduction

Plant, algal, and fungal cells have engineered unique solutions to obstacles encountered in their microenvironment. A paramount and fundamental challenge for all cells is the acquisition of energy to run and maintain the many complex chemical and biological processes that are necessary to sustain, maintain, and procreate life. Plant and algal cells have developed and honed unique biological structures (e.g., chloroplasts) and unique chemical processes (e.g., photosynthesis) that capture and convert solar energy to chemical energy in the form of organic compounds that sustain essentially all life on earth. Also, plant and algal cells have developed chemical reactions that convert the carbon dioxide in the atmosphere to oxygen, a molecule that is required for all respiring cells. Fungal cells have developed chemical reactions that decompose complex organic compounds produced by respiring organisms so that the molecules of life can be recycled in photosynthesis and respiration.

Another biological process that is unique to plant, algal, and fungal cells is expansive growth. Some plant, algal, and fungal cells can enlarge more than 30,000 times their initial volume (e.g., xylem vessels in plants) and achieve lengths of greater than 10 cm (e.g., fungal sporangiophores). Importantly expansive growth and its regulation are fundamental to development, morphogenesis, and environmental (sensory) responses of plant, algal and fungal cells.

### 15.1.1 Expansive Growth and the Cell Wall

It is interesting and somewhat surprising that plant, algal, and fungal cells have developed the ability to enlarge to such large volumes, because all of these cells are encased in a strong and somewhat rigid cell wall. Walls of cells undergoing expansive growth range between 0.1 and 1.0 μm in thickness. The cell wall is an exoskeleton that gives the cell shape and support while providing both physical and chemical protection from the external environment. The cell wall is a composite of networks composed of complex polysaccharides and proteins. The general composition of plant, algal, and fungal cell walls may be viewed as relatively large microfibrils made of complex polysaccharides embedded in an amorphous matrix composed of networks of complex polysaccharides and proteins (Fig. 15.1).

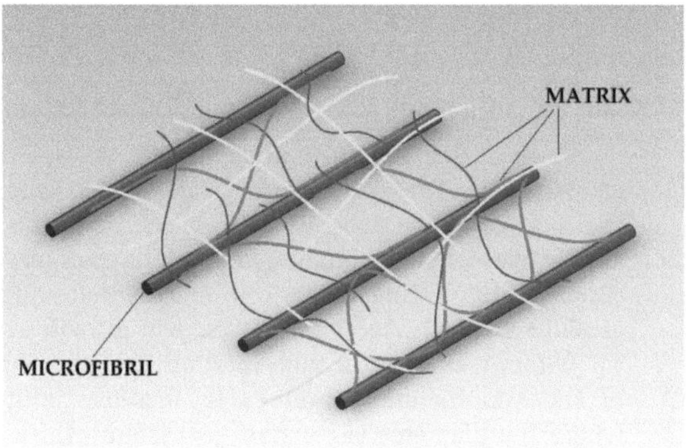


**Figure 15.1** Schematic illustration of microfibrils embedded in matrix networks. The wall is composed of relatively large microfibrils made of complex polysaccharides embedded in an amorphous matrix made of networks of complex polysaccharides and proteins.

Typical walls of plant and algal cells are composed of cellulose microfibrils embedded in an amorphous matrix of hemicelluloses and pectins, with small amounts of structural proteins [12]. Similarly, typical walls of fungi are composed of chitin and β-glucan microfibrils embedded in an amorphous matrix of chitosan, glycoproteins, and lipids [8,20,54]. The polymers in plant, algal, and fungal walls are linked together by covalent bonds, hydrogen bonds, hydrophobic interaction, and ionic associations [54].

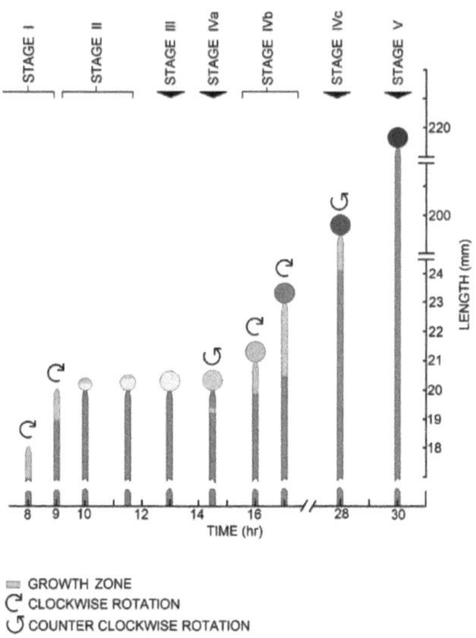


**Figure 15.2** Schematic illustration showing the developmental stages of the large, single-celled, sporangiophore of *Phycomyces blakesleeanus*. The stage I sporangiophore appears as a single pointed cell that grows longitudinally at the apical tip in the growth zone. The growth zone is shown as light green and the non-growing stalk is dark green. Clockwise rotation (when viewed from above) and elongation growth occur concurrently during stage I producing left-handed helical growth. Stage II begins with spherical growth at the apical tip without elongation and rotation growth. The diameter of the sporangium continues to increase until stage III, where the diameter is constant (≈0.5 mm). During stage III, there is no visible growth. Stage IVa begins with elongation growth concurrent with counter-clockwise rotation growth (right-handed helical growth) in a short growth zone located approximately 0.6 mm below the sporangium. The sporangium begins to darken and elongation growth, rotation growth, and growth zone length continue to increase. The right-handed helical growth continues for approximately 1h before the rotation rate gradually decreases to zero and clockwise rotation begins without interruption of elongation growth. Stage IVb begins with the initiation of left-handed helical growth. Stage IVb exhibits nearly constant elongation and rotation growth rates for many hours. Stage IVc is initiated by counter-clockwise rotation and right-handed helical growth. Stage V is the last stage and does not exhibit visible expansive growth. Reprinted with permission from [42].

The physical properties and strength of dead plant cell walls, i.e., wood, are well known. The strong and somewhat rigid living cell wall represents an obstacle in the microenvironment of these cells with which they must contend during expansive growth. A walk through the country or in the mountains can reveal the wonderful diversity of plants, algae, and fungi. How the cells of these organisms enlarge to such large volumes and regulate their growth to form the variety of shapes found in nature has long been of general interest and the focus of scientific studies.

Equally interesting is that cells with walls regulate their growth rate in response to environmental signals in order to improve the probability of their survival. For example, the large cylindrical single-celled sporangiophores of the fungus *Phycomyces blakesleeanus* are approximately 150 μm in diameter and grow in length to ten or more centimeters (Fig. 15.2). The sporangiophores detect gravitational acceleration, ethylene, mechanical stretch, gases, temperature, wind, light intensity, spatially asymmetric distribution of light, the presence of solid objects and turgor pressure. They respond to these stimuli with symmetric and asymmetric changes in expansive growth rate, i.e., growth responses and tropic responses, respectively [8].

It has become apparent that enlarging the cell wall is central to expansive growth and its regulation. However, simply synthesizing and adding more wall material cannot create the large thin exoskeleton that is observed in plant, algal, and fungal cells. Some force must be employed to deform the wall during expansive growth. Interestingly all plant, algal, and fungal cells employ the same force to enlarge the wall during expansive growth, pressure force generated by turgor pressure.

### 15.1.2 Turgor Pressure

Plant, algal, and fungal cells produce and transport active solutes inside their plasma membranes. The plasma membrane is semi-permeable and allows water to pass through it, but not the active solutes. This develops an osmotic pressure gradient across the plasma membrane that drives the absorption of water from its surroundings (osmosis). The absorption of water increases the volume of the cell and stretches its wall chamber. As the wall stretches (deforms), stresses develop within the wall and the wall resists the deformation. This wall resistance to deformation results in the production of pressure within the cell, Pi, and turgor pressure (defined as the difference of pressure inside and outside of the cell, i.e., $P = P_i - P_o$). As the magnitude of the turgor pressure increases, it produces an adverse pressure gradient that reduces the water flow into the cell, until equilibrium is achieved. The turgor pressure can be measured directly with a pressure probe (Fig. 15.3). Many different pressure probes have been developed to directly measure the turgor pressure of single plant, algal, and fungal cells, [9,10,30,38,39,48,49,58] and to measure the pressure in various plant structures (e.g., xylem) [1] and organs (stems, roots, leaves, etc.) [55].

Cells with walls generate and employ turgor pressure to produce a pressure force (pressure force is the pressure times the area) that acts on the inside surface of the wall perpendicular to its surface (Fig. 15.4). The wall essentially converts the pressure force (created by the turgor pressure) to stresses within the wall (stress is the force divided by the area) that are predominately tangent to the wall surface, thus putting the wall in tension longitudinally and circumferentially (Fig. 15.4). The stresses within the wall produce wall deformation in three orthogonal directions, generally stretching the wall in the two directions parallel to the wall surface (longitudinally and circumferentially) and contracting the wall in the direction perpendicular to the surface, making the wall thinner.

Overall, the direction of maximum wall deformation does not correspond to the direction of maximum wall stress. The direction and magnitude of the deformations are strongly dependent on the composition, structure, and architecture of the wall, and the chemical reactions within the wall, all of which modify and regulate the mechanical properties of the wall. In general, research has shown that the cell regulates expansive growth by controlling the magnitude of the turgor pressure (by increasing or decreasing the amount of active solutes in the cell) or by modifying the mechanical properties of the wall. Typically, the turgor pressure is sometimes regulated to control the magnitude of expansive growth rate, but regulating the wall's mechanical properties is used to control both the direction and magnitude of expansive growth during morphogenesis and tropic responses [2,6,12,16,18, 34,36,37,56].

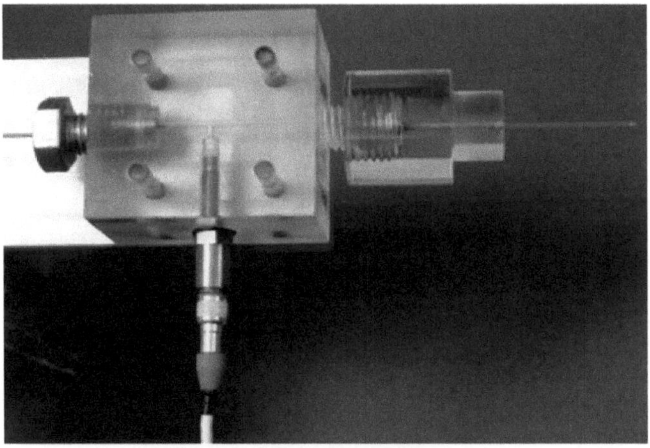

**Figure 15.3** Photograph of a cellular pressure probe. The pressure probe is an instrument that measures the turgor pressure of single plant, algal and fungal cells. The pressure probe is also used to monitor and produce changes in turgor pressure. The pressure probe consists of a pipette that tapers to a small tip (1–50 μm, depending on the size of the cell), a chamber, and a control rod attached to a micrometer, all of which are arranged in a block of clear acrylic. The chamber and pipette are filled with low-viscosity incompressible oil. In operation, the pressure probe is manipulated so that the tip of the pipette impales a cell. The tip passes through the cell wall and the cell membrane. Because plant and fungal cell walls are constructed of microfibrils that are essentially woven together like a fabric, the small tip can penetrate the wall without tearing it. The fluidic membrane provides a seal around the tip of the pipette to prevent leaking. Once the tip is in the cell, the system is pressurized and the cell sap will be pushed into the pipette producing an interface with the incompressible oil. Typically, the interface is returned to the surface of the cell by advancing the control rod. The incompressible oil transmits the pressure in the cell to the pressure transducer that measures turgor pressure ($P_i$ – $P_o$). The pressure transducer is connected to a voltmeter, which in turn is connected to a script chart recorder that will continually plot the turgor pressure as a function of time. The probe can also be used to manipulate the turgor pressure of a cell to gain insight to the dynamic growth responses that result from changes in turgor pressure. To be able to manipulate the turgor pressure of a cell, the pressure probe has to be able to move oil into the cell or move the cell sap out of the cell. This can be done by advancing or retracting the control rod.

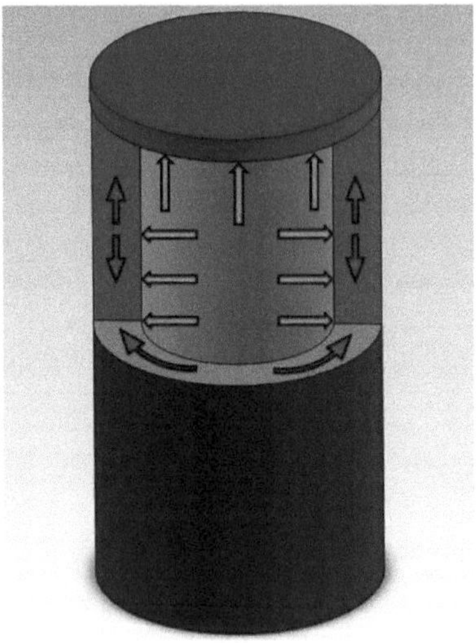

**Figure 15.4** Wall stresses in a thin cylindrical cell wall produced by turgor pressure. Turgor pressure (light blue arrows) generated within the cell acts perpendicular to the inner wall surface. The pressure force is converted to wall stresses (red arrows) that are parallel to the wall surface. The stresses create wall tensions that stretch (deforms) the wall.

### 15.1.3 Mechanical Behavior of the Wall and Relevant Molecular Mechanisms

The wall deformation behavior during expansive growth depends on the wall's composition, structure, architecture, and the chemical reactions within it, thus the wall properties are biochemically mediated. In all cases, the wall undergoes both reversible (elastic) and irreversible deformation during expansive growth. The irreversible deformation produces a permanent increase in wall surface area and ultimately determines the shape of the cell. The elastic deformation provides most of the wall resistance that produces turgor pressure. At a molecular level, elastic deformations are reversible rearrangements of the polymer networks within the wall that ultimately load molecular bonds within and between polymers.

Generally, the microfibrils are cross-linked by matrix polymers (Fig. 15.1). Experimental evidence indicates that wall's mechanical properties and deformation behavior is modified by molecular agents that affect load-bearing cross-links between microfibrils [11,12]. Breaking the cross links allows the microfibrils to separate [28], producing a controlled polymer creep, or wall loosening. The wall loosening produces wall stress relaxation and turgor pressure relaxation that in turn increases water uptake, wall deformation, and an increase in cell volume.

Two molecular agents have been identified that may regulate wall loosening during expansive growth: expansins and calcium. Expansins are pH-dependent wall-loosening proteins that are shown to catalyze wall stress relaxation and wall extension directly [11,12]. Interestingly, these proteins are not enzymes. The mechanism by which expansins produce wall loosening is unknown, but experimental and theoretical evidence [60] indicate that expansins break hydrogen bonds within the hemicellulose network that cross-link cellulose microfibrils. Expansins are candidates to be the primary wall-loosening agent for cells in organs of higher plants (e.g., stems, roots, and leaves).

Calcium is involved in cross-linking matrix polymers within the pectin network [3,12,44,47]. The carboxyl groups of newly synthesized pectin are mostly methyl esterified which prevents calcium cross-links from forming between pectin polymers. An enzyme (pectin methylesterases) can remove the methyl group allowing the carboxyl groups to form cross-links between pectin polymers. So when new pectin is delivered to the inner wall via exocytosis, it is hypothesized that it is de-esterified and loosens the wall by competing for and removing calcium from load-bearing cross-links. Regulating calcium-mediated pectin cross-links is hypothesized to be a wall-loosening mechanism used by some algal cells [3,47] and pollen tubes [44].

### 15.1.4 Expansive Growth Process and Growth Zones

The overall process of expansive growth occurs as follows. The wall is biochemically loosened, which reduces both wall stresses (stress relaxation) and turgor pressure (turgor pressure relaxation). Water absorption is increased in response to the decrease in turgor pressure and expands (stretches and deforms) the loosened wall. As water is absorbed, the turgor pressure begins to increase and wall deformation increases the wall surface area (expansion) and decreases the wall thickness. Concomitantly, new wall polymers and other wall materials are added to the inside surface of the wall, generally to maintain a nearly constant wall thickness. This series of events (i.e., wall loosening, wall stress relaxation, turgor pressure relaxation, water uptake, an increase in turgor pressure, wall expansion, an increase in wall stresses, and wall loosening again) is repeated continually during expansive growth.

In most plant and algal cells, the entire wall chamber undergoes irreversible and elastic deformation during expansive growth, exhibiting "diffuse" growth [36,37,42]. Typically cells in plant organs (e.g., stems, roots, and leaves) and algal internode cells exhibit diffuse growth. However, some cells restrict the region of wall that undergoes irreversible and elastic deformation during expansive growth to a local region called the "growth zone." Plant cells (e.g., pollen tubes and root hairs), fungal hyphae, and some fungal sporangiophores exhibit "tip" growth [20] where the growth zone is at the apical tip. Some sporangiophores exhibit "intercalary" growth where the growth zone is located immediately below a spherical sporangium that store spores [8]. Examples of tip growth and intercalary growth can be found in the development of the sporangiophores of *P. blakesleeanus* (Fig. 15.2) [6,8,42]. Tip growth occurs only in the first stage of development (stage I) before the sporangium is formed. Intercalary growth occurs during stage IV, after the sporangium and its spores have formed at the tip.

## 15.2 Mathematical Models

The qualitative concepts and processes described in the previous sections have been formalized as quantitative biophysical equations. The first set of biophysical equations were derived by Lockhart (1965) [27] and termed the "Growth Equations" [56]. The Growth Equations are *global* equations, i.e., they describe the average behavior of the entire cell and the magnitudes of the inclusive biophysical variables are average values for the entire cell. Importantly, the Growth Equations do not correctly describe and model the wall deformations that occur during expansive growth [9,10,16,34–37,48,49]. Subsequently, the Growth Equations were augmented with additional terms, by Ortega (1985) [32] and Ortega et al. (1988) [38] and termed the Augmented Growth Equations [32–37]. Importantly, the *global* Augmented Growth Equations correctly describe and model the wall deformation observed experimentally [9,10,34,40,43,48,49] and also account for transpiration in cells that are exposed and lose water to the atmosphere [34,38].

### 15.2.1 Augmented Growth Equations

The global Augmented Growth Equations describe and model two simultaneous and interrelated processes that produce expansive growth of cells with walls; water uptake and wall deformation. The global Augmented Growth Equations represents a small *system* model because they mathematically describe multiple underlying processes involved in expansive growth, *net* water uptake rate and wall deformation rate, and the turgor pressure behavior. Eq. 15.1 describes the relative rate of change in water volume, $(\mathrm{d}V_w/\mathrm{d}t)/V$, of the cell as the difference between the rate of water uptake and transpiration rate [33–38].

$$\underset{\text{(rate of change in water volume)}}{(\mathrm{d}V_w/\mathrm{d}t)/V} = \underset{\text{(rate of water uptake)}}{L(\Delta\Pi - P)} - \underset{\text{(transpiration rate)}}{T}, \tag{15.1}$$

where $V$ is the volume, $t$ is the time, $L$ is the relative membrane hydraulic conductance ($L = L_p A/V$), $L_p$ is the membrane hydraulic conductivity, $A$ is the membrane area, $\Delta\Pi$ is the osmotic pressure difference across the membrane (the solute reflection coefficient is assumed to be unity and omitted from these equations), $P$ is the turgor pressure (a gauge pressure relative to atmospheric pressure as measured with a pressure probe, $P_I - P_{atm}$, where $P_I$ is the pressure inside the plasma membrane and $P_{atm}$ is defined to be zero), and $T$ is the relative rate of change in water volume lost via transpiration (relative volumetric transpiration rate), i.e., $T = (\mathrm{d}V_T/\mathrm{d}t)/V$. In the case when the cell is not exposed to the atmosphere, $T = 0$.

Equation 15.2 describes the relative rate of change in volume of the cell wall chamber, $(\mathrm{d}V_{cwc}/\mathrm{d}t)/V$, as the sum of irreversible deformation rate and reversible (elastic) deformation rate of the wall [32–37].

$$\underset{\text{(wall expansion rate)}}{(\mathrm{d}V_{\mathrm{cwc}}/\mathrm{d}t)/V} \underset{=}{=} \underset{\text{(irreversible deformation rate)}}{\phi(P-P_c)} \underset{+}{+} \underset{\text{(elastic deformation rate)}}{(1/\varepsilon)\,\mathrm{d}P/\mathrm{d}t} \quad (15.2)$$

The biophysical variable, $\phi$, is the relative irreversible extensibility of the wall, $P_C$ is the critical turgor pressure, and $\varepsilon$ is the volumetric elastic modulus. Because the relative rate of change in volume of the cell contents (mostly water) and cell wall chamber are approximately equal, a biophysical equation describing the rate of change of turgor pressure, Eq. 15.3, can be obtained by combining Eqs. 15.1 and 15.2 with the elimination of $(\mathrm{d}V_W/\mathrm{d}t)/V$ and $(\mathrm{d}V_{cwc}/\mathrm{d}t)/V$, because $(\mathrm{d}V_W/\mathrm{d}t)/V \approx (\mathrm{d}V_{cwc}/\mathrm{d}t)/V$ [34–37].

$$\underset{\text{(rate of change of turgor pressure)}}{dP/dt} \underset{\propto}{=} \varepsilon\Big[\underset{\text{\{(water uptake rate)}}{L(\Delta\Pi-P)} \underset{-}{-} \underset{\text{(irreversible deformation rate)}}{\phi(P-P_c)} \underset{-}{-} \underset{\text{(transpiration rate)\}}}{T}\Big] \quad (15.3)$$

Equation 15.3 states that the rate of change of turgor pressure is proportional to the difference in magnitudes of water uptake rate and the sum of irreversible deformation rate of the wall and volumetric transpiration rate. Equation 15.3 may be solved to determine the turgor pressure change after an instantaneous change in the magnitude of the one or more of the inclusive biophysical variables and/or the relative volumetric transpiration rate, i.e., Eqs. 15.4 and 15.5.

$$P(t) = (P_o - P_{eq})\exp[-\varepsilon(\phi + L)t] + P_{eq}, \quad (15.4)$$

where

$$P_{eq} = (L\Delta\Pi + \phi P_c - T)/(\phi + L) \quad (15.5)$$

Equation 15.4 describes an exponential decay of the turgor pressure from $P_o$ (turgor pressure at $t$ = 0) to $P_{eq}$ (equilibrium turgor pressure) that occurs subsequent to an instantaneous change in the magnitude of one or more of the inclusive biophysical variables, and/or the relative volumetric transpiration rate. The exponential decay has a time constant, $t_c = [\varepsilon\ (\phi + L)]^{-1}$. The expression for $P_{eq}$ (Eq. 15.5) describes the equilibrium turgor pressure for a growing cell that is transpiring [35]. Equation 15.5 establishes a relationship between the equilibrium turgor pressure and the transpiration rate. It can be seen that an increase or decrease in transpiration rate will decrease or increase the magnitude of the equilibrium turgor pressure, respectively.

### 15.2.2 Validation of the Augmented Growth Equations

Importantly, the Augmented Growth Equations (Eqs. 15.1–15.3) describe both the net rate of water uptake and wall deformation rate. Furthermore, the equations are coupled by turgor pressure, thus explicitly demonstrating that they are interrelated processes. This is important because modeling expansive growth requires modeling both processes of water uptake and wall deformation. Because turgor pressure, $P$, appears in all of the Augmented Growth Equations and because the turgor pressure can be measured directly with the pressure probe, experimental protocols have been designed to employ the pressure probe to validate the Augmented Growth Equations. Importantly, it was demonstrated that the Augmented Growth Equations accurately model the results of in vivo stress relaxation and in vivo creep pressure probe experiments [9,10,34–37,40,48,49], which in turn demonstrates that these equations accurately model the measured mechanical properties and behavior of the cell wall.

Good agreement has been demonstrated between theoretical results obtained from the Augmented Growth Equations and experimental results obtained from pressure probe experiments on growing plant cells in pea stems [9,10], fungal sporangiophores [34–43], and algal internode cells [48,49]. Pressure probe experiments have been conducted to measure the magnitude and behavior of inclusive biophysical variables within the Augmented Growth Equations during development [43], during environmental responses [39], and with growth mutants [42]. Also, the Augmented Growth Equations have been extended to study the effect of wall loosening agents on growth rate behavior [46] and whole plants growth behavior over long periods of time [25].

The Augmented Growth Equations have been modified to model expansive growth and water uptake of plant cells embedded in tissue and organs, where pressures within the apoplasm (cell walls and xylem) are frequently different from atmospheric pressure [35]. These modified Augmented Growth Equations for cells in tissue predict growth-induced water potentials [29,31] and can model growth behavior obtained experimentally [57]. Ortega

[34,36], Geitmann and Ortega [16], and Ortega and Welch [37] have reviewed these equations. Currently, the Augmented Growth Equations have been validated by more experimental results than any other mathematical model.

### 15.2.3 Growth Rate Regulation

The results of pressure probe experiments, together with the Augmented Growth Equations, provide insight into how these cells with walls regulate expansive growth rate. For example, during development the elongation growth rates of the sporangiophores of *P. blakesleeanus* increase from stage I (tip growth) to stage IVb (intercalary growth). The results of pressure probe experiments demonstrate that the observed increase in elongation growth rate is predominately the result of an increase in the relative irreversible extensibility, $\phi$, and the growth zone length, $L_g$ [34,43]. A similar result is obtained for the transient increases in elongation growth rate that occur during the light growth responses and the avoidance growth responses; transient increases in elongation growth rates are predominately the result of increases in $\phi$ and $L_g$ [36,39]. Also, decreases in elongation growth rates observed during the pressure response are predominately the result of decreases in $\phi$ and $L_g$ [41]. These results for the sporangiophores of *P. blakesleenaus* are similar to those obtained from internode algal cells of *Chara corallina* (diffuse growth) [48,49] and plant cells within pea stems (diffuse growth) [9,10], both of which demonstrate that the expansive growth rate is predominately regulated by the magnitude of $\phi$. Overall, these results demonstrate that growth rate of cells with walls is predominately regulated by controlling the mechanical properties of the wall to undergo irreversible deformation. These results provide both impetus and justification for molecular mechanisms of wall growth that involve breaking cross-links between microfibrils to produce controlled polymer creep and wall loosening. Although experimental evidence indicates that controlling the wall's mechanical properties is the primary regulator of expansive growth of cells with wall, there is also evidence that turgor pressure is involved in growth rate regulation [19,40–43,48,49]. In addition, recent studies demonstrate that growth mutant sporangiophores of *P. blakesleeanus* regulate their elongation growth rate by controlling the magnitudes of $\phi$, $L_g$, and the effective turgor pressure ($P - P_c$) [42].

### 15.2.4 Utility and Limitations of the Augmented Growth Equations

Overall, the global Augmented Growth Equations provide a qualitative and quantitative system model for understanding expansive growth rate and its regulation, and their relationship with net water uptake, wall's mechanical properties and turgor pressure behavior. However, these biophysical equations have inherited limitations because they are *global* and not *local*. *Local* equations describe the behavior of a quantity at a point in space (actually a small volume in space because the assumption of a continuum is employed) and values for inclusive variables are defined for that point (small volume) in space. Local equations are needed to describe or model local spatial or temporal changes in the wall in order to describe and understand morphogenesis and the underlying molecular mechanism(s) of wall loosening. New local models and larger system models (some employing global equations and others employing local equations) are being developed to investigate other processes involved in expansive growth (e.g., wall thinning rate and wall deposition rate) and the relationship between wall structure, wall chemistry, and mechanical properties.

### 15.2.5 System Models

The underlying local constitutive equation for the global Augmented Growth Equation, Eq. 15.2, has been used to develop mathematical system models for expansive growth and growth regulation of pollen tubes. Interestingly, pollen tubes exhibit growth oscillations during elongation growth and both the magnitude and frequency of the growth oscillations are related the overall average elongation growth rate. Kroeger et al. [22–24] have developed mathematical system models that employ feedback loops and predict the growth oscillations of pollen tubes consistent with experimental results. These models and others like them are important because mathematical equations are provided for the known biological and chemical processes that are hypothesized to be associated with manipulating the wall's mechanical properties and regulating elongation growth rate. As an example, an equation is used to model the rate of change in wall thickness as the difference in the rate of wall deposition and the rate of decrease of wall thickness due to wall stretching. Also, the relationship between the wall deposition rate and the calcium concentration inside the pollen tube apex is described mathematically. Another equation describes the rate of change of calcium concentration in terms of influx through stretch-activated channels in the plasma membrane and binding of calcium ions with various components of the cytoplasm. And yet another equation describes the rate of change of wall viscosity as a function of wall deposition rate and rate of de-esterification of pectins.

Other investigators have developed mathematical system models that employ feedback loops and relate calcium dynamics with growth behavior [26,61]. Another system model by Rojas et al. [53] is especially interesting because a molecular mechanism is used to obtain a local constitutive equation that explores the relationship between the deposition of wall polymers and wall deformation. In this model, the wall deformation is considered to be the result of a chemo-rheological process where the effective viscosity of the wall is determined by a chemical reaction. They formulate a mechanism for expansive growth of pollen tubes in which deposition of wall material causes turnover of cross-links between wall polymers and wall deformation is related to the rate of turnover of cross-links.

Many assumptions are made in all of these system models. In the near future, the assumptions and inclusive equations must be validated experimentally. Also, the inclusion and sequencing of biological and chemical processes in the hypothesized system model also need to be validated experimentally. However, even though these recent models require a significant amount of experimental validation, they are important because they provide the foundation for future models that can explicitly address biological, chemical, and physical processes. In this respect, these new system models represent a significant advancement in obtaining a qualitative and quantitative mathematical framework for a better and more detailed understanding of expansive growth and its regulation.

### 15.2.6 Models for Morphogenesis

The shapes of cells with walls are obtained by controlling anisotropic mechanical deformation of the wall [2,12,18,37,56]. Some experimental evidence indicates that the orientations of microfibrils within the wall matrix affect the anisotropic mechanical properties of the wall [2,5,16,50–52]. However, other evidence suggests that morphogenesis is more complicated and depends on dynamic processes involving the deposition rates of new wall polymers at specific locations, and the action of biologically active molecules, ions, proteins, and enzymes within the wall [2,12,16,20]. At present, it is thought that anisotropic mechanical deformation is produced by temporal and spatial regulation of wall mechanical properties, which is produced by regulating the processes of wall deposition, wall assembly, wall loosening, and wall stiffening. It is apparent that mathematical models for morphogenesis require mathematical constitutive equations for multidimensional wall deformations.

Recent mathematical models exhibit diverse approaches to the challenges generated by this complicated multi-scale problem. Some models employ a modified local version of the global biophysical equations [45], others employ modified versions of constitutive equations and methods developed in fluid mechanics [4,13–15,17], elasticity [7,59], and continuum mechanics [21]. It is noted, that all of these models are in the early stages of development and require a significant amount of experimental validation. One model by Huang et al. [21] is noteworthy, because it was demonstrated that this model is able to describe the experimental results obtained from an in vivo creep experiment [48].

## 15.3 Summary, Future Research, and Conclusions

Cells with walls (plant, algal, and fungal cells) accumulate active solutes inside their plasma membrane to generate turgor pressure. The turgor pressure produces the pressure force needed to stretch (deform) its wall so that the cell can enlarge (grow) in volume, i.e., expansive growth. Two simultaneous and interrelated processes, rate of net water uptake and rate of wall deformation, are fundamental to expansive growth of cells with walls. Much of what is known about expansive growth and its relationship to turgor pressure behavior, net rate of water uptake and wall deformation rate is obtained from experimental and theoretical investigations conducted in concert. Some of the experimental results from pressure probe experiments conducted on fungal, plant, and algal cells in concert with analyses employing the global Augmented Growth Equations are summarized so they may be used to guide future global, local, and system models. (1) The wall deformation during expansive growth is both irreversible and reversible (elastic). (2) Elastic deformation rate, irreversible deformation rate, and expansive growth rate all change when the magnitude of the turgor pressure is changed. (3) The wall's mechanical properties and/or the turgor pressure are regulated by the cell to control expansive growth rate. (4) The wall mechanical properties and turgor pressure change during development. (5) Regulation of the wall's mechanical properties is used to control the overall magnitude of expansive growth rate and the differential growth rate during sensory responses, i.e., growth responses and tropic responses, respectively. (6) Morphogenesis can only be produced and controlled by temporal and spatial regulation of wall mechanical properties.

Future investigations will require *experimentally validated* local models of the wall (that can describe temporal and spatial changes in wall mechanical properties) to investigate morphogenesis and some sensory (tropic) responses of cells with walls. *Experimentally validated* local, global, and system models will all be needed to investigate and organize the many associated biological, chemical, and physical processes, as well as the underlying mechanisms used to control expansive growth. It is concluded that designing and conducting relevant experiments simultaneous to and in concert with the development of future local, global, and system mathematical models (in order to provide validation and insight) will be a major challenge that must be addressed in future research.

## Acknowledgment

The authors gratefully acknowledge financial support by the National Science Foundation Grant MCB-0948921 to Joseph K. E. Ortega.

## References


1. Balling, A., Zimmermann, U. (1990). Comparative measurement of xylem pressure of Nicotiana plants by means of the pressure bomb and pressure probe, *Planta*, **182**, 325–338.
2. Baskin, T. I. (2005). Anisotropic expansion of the plant cell wall, *Annu. Rev. Cell Dev. Biol.*, **21**, 203–222.
3. Boyer, J. S. (2009). Cell wall biosynthesis and the molecular mechanism of plant enlargement, *Funct. Plant Biol.*, **36**, 383–394.
4. Campas, O., Mahadevan, L. (2009). Shape and dynamics of tip-growing cells, *Curr. Biol.*, **19**, 2102–2107.
5. Carpita, N. C., Gibeaut, D. M. (1993). Structural models of primary cell wall in flowering plants: Consistency of molecular structure with the physical properties of the walls during growth. *Plant J.*, **3**, 1–30.
6. Castle, E. S. (1942). Spiral growth and the reversal of spiraling in Phycomyces, and their bearing on primary wall structure. *Am. J. Bot.*, **29**, 664–672.
7. Chaplain, M. A. J. (1993). The strain energy function of an ideal plant cell wall, *J. Theor. Biol.*, **163**, 77–97.
8. Cerda-Olmedo, E., Lipson, E. D., *Phycomyces.* Cold Spring Harbor Laboratory Press, Cold Spring Harbor, NY (1987).
9. Cosgrove, D. J. (1985). Cell wall yield properties of growing tissue evaluation by in vivo stress relaxation, *Plant Physiol.*, **78**, 347–356.
10. Cosgrove, D. J. (1987). Wall relaxation in growing stems: Comparison of four species and assessment of measurement techniques, *Planta*, **171**, 266–278.
11. Cosgrove, D. J. (2000). Loosening of plant cell walls by expansions, *Nature*, **407**, 321–326.
12. Cosgrove, D. J. (2005). Growth of the plant cell wall, *Nat. Rev. Mol. Cell Biol.*, **6**, 850–861.
13. Dumais, J., Shaw, S. L., Steele, C. R., Long, S. R., Ray, P. M. (2006). An anisotropic-viscoplastic model of plant cell morphogenesis by tip growth, *Int. J. Dev. Biol.*, **50**, 209–222.
14. Dyson, R. J., Jensen, O. E. (2010). A fibre-reinforced fluid model of anisotropic plant cell growth, *J. Fluid Mech.*, **655**, 472–503.

15. Dyson, R. J., Band, L. R., Jensen, O. E. (2012). A model of crosslink kinetics in the expanding plant cell wall: Yield stress and enzyme action, *J. Theor. Biol.*, **307**, 125–136.

16. Geitmann, A., Ortega, J. K. E. (2009). Mechanics and modeling of plant cell growth, *Trends Plant Sci.*, **14**, 467–478.

17. Green, J. E. F., Friedman, A. (2008). The extensional flow of a thin sheet of incompressible, transversely isotropic fluid. *Euro. J. Appl. Math.*, **3**(19), 225–257.

18. Green, P. (1969). Cell Morphogenesis, *Ann. Rev. Plant Physiol.*, **20**, 365–394.

19. Green, P. B., Erickson, R. O., Buggy, J. (1971). Metabolic and physical control of cell elongation rate: In vivo studies in Nitella, *Plant Physiol.*, **47**, 423–430.

20. Heath, I. B. (1990). *Tip Growth in Plant and Fungal Cells.* Academic Press, Inc., San Diego, CA.

21. Huang, R., Becker, A. A., Jones, I. A. (2012). Modelling cell wall growth using a fibre-reinforced hyperelastic-viscoplastic constitutive law, *J. Mech. Phys. Solids*, **60**, 750–783.

22. Kroeger, J. H., Geitmann, A., Grant, M. (2008). Model for calcium dependent oscillatory growth in pollen tubes, *J. Theo. Biol.*, **253**, 363–374.

23. Kroeger, J. H., Geitmann, A. (2013). Pollen tubes with more viscous cell walls oscillate at lower frequencies, *Math. Model Nat. Phenom.*, **8**(4), 25–34.

24. Kroeger, J. H., Zerzour, R., Geitmann, A. (2011). Regulator or driving force? The role of turgor pressure in oscillatory plant cell growth, *PLoS One*, **6**, e18549.

25. Lewicka, S. (2006). General and analytic equations of the Ortega equation, *Plant Physiol.*, **142**, 1493–1510.

26. Liu, J., Piette, B. M. A. G., Deeks, M. J., Franklin-Tong, V. E., Hussey, P. J. (2010). A compartmental model analysis of integrative and self-regulatory ion dynamics in pollen tube growth, *PloS One*, **5**, e13157.

27. Lockhart, J. A. (1965). An analysis of irreversible plant cell elongation, *J. Theor. Biol.*, **8**, 264–275.

28. Marga, F., Grandbois, M., Cosgrove, D. J., Baskin, T. I. (2005). Cell wall extension results in the coordinate separation of parallel microfibrils: Evidence from scanning electron microscopy and atomic force microscopy, *Plant J.*, **43**, 181–190.

29. Molz, F. J., Boyer, J. S. (1978). Growth-induced water potential in plant cells and tissue, *Plant Physiol.*, **62**, 423–429.
30. Murphy, R., Ortega, J. K. E. (1995). A new pressure probe method to determine the average volumetric elastic modulus of cells in plant tissue, *Plant Physiol.*, **107**, 995–1005.
31. Nonami, H., Boyer, J. S. (1993). Direct demonstration of a growth-induced water potential gradient, *Plant Physiol.*, **102**, 13–19.
32. Ortega, J. K. E. (1985). Augmented growth equation for cell wall expansion, *Plant Physiol.*, **79**, 318–320.
33. Ortega, J. K. E. (1990). Governing equations for plant cell growth, *Physiol. Plant*, **79**, 116–121.
34. Ortega, J. K. E. (2004). *A quantitative biophysical perspective of expansive growth for cells with walls*. Pandalai, S. G., ed., *Rec. Res. Dev. Biophys.*, Transworld Research Network, Kerala, India, **3**, 297–324.
35. Ortega, J. K. E. (2010). Plant cell growth in tissue, *Plant Physiol.*, **154**, 1244–1253.
36. Ortega, J. K. E. (2012). *Growth rate regulation of cells with walls: The sporangiophores of Phycomyces blakesleeanus used as a model system*, Pandalai, S. G., ed., *Rec. Res. Dev. Plant Physiol.*, Transworld Research Network, Kerala, India, 5, 1–19. ISBN: 978-81-308-0497-2.
37. Ortega, J. K. E., Welch, S. W. J. (2013). Mathematical models for expansive growth of cells with walls, *Math. Model Nat. Phenom.*, **8**(4), 35–61.
38. Ortega, J. K. E., Keanini, R. G., Manica, K. J. (1988). Pressure probe technique to study transpiration in Phycomyces sporangiophores, *Plant Physiol.*, **87**, 11–14.
39. Ortega, J. K. E., Manica, K. J., Keanini, R. G. (1988). Phycomyces: Turgor pressure behavior during the light and avoidance growth response, *Photochem. Photobiol.*, **48**, 697–703.
40. Ortega, J. K. E., Zehr, E. G., Keanini, R. G. (1989). In vivo creep and stress relaxation experiments to determine the wall extensibility and yield threshold for the sporangiophores of Phycomyces, *Biophys. J.*, **56**, 465–475.
41. Ortega, J. K. E., Lesh-Laurie, G. E., Espinosa, M. A., Ortega, E. L., Manos, S. M., Cunning, M. D., Olson, J. E. C. (2003). Helical growth of stage-IVb sporangiophores Phycomyces blakesleeanus: The relationship between rotation and elongation growth rates, *Planta*, **216**, 716–722.

42. Ortega, J. K. E., Munoz, C. M., Blakley, S. E., Truong, J. T., Ortega, E. L. (2012). Stiff mutant genes of Phycomyces affect turgor pressure and wall mechanical properties to regulate elongation growth rate, *Frontiers Plant Sci.* (*Plant Biophys. Model.*), **3** (Art 99), 1–12. doi: 10.3389/fpls.2012.00099.
43. Ortega, J. K. E., Smith, M. E., Erazo, A. J., Espinosa, M. A., Bell, S. A., Zehr, E. G. (1991). A comparison of cell-wall-yielding properties for two developmental stages of Phycomyces sporangiophores: Determination by in-vivo creep experiments, *Planta*, **183**, 613–619.
44. Parre, E., Geitmann, A. (2005). Pectin and the role of the physical properties of the cell wall in pollen tube growth of Solanum chacoense, *Planta*, **220**, 582–592.
45. Pietruszka, M. (2011). Solutions for a local equation of anisotropic plant cell growth: An analytical study of expansin activity. *J. Roy. Soc. Int.*, **8**, 975–987.
46. Pietruszka, M. (2012). Special solutions to the Ortega equation, *J. Plant Growth Regul.*, **32**(1), 102–107.
47. Proseus, T. E., Boyer, J. S. (2012). Calcium deprivation disrupts enlargement of Chara corallina cells: Further evidence for the calcium pectate cycle, *J. Exp. Bot.*, **63**(10), 3953–3958.
48. Proseus, T. E., Ortega, J. K. E., Boyer, J. S. (1999). Separating growth from elastic deformation during cell enlargement, *Plant Physiol.*, **119**, 775–784.
49. Proseus, T. E., Zhu, G. L., Boyer, J. S. (2000). Turgor, temperature and the growth of plant cells using Chara corallina as a model system. *J. Exp. Bot.*, **51**, 1481–1494.
50. Richmond, P. A., Métraux, J.-P., Taiz, L. (1980). Cell expansion patterns and directionality of wall mechanical properties in Nitella, *Plant Physiol.*, **65**, 211–217.
51. Roelofsen, P. (1950). The origin of spiral growth in Phycomyces sporangiophores, *Rec. Trav. Bot. Neerl.*, **42**, 72–110.
52. Roelofsen, P. (1951). Cell wall structure in the growth zone of Phycomyces sporangiophores. II. Double refraction and electron microscopy. The origin of spiral growth in Phycomyces sporangiophores, *Biochem. Biophys. Acta*, **6**, 357–373.
53. Rojas, E. R., Hotton, S., Dumais, J. (2011). Chemically-mediated Mechanical expansion of the pollen tube cell wall, *Biophys. J.*, **101**, 1844–1853.

54. Ruiz-Herrera, J. (2012). *Fungal Cell Wall: Structure, Synthesis, and Assembly,* CRC Press, New York.

55. Steudle, E. (1993). *Pressure probe techniques: basic principles and application to studies of water and solute relations at the cell, tissue, and organ level.* In Water Deficits: Plant Responses from Cell to Community. (Smith, J.A.C. and Griffiths, H., eds). Oxford: Bios Scientific Publishers Ltd, pp. 5–36.

56. Taiz, L. (1984). Plant cell expansion: Regulation of cell wall mechanical properties, *Ann. Rev. Plant Physiol.*, **35**, 585–657.

57. Tang, A.-C., Boyer, J. S. (2008). Xylem tension affect growth-induced water potential and daily elongation of maize leaves, *J. Exp. Bot.*, **59**, 753–764.

58. Tomos, D. A., Roger, A. L. (1999). The pressure probe: A versatile tool in plant cell physiology. *Ann. Rev. Plant Physiol. Plant Mol. Biol.*, **50**, 447–472.

59. Vandiver, R., Goriely, A. (2008). Tissue tension and axial growth of cylindrical structures in plants and elastic tissues. *Europhys. Lett.*, **84**, 58004.

60. Veytsmann, B., Cosgrove, D. J. (1998). A model of cell wall expansion based on thermodynamics of polymer networks. *Biophys. J.*, **75**, 2240–2250.

61. Yan, A., Xu, G., Yang, Z.-B. (2009). Calcium participates in feedback regulation of the oscillating ROP1 Rho GTPase in pollen tubes, *PNAS*, **106**, 22002–22007.

# Index